Schnetzer Puskas Engineers

Design Structure Experience

Aita Flury (ed.)

gta Verlag

Prologue

Conversations – Prologue

Aita Flury All three of you are different ages and have had different careers. Heinrich and Tivadar, each of you initially completed an apprenticeship as a draftsman before studying civil and structural engineering at ETH Zurich. Stefan, you're considerably younger and followed a direct path from your school-leaving exam to the ETH.

Career
Influences
Characters

Heinrich Schnetzer My career has been influenced by the interaction between formal training and practical experience: After my apprenticeship as a structural steel draftsman, I worked in that profession for a while. Then came my formal training at a technical college and then again a practical experience, until I ultimately graduated from the ETH and completed my doctoral studies there – but the latter was again only after gaining more practical experience.

Tivadar Puskas Even as a small boy, I was influenced by my father's engineering work – I remember very well my impressions from visits to highway construction sites in the 1960s. My career path and Heinrich's are similar: after my apprenticeship, I got some brief practical experience, then went to a technical college and finally attended the ETH. That's where I got to know Heinrich, and it soon became clear that we eventually wanted to start a firm together.

Stefan Bänziger I studied at the ETH directly after passing my school-leaving exam, but nevertheless I also have a practical background: my parents ran a roofing business, so as a schoolboy I often spent time at construction sites and was able to gain practical experience, albeit in construction-related fields instead of the main trades. As a teenager I even thought about pursuing an apprenticeship as a craftsman. In the end, I studied civil and structural engineering because I wanted to learn a specific profession that clearly targets a practical activity.

AF What influenced you most in your education? How do you assess the significance of your choice of professor and the theme of your diploma thesis or dissertation for your work today? What or who was important to developing your technical and intellectual skills? Were there any formative work experiences in offices or engineers as role models?

HS It was clear to me that I wanted to get a degree in construction, and of course, as my professor, Christian Menn played a central role. He was *the* authority and doing my diploma thesis under him was very demanding. The actual topic of my thesis played a secondary role – the professor's temperament and attitudes were far more influential. After all, he also convinced me to work with him as an assistant, even though I had already planned on working in tunnel construction. So he greatly influenced my subsequent career.

TP As Heinrich has already indicated, our professor, Christian Menn, was very important. As a bridge builder, he opened up new worlds to us in structural analysis, in construction, and especially in bridge construction. Professor Dubas and Professor Lang, who taught steel construction and geotechnical engineering, were also important. I followed up my thesis on bridge construction with a degree in geotechnics. Professor Menn had offered me a job as an assistant, too, but by that time, after having spent a long time studying, it was very important to me to pursue a practical occupation again. That's when I began working at WGG, our predecessor firm and successor to Heinz Hossdorf's practice. There, I was particularly impressed by the work on the project for Basel's municipal theater.

SB My diploma thesis, a pedestrian bridge, already pointed the way forward, presaging my current work. Right after graduating, I took a position at WGG Schnetzer Puskas, where, during the first three years, I was heavily influenced by my "apprenticeship" with René Guillod and learned the proper "handicraft" needed for the job. Later, a key experience came while working on the Messeturm [Trade Fair Tower] in Basel, when – under Tivadar's steady guidance – I was granted the chance to demonstrate my ability as a project manager for the first time.

TP Looking back, I was very impressed by the Centre Pompidou in Paris and the John Hancock Center in Chicago, with its *tube system.* I got to know the skyscrapers in

Chicago, by the way, on an educational trip with a colleague who was studying architecture – and who also introduced me to the work of Mies van der Rohe and Frank Lloyd Wright.

HS I was fascinated by Professor Menn's arched bridge in Reichenau, and while working toward my degree, I studied it in depth. Ultimately, though, it was more the professors' personalities that greatly impressed and influenced me. Besides Professor Menn, that also includes Professor Peter Marti, with whom I did my doctorate. They both showed me how important it is to be able to separate the essential from the inessential, to be able to single out the heart of the matter. Analytical thinking is one of the primary qualities of a good engineer.

AF How is it that the three of you came together?

TP As I said, Heinrich and I got to know each other during our studies because our training happened to follow the same path. Stefan was one of our employees and someone whom we saw as complementing us perfectly to help establish an office in Zurich. That's how our collaboration has come about, and in the case of me and Heinrich, we have now been working together for twenty years.

HS As in life generally, chance plays an important role!

AF Do you pursue other interests – independent of your engineering work – that implicitly impact upon your vocation? That is, personal backgrounds or activities that have a strong influence on your work?

TP For me, significant inspiration comes from study trips. In April 2013, I followed the traces of Mendes da Rocha and Niemeyer in Brazil. Their works interest me because they oscillate between architecture and engineered structures.

SB Getting close to nature by fishing and panning for gold helps me find a balance, and the natural sciences serve as a foundation; these are fundamental for my work as an engineer. Physical principles and properties interest me – and that comes into play, for example, when panning for gold: Gold's high specific gravity is what makes it possible to separate it – that's fascinating!

HS My passion is gliding. The bird's-eye perspective and the beauty of the topography impress me every time as if it were the first. At the same time, it's a big challenge: you must be able to abstract details and focus on a few important things in order to make a safe landing.

AF Are there certain landscapes that have influenced you? Switzerland is, after all, renowned for its outstanding works of civil engineering in the Alpine region, but you are based in the Central Plateau. To what extent is the local topography, or even the mountains, part of the sphere of influence for your work?

HS There's more than just the Alps and the Central Plateau; here in Basel, for instance, we live on the other side of the Jura! The Jura mountain range has immense topographical beauty – but the people living in Zurich simply haven't noticed yet!

SB As a Zurich resident, I'm impressed by the strongly perceptible spatial proximity between the urban center and the mountains. To gaze upon the Glarus Alps from Lake Zurich, to feel their presence, is sheer pleasure! Professionally, I'm interested in Switzerland's different regions because of their subsoil: the foundation of every built structure must always be adjusted to completely different soil properties. As engineers, we thus have the fortune of dealing with the soil's geology and with the whole story of its evolution.

TP The mountains, and especially the light in the mountains, are very important to me. Every August, I climb a four-thousand-meter peak in Valais, and in the winter, I go cross-country skiing in the Engadine, one of the most beautiful valleys I know. However, the unplanned and speculative construction there has left behind such enormous traces that, unfortunately, the magnificent engineered structures often fade into the background in people's perceptions. These political and social issues are meanwhile of more importance for politics and society than for engineers or architects.

AF Your firm has two geographically and culturally distinct locations in Zurich and Basel – to what extent do they play a decisive role in your work? How do you recruit your employees and how do you go about working on international projects?

TP Our main office is located in Basel, there's a branch office in Zurich and, as of late, another one in Berne – that's a result of the fortuitous conjuncture in our biographies, which we discussed earlier. We work on international projects from our headquarters in Basel, since the employees with the needed skills work there.

HS Naturally, we want the best employees – and usually, they cannot be found through want ads; it takes word of mouth, like when our younger employees inspire acquaintances to work for us. But unfortunately there's really a shortage of good engineers. We also employ foreign professionals – e.g. Germans, Italians, Spaniards and Japanese – although it should be pointed out that today Switzerland still offers a very good education in engineering.

SB There's a lively exchange between the offices in Basel, Zurich, and Berne. By now, we also work with Zurich architects in Zurich on projects in Basel – and vice versa. So in that respect, there's a good mix – the border between Basel and Zurich is becoming blurred.

HS We see ourselves as a firm with various locations. In terms of our philosophy, it makes no difference whether the project is located in Basel, Zurich or abroad. Our work abroad is limited to developing concepts. The design is developed in cooperation with local offices, with our technical support.

Design Structure Experience

Heinrich Schnetzer, Tivadar Puskas,
Stefan Bänziger

Our mission is to develop load-bearing structures in collaboration with everyone involved in the planning and construction process: along with the technical and economic aspects, what interests us when designing a structure is having a constructive interdisciplinary dialogue. We are wholeheartedly committed to strengthening architectural expression in an integral way by applying our theoretical and practical knowledge about construction and our experience in the design of load-bearing structures.

This book is an attempt to give the reader an understanding of our passion for structural engineering – to make its essence tangible and visible. In the form of short essays looking at twenty-seven representative projects from the last fifteen years that have pointed the way for our work, we examine issues that are important to us. Our conversations with the editor, Aita Flury, constitute a complementary, discursive approach orbiting these topics. Additionally, the publication is enriched by views on our work from people outside our firm: text contributions by Marcel Baumgartner, Jürg Conzett, Harry Gugger, Christian Menn, Daniel Meyer and Quintus Miller illuminate our work, the work of engineers in general or our collaboration with architects, each as seen from a different perspective.

Design Structure Experience is a snapshot in time: Our engineering firm was founded in 1953 by Heinz Hossdorf and, after his retirement in 1980, the business was continued by Kilian Weiss, René Guillod and Rudolf Gisi (WGG). We regard ourselves as the third generation, as a link in a continuous chain. Thanks to our younger partners, who share our passion for building and for tackling challenging structures, we look forward with great confidence to our firm's continued development. This publication and the accompanying exhibition coincide with the sixtieth anniversary of our firm – a nice coincidence, but also a good occasion to pause (briefly) along our journey and reflect on what we do.

Coming from the Institute for the History and Theory of Architecture at the ETH Zurich Department of Architecture, the invitation to make this book and the accompanying exhibition has delighted us, indeed touched us. We have sought to use this opportunity to provide insight into engineering-specific ways of thinking and working, so as to give others an understanding of the design of load-bearing structures and open up a view of what the work of engineers can be today.

Constructive Intelligence

Aita Flury

Albeit slowly, recent years have seen a renewed worldwide appreciation of works of structural engineering: indeed, functional buildings and their design are once more forming the focal point for cultural encounters, as the Swiss contribution to the Biennale 2010 has demonstrated. This comes after decades in which structural engineering, which had become synonymous with the massive structural transformations of our landscape, had received little appreciation or interested attention on the part of society. In fact, since the middle of the nineteenth century Switzerland in particular has, time and again, provided brilliant examples of the art of structural engineering, which to a significant extent is attributable to the training in engineering provided at the Swiss Federal Institutes of Technology (ETH + EPFL). So to exhibit and reflect on the work of structural engineers at a place like the ETH Zurich, the cradle and heart of engineering and architectural training in Switzerland, seems logical. Nonetheless, the exhibition showcasing the works of Schnetzer Puskas Engineers is, remarkably, one of still only a few monographic presentations of contemporary structural engineering at the ETH Zurich. Against the background of our media-based culture which – in contrast to the construction-based culture that prevailed into the late twentieth century – is only interested in technical aspects on the user level, this publication accompanying the exhibition therefore offers a unique opportunity to raise some general questions about the professional standing of the structural engineer, and come up with some possible answers: the outstanding works by Schnetzer Puskas Engineers are an exemplary illustration of the skills and expertise that can be associated with this discipline today, as well as the nature of the challenges involved and possible solutions. Naturally, addressing these aspects means going beyond simple consideration of the built product: the world of technology – which is unquestionably implicit in engineering – only begins to unfold itself in new ways and reveal its secrets if one steps back from the end product to consider the process through which it came into existence.[1] In this respect, we therefore need to pay attention to the nature and structure of production, to the rules and conditions that govern this process.

Finding a Language

Analyzing the different aspects involved in the creative act, in particular understanding which identical, complementary or differentiating factors relate engineering and architectural thinking, has long fascinated me: the "constructors' dialogue,"[2] in the sense of cultivating the discussion between two kindred disciplines with differing perspectives, forms my personal basis for this. This discursive analysis has always been accompanied on my part by an admiration for the achievements and skills of engineers, coupled at the same time with a sense of regret at the generally very marked reticence of a discipline not exactly noted for self-promotion.

This book therefore seemed to offer a marvelous opportunity to allow engineers to speak in their own words, to animate them to comment on their extensive and diverse body of work in an autobiographical way. Naturally, this also places the book in the tradition of Heinz Hossdorf, the founder of the original predecessor firm which later evolved into Schnetzer Puskas, who with his book *Das Erlebnis Ingenieur zu sein* [The experience of being an engineer] presented, in exemplary manner, a broad and enduring commentary on his own work.[3] In view of the very extensive list of works and the fact that the three busy partners Heinrich Schnetzer, Tivadar Puskas, and Stefan Bänziger were under pressure with current projects, this ambition represented no small challenge. The partners' own texts, in which they comment on a selection of their works from a thematic viewpoint, are supplemented by so-called "conversations" – discussions about building, about the profession of the engineer, about engineering itself. These "Q+A sessions" share elements of both a public interview and a private conversation. They offer space for general reflections and considerations, but also elicit private opinions and preferences of the three participating partners: a mixture which offers the spontaneity and the possibility of dialectic confrontation of an active interview and which allows the three different personalities to come across – in the exhibition, the film performs the purpose of showing the people behind the work. Finally, the exemplary quality of the firm's work, which is "at the cutting edge," is summarized through observations from outside, in which engineers and architects who work together with Schnetzer Puskas Engineers in different ways or who have friendly connections with the firm illuminate individual aspects from their own viewpoints.

Participative Congeniality

In terms of responsibilities, an unfortunate process of cell-division has become prevalent in the sphere of building over the past twenty years: the continuously increasing level of complexity in terms of technical and energy-related requirements has led to a culture of specialization which brings with it more dialogue-based, more discursive design processes – the growing number of critical components are increasingly shifting design and planning in the direction of what are in effect processes of "negotiation" between structural and spatial requirements: cooperation between the different disciplines has become essential. In this era of specialization, the importance of the engineer should not be underestimated – ideally, engineers are not only skilled on the level of structural engineering, they also embody a wealth of interdisciplinary technical expertise. This is one of the reasons why a growing number of contemporary architects favor an intensive collaboration with the structural engineer right from the beginning of the conception of the building. This approach makes it possible to develop core design ideas in participative processes involving a small number of participants (architect and engineer) who make contributions of equal importance. In close, established teams, dynamic processes of creative debate take place in which space, form and structural design are mutually negotiated in a fluctuating, reciprocal process. In this case, the actual creative act of conceptual design is to a large extent actually performed as a team.

Essential as cooperation may be, it also has its own preconditions. The originality of Schnetzer Puskas Engineers is clearly expressed in dialogue-based collaboration of this sort with architects from varying backgrounds. The creativity of the engineers is invisible in this context and, when viewed superficially, is overlaid by the architectural image, since the focus here is not primarily on their own formal intentions or visual expression. In their design processes, the analysis of the conditions follows a logic of possibilities and decisions: initially vague, sometimes audacious spatial ideas which the partner architects formulate from an architectural perspective are, from the outset, examined synchronously and on an equal level according to engineering criteria. The spatial is incorporated in the technical as a premise: spatial concepts are integrated into the design brief as "conditions" which, while central, are subject to observation and adjustment (for example, the interpretation of the details of the program can be questioned). The aim is to make possible, but also to enhance, the realization of architectural visions in that the structure and design are conceived integrally and consistently – the structural design should participate in the space as subtly as possible.

In several of the projects, the designers' efforts are founded on the architectural wish to overcome the rigid adhesion of the architectural mass to the ground, to set it floating in suspended motion – a theme which has long inspired the history of architecture, in continually new ways. Spatial concepts such as "levitation" and "defying gravity/weightiness" are starting points for projects such as the Elbphilharmonie Hamburg, the CaixaForum Madrid, the Forum 2004 in Barcelona or the Basel Trade Fair Tower with its overhanging base. Logically, in most cases the reduction in the vertical load-bearing elements is implicit in the structural concepts – the minimization of the bearing points on the ground or on an existing structure is a fundamental prerequisite of the architectural image. The challenges in terms of structural design focus on the relationship between the geometry of the horizontal load-bearing elements and the position of their vertical supporting points. The theme of the deflection of forces from one spatial plane into another, the indirect flow of force, is also promoted by the current trend of stacking different load-bearing structures on top of one another: in such cases of three-dimensional structural engineering, as manifested for example in the project for the Actelion Business Center, the structural design is based on a spatial suggestion: the destruction of the traditional vertical fixed order. However, the contribution of Schnetzer Puskas Engineers is also particularly interesting in cases which involve machinery – equipment, rather than architecture: in the case of the St. Prex balloon, the inconvenience of architecture's immovability and earthbound nature is overcome in real terms in that a temporary stage covering can rise into the sky as a movable, hemispherical balloon. In this case, it is the boundary condition of movement, and thus the different positions of the structure in space, which inspire the clever and inventive elements in the structural design.

Even though in some projects it is the architectural image that strikes one at first glance, it quickly becomes clear on closer examination that the final form of the buildings is largely determined by the artful combination of architectural and engineering criteria. The conceptual and technical sophistication, its constructive intelligence, therefore elevates the engineers' contribution far beyond a mere "enabling" function, making it a creative one.

Design Thinking

The engineer's contribution to the design of buildings which transcend our familiar expectations and in some respects exceed the everyday conceptual logic of the discipline requires a high degree of open-mindedness. Schnetzer Puskas Engineers are distinguished by their readiness to challenge the conventions of structural design and think up new possibilities – to sound out the limits, as they themselves put it. One key basic requirement for this is the ability to think hierarchically. This makes it possible for them to cope with complexity and the integration of numerous contradictory boundary conditions: design processes are open processes, not linear ones, in which the participants are confronted with so-called *wicked problems*[4]. Hierarchical thinking makes it possible for the designers (architect and engineer) to accept, in each phase, certain problems which are at present still unsolved. At the same time, they must be able to judge whether they will be able to develop and solve these outstanding problems during a later phase through further mental effort.

Design thinking can also be described as a synthesis of analysis and empiricism, which also applies to both architect and engineer.[5] Whereas the architect focuses on the spatial aspects, the priority for the engineer is an understanding of the internal forces involved. The engineer thus translates the physical reality into an analytical model and thinks in an abstract mathematical language. The "substance" is a material acted on by a linear system of forces, an abstract network of relationships which, divided into tension and pressure, can be ordered and calculated. The young engineer can develop a feeling for the behavior of the internal forces by gaining experience through calculation work and developing a sense of the assumptions which need to be made and the properties of materials. This intuitive feel for the internal forces has a key impact on a design from the outset in terms of weighing up and appraising the available options: during the conceptual phase, it is decisive and enables the engineer initially to follow an intuitive approach without placing too much emphasis on theoretical calculation at too early a stage. Schnetzer Puskas methodical approach is characterized by the skilled, chronologically-harmonized combination of estimation, mental and manual calculation and subsequent verification by means of computer processing. To this is added a great openness when it comes to using the computer: the digital possibilities of simulating models of load-bearing structures through computer animation are fully exploited.

Certainly, Schnetzer Puskas' effective use of the computer can also be viewed in the context of the general digital *zeitgeist;* nonetheless, in this regard in particular, it is scarcely by chance that they find themselves following in the footsteps of Heinz Hossdorf: he was among the first (at least in Switzerland) to recognize the importance of the virtual model reflecting the spatial-material reality as an indispensable, because absolutely unequivocal, basis for any computer-supported work of the engineer.[6] Digital models do not provide absolutely reliable information on the appearance of the real building in the same way as physical models. However, the digital simulations and models created for the purpose of calculation produce, as by-products, images with their own unique aesthetic which present calculation data in a fascinating way as a distinct visual art form.

Experiences

The key to the art of structural engineering is, as David P. Billington observes, the way in which mechanics and aesthetics are brought together in practice.[7] In other words: theoretical knowledge and familiarity with engineering principles can only be developed and deepened through the connection with the building site, with building itself. As the engineers themselves emphasize, experience in execution is essential, because without this an important source of feedback for subsequent designs is lacking. Direct, local observation and assessment – whether in the continuous study of building processes or in the observation of structures in use – is thus of central importance. Only this practical orientation makes it possible for the engineers to develop structural design proposals, also for complex structures, which it is economically feasible for the contractor to execute – that is to say efficient structures.

Efficiency in this respect does not focus simply on the economic or monetary aspect, but also, above all, on the efficient use of resources and an intelligent flow of force. Accordingly, an indirect flow of force resulting from a stacking principle is not simply optimized. Rather, efforts are focused on the search for a flow of force which is once again more or less direct and proportionate. Economy, feasibility and efficiency are part of the discipline of engineering and are not incompatible with artistic considerations. For Schnetzer Puskas Engineers, too, the concept of technical and economic optimization is an important reference value in any project – like all other responsible engineers they respect the imperative of economic efficiency as a "taskmaster." Having to deliver ingenious results under constraining economic conditions encourages the engineer to seek out new solutions. These can influence the actual structural concept, focusing on a transition, a bearing detail, or they can involve an idea relating to the sequence of construction work, as became of key importance in the case of the Puna Tsang Chhu Bridge in Bhutan for example: an intelligent development of the load-bearing elements made it possible to re-use these several times during the construction process, in different functions – as falsework beams and later as composite girders in the deck slab.

Also, it is only through continuous practical experience that it becomes possible to evolve typological concepts: structural ideas or geometrical principles endure because they are transferred into other projects or collaborations, where they undergo optimizations and formatting. In contrast to architects, who use the type to describe the layout, organization or function of a building, the engineer interprets the type in terms of the load-bearing element or the load-bearing structure – that is to say simple beams, trusses, folded structures, etc. Only a profound knowledge of the basic types and their material correlations make further interpretations such as Prada Aoyama in Tokyo, or the elaborate mixing of typologies as in the Actelion Business Center possible in the first place. Incidentally, the hybrid Vierendeel truss developed for this project highlights the importance of typological classification: when conventional typologies are transcended, new terms are created.

The Actelion Business Center is also a good example of a structure which combines building components with differing load-bearing effects. Schnetzer Puskas Engineers are interested in systems in which the interaction of all elements creates static indetermination or over-determination – so-called redundant systems. In such structures, which tend to isotropy, the forces are distributed between the different load-bearing systems as required, providing mutual support in a democratic and natural regime of structural solidarity – if one load-bearing element fails, another takes on its function. Nowadays, the actual dimensioning of such efficient but complex systems has become relatively simple through the introduction of new computing capacities using finite elements. However, plausible assumed values, which can only be based on experience gained from prolonged and persistent research, remain a fundamental prerequisite for the conception of an economically sound redundancy.

Conclusion

The works of Schnetzer Puskas Engineers do not fit into a clearly readable line of aesthetic or technical development. Their structural designs are particular, just as different as the partners themselves, and can in no way be forced into a single formula or concept. The distinctly individual projects reflect both the influence of the three author-partners and the collaborations with architects bringing in different perspectives. What gives the firm its identity is the broad spectrum of technical expertise which is applied intelligently in creative interplay with the architects. The engineers' strength lies in the definition and exploitation of the scope available for experimentation. Schnetzer Puskas Engineers are a model, indeed an advocacy, for the logic of exploiting the profession's particular perspectives in all their different facets, and establishing a joyful, creative approach to working with these.

1 See Ernst Cassirer, "Form and Technology," in: Peter Fischer (ed.), *Technical Philosophy,* Leipzig 1996, p. 161.

2 The "Constructors' Dialogue" is a sequence of exhibitions and publications which started at the beginning of 2006 with an exhibition in the Architekturforum in Zurich curated by the author. Its provisional conclusion is the publication: Aita Flury (ed.), *Cooperation: The Engineer and the Architect,* Basel 2012.

3 Heinz Hossdorf, *Das Erlebnis Ingenieur zu sein,* Basel 2003.

4 The concept of "wicked problems" originated with the philosopher Karl Popper and was used in the 1960s by Horst Rittel, professor at the Hochschule für Gestaltung in Ulm (Ulm School of Design) as an approach to a design methodology. In 1992, Richard Buchanan seized on the term again in his essay "Wicked Problems in Design Thinking" (in: *Design Issues* (1992), vol. 8, no. 2, pp. 5–22). In his essay, he hypothesizes that every designer has a personal set of positions that are developed and tested through experience. The inventiveness of a designer lies in his natural or cultivated artistic ability to be able, at all times, to draw upon these positions in order to apply them to a new situation or to discover aspects of the situation that have impact on the design.

5 See Tom F. Peters, "Bridging the Gap: Points of Contact Between the Architect and Engineer," in: *Architronic* (Dec. 1999), vol. 1, no. 1, ISSN 1066-6516, avail. online: http://corbu2.caed.kent.edu/architronic/v1n1/v1n1.09.html (September 2, 2013).

6 Peter Dietz, "Der Schritt in die reine Virtualität" [The step into pure virtuality], in: Heinz Hossdorf, *Das Erlebnis Ingenieur zu sein,* Basel 2003, pp. 249–260, here p. 249.

7 David P. Billington, *The Art of Structural Design: A Swiss Legacy,* Princeton 2003, p. 15.

1 On Collaboration of Engineers and Architects

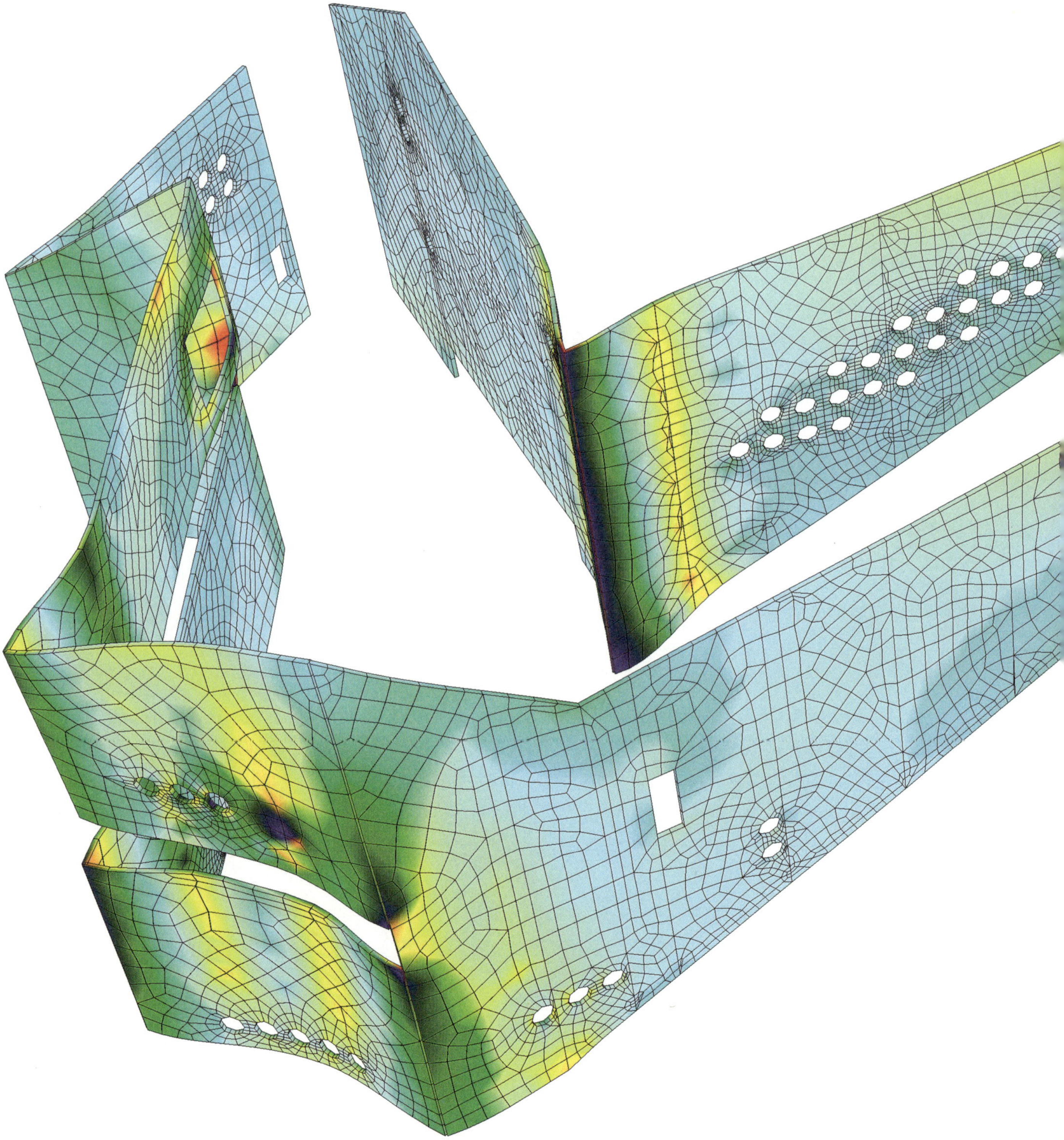

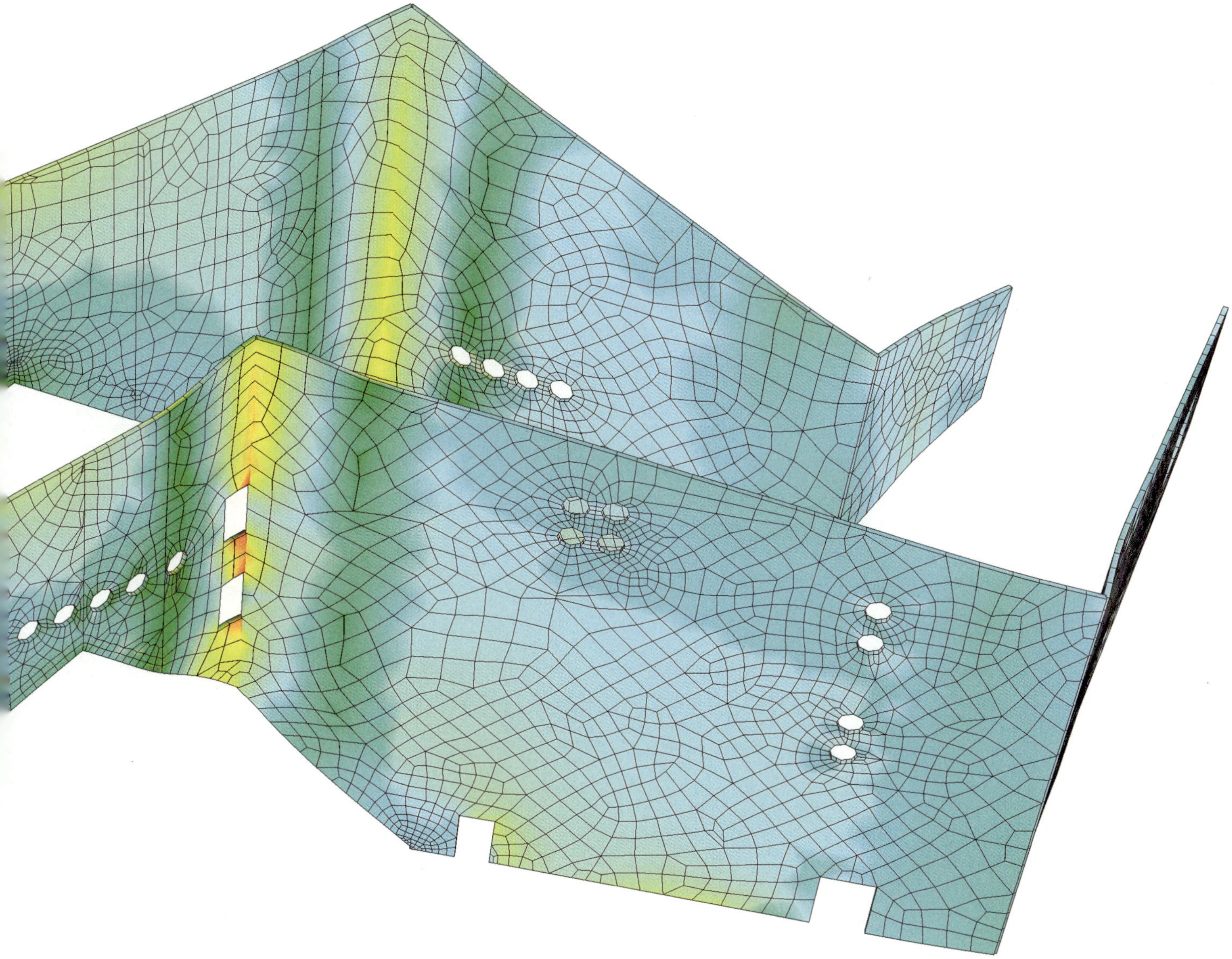

← Stress distribution on the joint-free facade; expansion of the Swiss National Museum Zurich (computer simulation)

Conversations I

AF How roles are perceived and shared between architect and engineer is always dependent upon the chronology of events. To what extent are you involved with participatory design between engineer and architect? Role Perception

TP For buildings, architects are usually approached to do a project, then they form a team accordingly and, if it's a competition, for example, they also take over the lead. On building projects, we see ourselves as participatory planners who seek to anticipate the ideas of the architect, to strengthen them or even to question them through our use of structural concepts. The goal is to establish a unity of space and structure.

AF So do you see yourselves as part of the orchestra, with the architect as the conductor? And if that's the case, then what is the role of the structural engineer?

TP It's true for building construction, but for bridges or other civil engineering works, the roles are different. But also on projects where we have the lead role, we often draw upon the design assistance of architects, and that usually pays off.

HS It's possible to think of the architect as a conductor, but for me the idea of a coordinator comes closer. Since architects are generally the first to come in contact with the client, they are also the first to assimilate the client's needs and translate them into a spatial program or even an initial design. The architect is, in this respect, closer to the client and conveys their needs on a technical level. For bridge construction, by contrast, it is the structural engineer who leads the team, and the architect is not needed to translate the program because the engineer understands the client's requirements. Civil engineering and bridge construction projects call for architects more for urban design considerations.

AF How definitive are your communications, your ingenious messages? Or, asked provocatively: How far do you go to sell your ideas to the architect? How far will you go to stretch the technology for them? Do you actually delight in architectural demands that create structural problems because then you can demonstrate your virtuosity in dealing with them? Ideology Chameleon

TP We're not necessarily known for saying no. We clearly view it as an exciting challenge for us to take what seems at first to be a difficult project and translate it into a structurally plausible concept. And the concept doesn't always have to follow the direct flow of forces if exciting spaces are made possible. Such architectural excursions are more interesting to us than purely commercial buildings with no real structural demands.

HS Behind your question lies a profound implication. The question suggests that we enable the architect to create a structure that doesn't make sense. But we presume that our partner architects are acting with a clear idea about their intentions. If we are enabling something that seems to be on the edge of sensibility, it's naturally important to carefully reflect on it. When exploring limits, the dialogue between architect and structural engineer is key. Sometimes we explore a limitation, only to realize in retrospect that we were operating along the margin. A good example is the Actelion project, where you might wonder whether it is still a building at all or whether it is, in fact, a sculpture. For the engineer, the question of whether and when to draw the line unquestionably arises again and again. At the same time, it's also simply exciting to take something to the edge.

SB Along with the engineer, there are other factors that determine whether a project that goes to the limit will get built. Most often it's the cost of construction, the building services or a design idea that is not feasible – for instance a steel and glass structure that cannot meet the thermal requirements or those of comfort. Personally, I often don't recognize spatial qualities that might, for instance, be achieved by omitting columns, until after the building exists. So with that in mind, I'm glad that architects exert a certain development pressure on us.

AF Overstating it perhaps, that means you rely on the spatial skills of your architect partners.

HS That's right. But we must also critically note that there are certainly architects who narcissistically just work toward that "one" special photograph and do not necessarily pursue the needs of the client.

AF Doesn't that contradict your previous statement?

HS It is a contradiction, but it's also the reality. We fundamentally trust our partners. But sometimes you simply realize in retrospect that the path was somewhat less than ideal.

TP You need to distinguish between various clients, architects and different programs. Designing a Prada flagship store in Tokyo or a tall residential building in Basel are two very different things. When designing a building, it's naturally important whether it is supposed to represent a brand or should simply have a facade that is as intelligent as possible.

AF What are the means that enable you as a firm to draw the line between artistic desire and what you view as irrational? How extreme can the structural efforts be when seeking to implement an architectural concept? Limits

HS As Stefan already mentioned, in that regard there are three points worth mentioning. To begin with there is of course physics itself – the statics and the mechanics that set limits, although we try to explore these and experiment within them. A second point is the contractibility and the third is the cost. The last two points are especially limiting, representing constraints that I wouldn't want to be without. As engineers, we are always looking to produce things rationally and cost-effectively. In recent years, vast changes have taken place inasmuch as very complex geometries can now be developed with CAD. In steel construction, it is even used to prefabricate the component parts. The actual process of assembling these parts, however, has not been simplified in the last fifty years – to this day, people or scaffolding must hold the parts in place so they can be put together! Those are inherent cost limitations that persist despite CAD, and about which I am quite happy.

AF Where do you see the opportunity for improvement in collaborations between engineers and architects?

SB To some extent, on both sides there's a lack of understanding for the other discipline. Engineers are not the only specialists – the architects are just as well, in their own way. Our education clearly lacks an involvement with architecture, and at the same time, architects often lack a rudimentary understanding of statics. To counteract that, both sides need to come closer together beginning in their training.

TP Students would need to attend lectures dealing with the other discipline: architects would learn about developing load-bearing structures and engineers would learn about urban design. Pier Luigi Nervi claimed spatial thinking for the engineer, too. We strive in that direction, although seen from the other side, it's important that the architect brings along a "sensibilità statica" – a sense for statics. The greater the common denominator, the better off we all are.

AF Your projects show that you can work together very well and very closely with different architects. What then are the key skills needed for collaborating with architects – what distinguishes you?

HS It's important to recognize the common goals from the beginning. To do so, the architect must first spatially interpret the client's requirements. Then an approach is needed that can in turn abstract and focus those ideas in order to arrive at a structure.

AF So you're good at listening.

HS That's key to good collaboration.

SB It also requires a very good understanding of the other disciplines that are involved. A structural concept that doesn't consider building physics or permit the building services to be integrated rationally is pointless. One of our strengths is certainly that we have a good knowledge of the important aspects of other disciplines, so we can keep an eye on complex interdependencies during the early phases, when the structure might initially be developed only between the engineer and the architect.

Architecture in a Crisis

AF Do you think architecture is in a crisis, and if so, what is its nature and what role does the engineer play? Building as art or the art of building?

SB It's certainly possible to observe that for many architects, individuality and novelty, "invention", is key. In this regard, sometimes a certain obsession really cannot be denied and, thus, tried and true solutions are often not used.

AF The problem of architect's narcissism?

SB That's perhaps a bit exaggerated, but not totally wrong.

AF But as engineers, you're also interested in innovations.

HS We're clearly interested in exploring limits and operating within the available methods. But in response to the fundamental question of whether architecture is currently in a crisis, I diagnose a social problem: These days, only the "cultural" aspect seems to matter; technology is taken for granted. Design seems to be very important, but the implementation, the craft, no longer has much significance, at least not as part of the training. The statement is definitely aimed at those architects who would like to be "star architects," but don't have the tools of the craft needed to do so. So I have to say yes: the crisis is a crisis of the architects' technical know-how.

TP The projects that are discussed within professional circles and appear in professional journals make up perhaps two to five percent of the total construction volume. Still, I believe these projects serve as models, which I would identify as building culture. They are important because they demonstrate that the act of building is not only, or not primarily, about monetary values and optimizing rates of return, but that above all, it also has societal, social or cultural relevance.

HS I would put a question mark at the end, asking whether such projects raise the level of quality. Just look at how the politicians in our country go about their business with blinders on! Our everyday built environment manifests a general spatial insensitivity that is becoming rampant!

AF In talks at engineering events, the same self-reflexive question repeatedly comes up: what are structural engineers? Is engineering in the grips of an identity crisis? Are you service providers for society? Identity Backdrop

SB Thus far we have spoken primarily about structural engineers involved with building design. But engineering is, of course, a highly diversified discipline. The spectrum is very broad, ranging from traffic planning and hydraulic engineering to construction management and environmental biology.

AF So the engineer's identity is diffuse because it is pluralistic, with many orientations? Is that the reason for the self-scrutiny and the unclear position of the engineer in society?

SB Our association, the USIC [Swiss Association of Consulting Engineers], has made various efforts in recent years to strengthen the engineer's position in society. Not without success, as shown by the increasing student numbers. We are in the midst of moving beyond an identity crisis.

HS The whole must be seen in the context of time: in the 1950s and 1960s, engineers created large bridges and other works of infrastructure – including, going further back, railroad structures (like the Gotthard Tunnel) – for which the technical aspects were very important. As said before, the so-called cultural aspects are more important today and the technology is simply taken for granted. Put bluntly, the technology is so complex today that people are no longer specifically interested in it. Architects realized that and focused their attention on the cultural aspects, but suddenly engineers, who took refuge in technology, no longer received society's attention. That's why it's important to start early and also raise awareness for the cultural aspects of engineering work as part of the training, since that's what is perceived and judged by society.

TP Elements of the transportation infrastructure once had decisive influence on the course of settlement. And engineers often played visionary roles – but that was lost later on. I see large engineering firms – ones with four to five hundred engineers who cover all the disciplines – as belonging to a category of "unknown service providers for society." Their projects must simply work on a technical level – they are not scrutinized by society. Another burden that engineers have to carry is the fact that they are usually only seen in the event of a disaster – only after something collapses, for instance. As long as the structures hold up, the architect is usually the only one in the limelight.

Generalists
Specialists
Interdisciplinarity

AF Where do you see the future for your discipline – where is it heading? Do you view engineers as generalists or specialists? Is the engineer's task primarily in developing the load-bearing structure, or to what extent is focusing on the interfaces with the other disciplines – interdisciplinarity – of significance?

HS If we limit your question to supporting structures, I certainly view myself as a generalist. It is key to keep in mind the overall context of design, costs and implementation. Without that knowledge, our firm would not last very long – this total package is precisely what's expected of us. At the same time, it makes sense to also have people in the firm who are proficient in different fields of specialization, such as earthquake engineering or material technology, for example.

TP In our field, we are definitely generalists. A key element is contextual thinking, especially in relation to the use of materials or resources. So we definitely look beyond the immediate "structure" or "structural system" and usually begin with an analysis of the surroundings.

HS Indeed, there are extremely diversified fields of specialization: all these specialists – for fire protection, wood construction, refurbishment, etc. – often lack the distance needed to see the bigger picture. Personally, I'm very skeptical of specialization – it's basically like being in a blind alley. It's an unnecessary fragmentation of our profession, often with the overtone of securing a market niche. Ultimately, though, the brains – who can bring everything together and take responsibility for the whole – are still needed every time.

AF So the training of engineers must be aimed at creating generalists.

HS Absolutely. But, of course, individual areas of specialization should continue to be taught.

Building with Ambiguity

Quintus Miller

Building is an early characteristic of civilized human activity and over the millennia, it represents a multifaceted archive of human ways of living. It is a collective process involving all participants in the construction, with the objective of creating a building that satisfies their respective needs in life. The productive interplay of various disciplines promotes the quality of the buildings and increases their cultural relevance. This multifaceted interconnection holds great potential for the future.

The Historical Shift in Building

As a communal effort, the act of building was a decisive prerequisite for the emergence of the early cultures along the Mediterranean. There are hardly any historical accounts from this period, but impressive built structures like the Bubastis Canal between the Nile Delta and the Red Sea demonstrate that engineering thought and meticulous planning were the basis for these accomplishments. The Romans also used technologies in masterly fashion, pushing them to the then-existing limits of feasibility. In his *Ten Books on Architecture,* Vitruvius describes the métier of the architect and defines the relationship of the categories within his architectural theory: the *firmitas,* the *utilitas* and the *venustas* must be taken into equal account.[1] Support, use, and beauty are prerequisites for a strong, holistic result.

With the fall of the Roman Empire, engineer-like thinking in architecture came to a halt. In the Renaissance, engineering thought returned to architecture with awareness of the composition of forces. In pre-revolutionary France, the spirit of the Enlightenment led to origination of the profession of military engineer, who – in contrast to the civil engineer – completed a course of studies predominantly shaped by mathematics. This change meant, however, that in the architecture of those times, the categories postulated by Vitruvius were no longer contemplated holistically. Entirely in keeping with the division of labor in industrial society, the professions of architect and structural engineer were understood and cultivated as different vocations. The mastery of problems through the fragmentation of complex issues also corresponded to the scientific opportunities available. Simultaneously accompanying this tendency was an increasing loss of the mutual cultural bond of buildings' structure and form, which, as separate systems, constitute the building as an artifact. This historically founded circumstance proves to be culturally fatal, as the separation of form and content leads to mutual alienation and interchangeability. Things are no longer what they are; they are just meaningless images.

Building as Cultural Layering

Building causes profound changes in the landscape and in urban space. It is accompanied by interventions that have an impact on our habitat for generations: Building can be understood as a material reflection of human civilization that literally imprints itself in the territory. Through the centuries, the cultural layers overlap and a multifaceted framework of traces emerges, forming the natural landscape into a cultural landscape. In this totality, the aggregate states of human societies can be deciphered and reconstructed over millennia. The territory is interwoven with human ways of life, much like a palimpsest in which the new arises through constant reshaping, but which implicitly always retains the antecedent and its meaning. The current meaning of the zeitgeist appears on the surface, while the entirety hides multitudinous meanings pointing to their cultural origins.

The cultural layering harbors substantive constancy that passes on socially relevant meanings beyond the now. For the city as a system, it becomes evolved programming with a deeper meaning, because social contents are thereby superimposed

1

2

1 Tardis Bridge, Landquart, competition entry 2001, 2nd place
2 Spirgarten Senior Residence, Zurich-Altstetten, completed 2006
3 Courtyard housing on Sempacherstrasse, Basel, in progress 2013
4–5 Alternating design of the vertical structural elements (columns, hangers, diagonals, and shear walls)

and their respective importance ranked. The city becomes ambiguous and legible on many layers; as a system, it can always re-adapt to the changing social conditions and draw upon the accumulated cultural content. Some of these semantic contents may appear to be redundant, but they represent different contexts. They are not identical – they are, on the contrary, analogous and thus, considered culturally, not redundant. The ambiguity permits the reinterpretation of the existing condition and, indeed, an enhanced mutability can also be recognized therein, which hinders succumbing to strategies of the tabula rasa when confronted with the loss of meaning. The stability of architecture lies in the simultaneity of change and constancy.

Building as a Communal Effort

To avoid the separation of form and content described above, close collaboration among the parties involved in the construction is fundamental. In particular, engineers and architects, whose professions are now so diversely oriented in terms of their complexity, must comprehend their work on the project as a joint effort. Our postmodern capabilities in fields like computer science have created the potential to also resolve complex matters pertaining to building, so our fragmented view of built structures can be interrupted and the holistic view of building can be restored.

Such a modus operandi can be illustrated by numerous works that we have confronted together with the engineers Heinrich Schnetzer and Tivadar Puskas: In the early work done in collaboration on the competition for the Tardis Bridge near Landquart (fig. 1), for example, the design is heavily based on dealing with the perception of the landscape. A fundamental decision was to not allow the supporting structure to be seen as a marker in the landscape – the spatial impact should instead unfold in the area along the river. As a consequence,

3

4

5

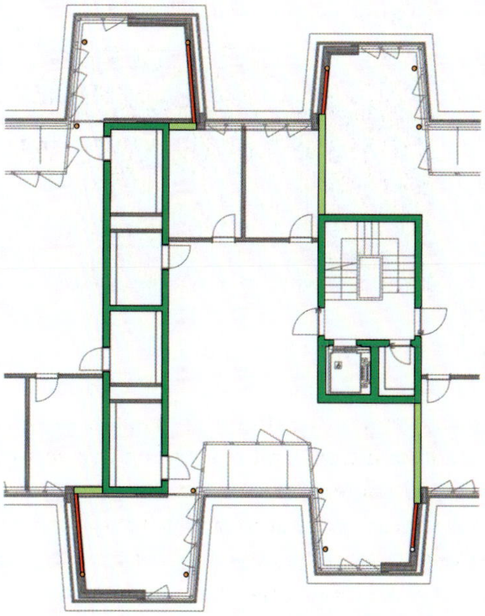

the bridge form featured a slightly arched roadway on delicate, V-shaped pillars – an appearance that respects the more important landscape and which was developed in a jointly conducted examination of form and structure.

A different approach was used for the design of the Spirgarten senior residence in Zurich-Altstetten (fig. 2). Urbanistically, the new building was intended to mediate between the rural structures along Spirgartenstrasse and the urban-scaled structures on Badenerstrasse. Not only the different granularity, however, but also the internal inconsistency of the rapidly growing agglomeration was to become a form-giving factor. Thus, the building's facade along Badenerstrasse features display windows at grade, while the form facing Spirgartenstrasse opens up to the entry court with a freely interpreted ground floor. On the upper floors, strip windows between solid spandrels dominate the facade's design, which reflects the retirement apartments' orientation to all sides. But at the same time, the large living room windows interlock with the concrete spandrels – they cut into them – and thus challenge the uninterrupted flow around the building form. Our interest in these formal contradictions comes from a reading of the situation that accepted no pure rules and revealed their inconsistency. These urbanistic and spatial intentions could only be realized, however, thanks to the close link between our design and the structural concept: whereas the facade along Badenerstrasse rests directly on the ground, shear walls balancing on columns allow the building to have an open lobby. The architecture and structure are thought of as an integral whole, with the facets mutually enhancing each other.

The basic idea of the Sempacherstrasse courtyard development in Basel (fig. 3–5) – the youngest example of our collaboration – is based on creating spacious park apartments that refer, in character and form, to the commercial buildings that are typical for these courtyards. A ground-floor *rue intérieure* provides a lengthwise circulation space for the building in a north-south direction, while the three residential stories above are oriented east-west. The logic of the supporting structure is derived from this arrangement: the load-bearing structure runs in a longitudinal direction on the ground floor, transitioning to three large, transverse frames on the upper floors, which incorporate the stairs and the wet rooms. The slabs are hung from these frames, creating a freely configurable ground floor. The dark, lightweight facade is attached to the slab edges and each story is slightly recessed in relation to the floor above, which, apart from constructive considerations, magnifies the industrial character of the building. In this project, too, the supporting structure and the architectural theme are mutually dependent on each other and form an inseparable whole – the structure, form, and content are congruent.

Building for the Future

The potentials of and the fascination for designs that intertwine architectural and structural facets with each other is far from exhausted. The guides for integrated design must be more than just the direct dependency of form and structure. As described in the introduction, architectural and substantive ambiguity sustainably increases a project's long-term cultural survivability. A similar effect can also be presumed in relation to the structural analysis. If enlightening and unambiguous load-bearing structures make way for ambiguous building structures, then buildings will gain robustness in the long term. Through their structural redundancy, they can more easily accommodate massive architectural changes. Structural redundancy, which is usually seen as superfluous, is thus favorable when viewed from the standpoint of sustainability. Could a path shared on one of our future projects lead there?

1 "Haec autem ita fieri debent, ut habeatur ratio firmitatis, utilitatis, venustatis. Firmitatis erit habita ratio, cum fuerit fundamentorum ad solidum depressio, quaque e materia, copiarum sine avaritia diligens electio; utilitas autem, <cum fuerit> emendata et sine inpeditione usus locorum dispositio et ad regiones sui cuiusque generis apta et commoda distributio; venustatis vero, cum fuerit operis species grata et elegans membrorumque commensus iustas habeat symmetriarum ratiocinationes."
"Now these should be so carried out that account is taken of strength, utility, grace. Account will be taken of strength when the foundations are carried down to the solid ground, and when from each material there is a choice of supplies without parsimony; of utility, when the sites are arranged without mistake and impediment to their use, and a fit and convenient disposition for the aspect of each kind; of grace, when the appearance of the work shall be pleasing and elegant, and the scale of the constituent parts is justly calculated for symmetry." *Vitruvius: On Architecture,* Volume I, Books I–V, edited and translated by Frank Granger, Cambridge, MA 1931, Book I, chapter III, p. 35.

On Collaboration of Engineers and Architects

Tivadar Puskas

"I would distinguish the difference between the engineer and the architect by saying the architect's response is primarily creative, whereas the engineer's is essentially inventive."[1]

In his book *An Engineer Imagines,* Peter Rice explains that the best buildings develop from a symbiotic relationship between architect and engineer – if and when the engineer's skills as a goal-oriented innovator come together with the architect's more subjective and creative skills. The Centre Pompidou in Paris is the most important work of collaboration between Peter Rice and the architects Renzo Piano / Richard Rogers. Its multifunctionality and the transparent yet structural external appearance have made it a cultural and architectural symbol of the 1970s: The cast-iron gerberettes add drama to the building's exterior, and their form follows function, so that the support reactions of the story-high trusses – which span forty-five meters within the building – can be guided without moments into the columns and counteracted by external restraints.

The notions of the (ideal) relationship between engineers and architects are as varied as the personalities who have expressed them. What follows is a small compilation:

Sigfried Giedion, a structural engineer who studied art history, wrote in 1946 of construction as the subconscious of architecture. Construction is not merely a product of reason, but also implies an instinctive, that is, artistic form of expression.[2]

Le Corbusier visualizes the collaboration between architect and engineer as the intersection of two interlocking circles. His symbolic drawing "Constructeurs,"[3] features "two clasped hands, the fingers enlaced horizontally, demonstrating the friendly solidarity of both architect and engineer engaged, on the same level, in building the civilization of the machine age."[4]

The engineer August E. Komendant, in turn, reflects on his collaboration with the architect Louis I. Kahn: "The present principles and trend in structural engineering and architecture are simplicity, well balanced masses and areas, and harmony of the structure with its surroundings. The structural system should agree with the particular material it is made of. The transmission of loads and forces to which the structural element is subjected should be clear and visible. The construction should be simple, rapid and economical."[5] The Kimbell Art Museum in Fort Worth, Texas (fig. 1), represents the culmination of the collaboration between Kahn and Komendant. Here, architecture and engineering coalesce into an inseparable unit.

In *Cooperation: The Engineer and the Architect,* Aita Flury describes working within a team and cultivating the harmony of the varied responsibilities as follows: "The prerequisite for this is an attentive inquisitive readiness to cross borders: architects can find the key to a fruitful dialogue if they rediscover the master builder in themselves, the building designer with a keen understanding of structure. The engineers, on the other hand, would refresh their approach if they combined 'sensibilità statica' (Pier Luigi Nervi) with a spatial sensibility [...]"[6]

The architect Rem Koolhaas and the engineer Cecil Balmond strive to free themselves from the logic of the grid. Balmond's book *Informal* repre-

1

sents the desire for a counterpoint to the rationalist, industrial thinking of modernism. About the Kunsthal in Rotterdam, he writes: "Trapped by the Cartesian cage I wanted to break out. The informal beckoned ..."[7] The bracing, the exception to the grid, the frame and the juxtaposition of load-bearing elements were the four construction strategies for the Kunsthal, in which architecture and possible structural elements mutually provoke and challenge.

Similar to the varied opinions expressed in these text fragments is how we experience collaborations with architects, for which there are no magic formulas. In the development of load-bearing structures, however, it is our declared intention to explore – together with the architects – new avenues that enable us, with the aid of our analytical knowledge and intuition nurtured by experience, to enhance architectural intentions. Sometimes it is emphasized language that stimulates the design process, and sometimes it is a vague formulation that opens one's eyes to surprisingly innovative concepts. It is rarely a scientifically linear increase in knowledge. Rather, it is usually a matter of iterative processes within an environment shaped by culture, emotions, training, and career. Architecture, and consequently the load-bearing structure, are always subject to cultural spheres of influence. They are subject to the spirit of the times and the willingness therein to sound out technical possibilities. But for a profound dialogue between engineer and architect, mutual understanding of the problems of the other discipline is essential. Only under this condition can a common language be developed. Not infrequently, such professional collaborations also lead to lifetime friendships.

From project to project, the work we do together with architects is shaped by different aspects, as illustrated by the following examples.

Laboratory Building WSJ 352, Novartis Campus, Basel 2010

The different cultural context and the sheer distance of nine thousand five hundred kilometers made the collaboration with Tadao Ando on the WSJ 352 laboratory building a great challenge. In the biweekly workshops held on the campus with the managing architects, the medium of sketches functioned well as a universally understandable language for development and planning.

The laboratory building is located on the northernmost site on the campus, on a triangular lot that borders French territory to the north and is flanked on the west by Richard Serra's impressive sculpture "Dirk's Pod." A seven-story, triangular building that derives its form from the shape of the site was designed for this special context. Alone due to the geometric constraints of the pie-shaped building, the clear organization of the spatial program and the development of an appropriate structure evolved an unusual complexity.

The structure consists of prestressed flat slabs, columns, and a core. The columns are laid out on a 7.2 meter grid, which ensures that the laboratory spaces can be used flexibly. The layout and configuration of the cores, which incorporate the stairs, elevators and service shafts, follow the architectural requirement aimed at obtaining the most possible open space and keeping to the column grid. During the design process, many different core typologies were studied: single cores, cores following the outer shape of the building, cores along the three facade planes, double cores, split cores, freestyle cores unrelated to the grid, and individual linear cores, and it was a linear core that was ultimately constructed. The core's walls as well as the columns and the other walls were all made with Ando's trademark silky smooth exposed concrete. That finish is achieved by using formwork panels lined with

2

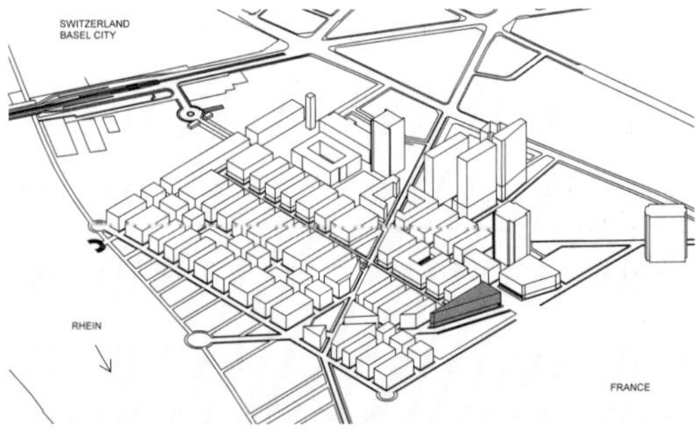

3

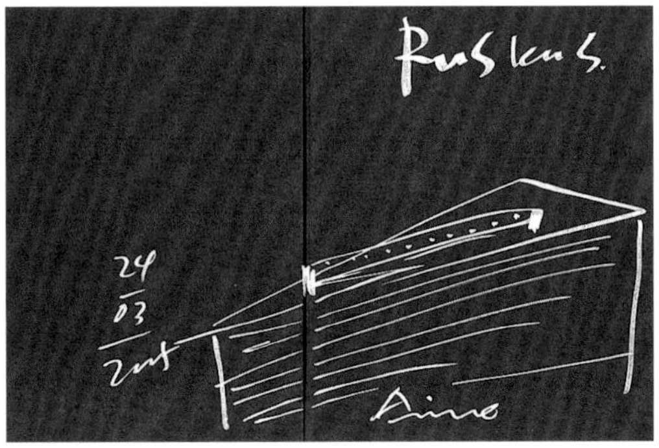

4

1 Kimbell Art Museum in Fort Worth, Texas (computer model)
2 Novartis Campus, Basel; master plan by Vittorio Magnago Lampugnani
3 Dedication in book: WSJ 352 Laboratory Building by Tadao Ando
4 Workshop in Tadao Ando's office in Osaka

coated plywood and tie-hole spacing that is specified to the millimeter. The core's remarkable slenderness ratio – a mere tenth of the building's height – and its eccentric position led to the deep foundation being subjected to tension under seismic loads.

The pursuit of precision reached a peak at the tip of the site, both literally and figuratively. The sharply pointed corner – *the tip of the building* – is less than 20 degrees and was supposed to float freely in the air. This required suspending the entire tip of the building, including the intermediate floors. A massive, six-ton steel wedge element assumes the accumulated loads and transmits them via prestressed upstand roof beams to the columns on the grid. What looks at first glance like an awkward way to route the forces results in an extremely light and transparent appearance that, in this project, almost symbolically represents the collaboration between engineer and architect.

Hertenstein House, Weggis, Switzerland, 2012

The private residence on Lake Lucerne depicts, in an exemplary way, our frequent collaboration with Buchner Bründler Architekten of Basel. From the first sketch onward, the typology of the building was designed in a fruitful collaboration.

Atria, niches, upstand beams and large openings set the tone for the house, which was designed for a pianist and her partner, a photographer. Interlocking horizontal slabs, projecting shear walls, external portal frames and columns – all made of light-colored exposed concrete – constitute the sculptural and refined building shell that both contains and transcends the volume: the building shell is a visible, load-bearing exterior that enters into a visual dialogue with the open space of the surrounding natural environment.

The overall structure, the upstand beams in the roof and the slabs suspended from this upper grid are, in an almost metaphorical way, evocative of musical rhythms. These elements are not just a means to fulfill structural requirements and resolve problematic thermal transitions – they are also components that have been consciously placed in relation to one another to create a rhythmical composition that refers to the activities of the client.

Expansion of the Swiss National Museum, Zurich 2013–2016

An intensive collaboration between architect and engineer is also necessitated for this project by the space-defining structure. The U-shaped museum, designed and built in 1898 by Gustav Gull, was renovated more than one hundred years later and overhauled to meet current standards and technical and architectural requirements. The museum is now moving forward with the construction of an extension that establishes a U-shaped counterpart to the existing building, based on Christ & Gantenbein Architects' winning submission to the 2002 competition. The annex building is joined to the existing building in such a way as to create a closed circuit linking the old and the new, and it forms a bridge – the central element of the project – that establishes a spatial and visual connection between park and museum.

The building, which has no visible joints, is a snake-like form of varying cross-section that is bent three times in plan and five times in section. What initially appears to be an architectural form that visualizes the visitor flow is at the same time a load-bearing structure that reveals its flow of forces: where the building meets the ground – at the feet of the bridge, as it were – the forces flow into the earth. These points of bearing are the building's fixed points. From here, the building will expand due to shrinkage of the concrete and differential temperature variations.

Where the building arches over the park like a portal, it is a bridge structure. This can be understood as a vector-active three-hinged truss, in which the inclined soffit surfaces defining the portal frame function as compression struts and the foundation slab is the tie rod. The prestressing of the foundation slab causes the three-hinged truss to rise – meaning the formwork and bracing are artificially relieved. The lateral, ascending perforated walls thereby stabilize the compression struts.

Where the annex joins the existing building, the girth of the new building is tapered. The structural requirements in this area are consciously reflected in the architectural design: the prospective docking site projects out to prevent forces from being transferred to the existing building, thus avoiding costly measures that would otherwise be needed to reinforce it. The concrete slabs wrapping around this projecting "beak" constitute a tube that is resistant to bending, shear, and torsion.

The concrete facade of the new building is intended to pick up on the character of the old building's stone facade and reinterpret it through the admixture of tuff as a medium-grain aggregate. Together with the concrete producer, research into a suitable grading curve with a balanced water-cement ratio is being made so as to develop a mix that guarantees a concrete with sufficient loading capacity and good workability for ease of placement on the construction site. The physical behavior of this concrete and the characteristic values resulting from the concrete testing are eminently suited for the design and execution of the joint-free facade. The flow of forces in the supporting shear walls is the decisive aspect defining the possible positions of the perforations, which were jointly developed in an iterative process between architect and engineer.

1 Peter Rice, *An Engineer Imagines,* London 1994, p. 17.
2 See Sigfried Giedion, *Space, Time and Architecture. The Growth of a New Tradition,* Cambridge 1946, p. 24.
3 Le Corbusier, "Le Corbusier et les constructeurs" in: *Science et Vie* 98 (August 1960), vol. 515, p. 3, transl. in: Olga Popovic Larsen, Andy Tyas, *Conceptual Structural Design: Bridging the Gap Between Architects and Engineers,* London 2003.
4 Ibid.
5 August E. Komendant, *Contemporary Concrete Structures,* New York 1972, p. xi.
6 Aita Flury, "Inquisitive Openness," in: Flury (ed.), *Cooperation: The Engineer and the Architect,* Basel 2012, p. 15.
7 Cecil Balmond, *Informal,* Munich 2007, p. 59.

Laboratory Building WSJ 352, Novartis Campus
Basel 2010

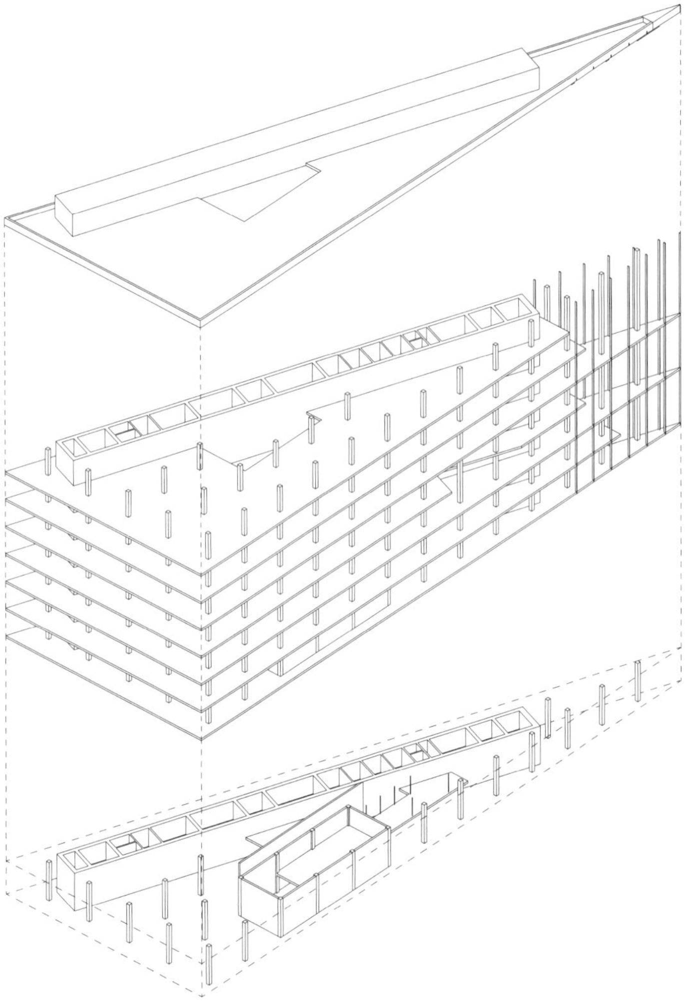

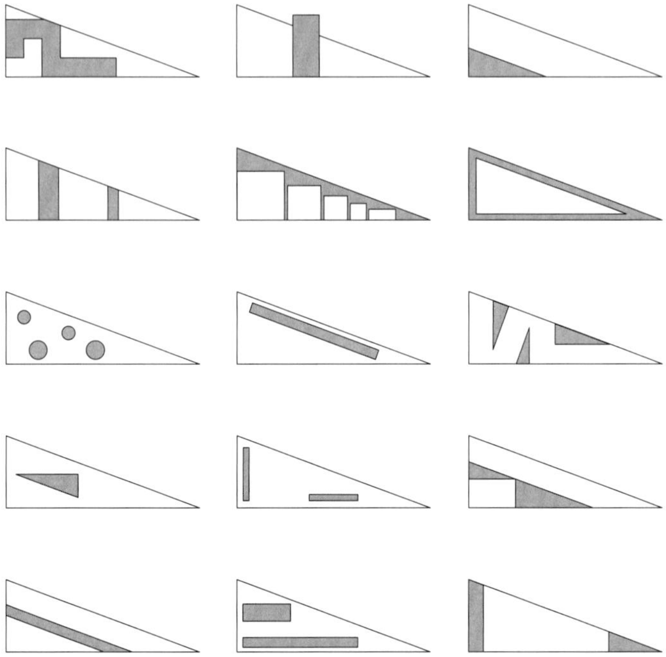

← Erecting the building tip
↑ Isometric view of the load-bearing elements
↗ Evaluation of possible core geometries
→ Slab reinforcement and prestressing of upstand beams at top floor
→→ Suspended structure at the building tip

1 On Collaboration of Engineers and Architects
Laboratory Building WSJ 352, Novartis Campus

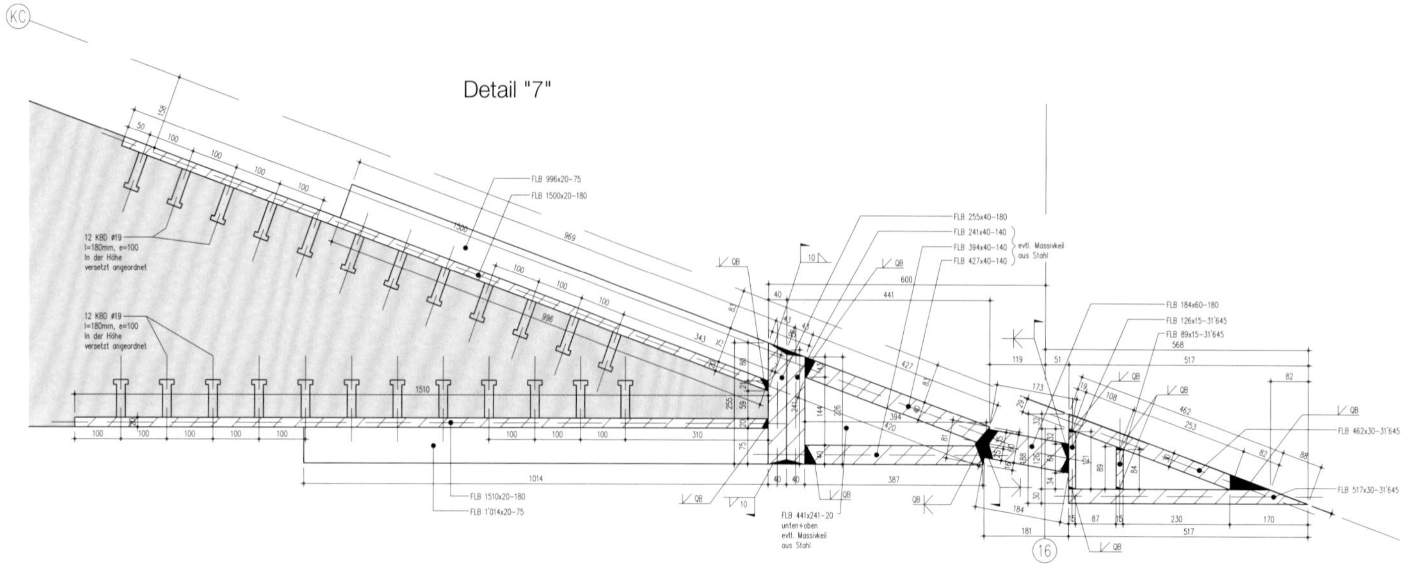
Detail "7"
FLB 996x20-75
FLB 1500x20-180
12 KBD #19
l=180mm, e=100
in der Höhe
versetzt angeordnet
FLB 255x40-180
FLB 241x40-140
FLB 394x40-140
FLB 427x40-140
evtl. Massivkeil
aus Stahl
FLB 184x60-180
FLB 126x15-31'645
FLB 89x15-31'645
FLB 462x30-31'645
FLB 517x30-31'645
FLB 1510x20-180
FLB 1'014x20-75
FLB 441x241-20
unten+oben
evtl. Massivkeil
aus Stahl
QB

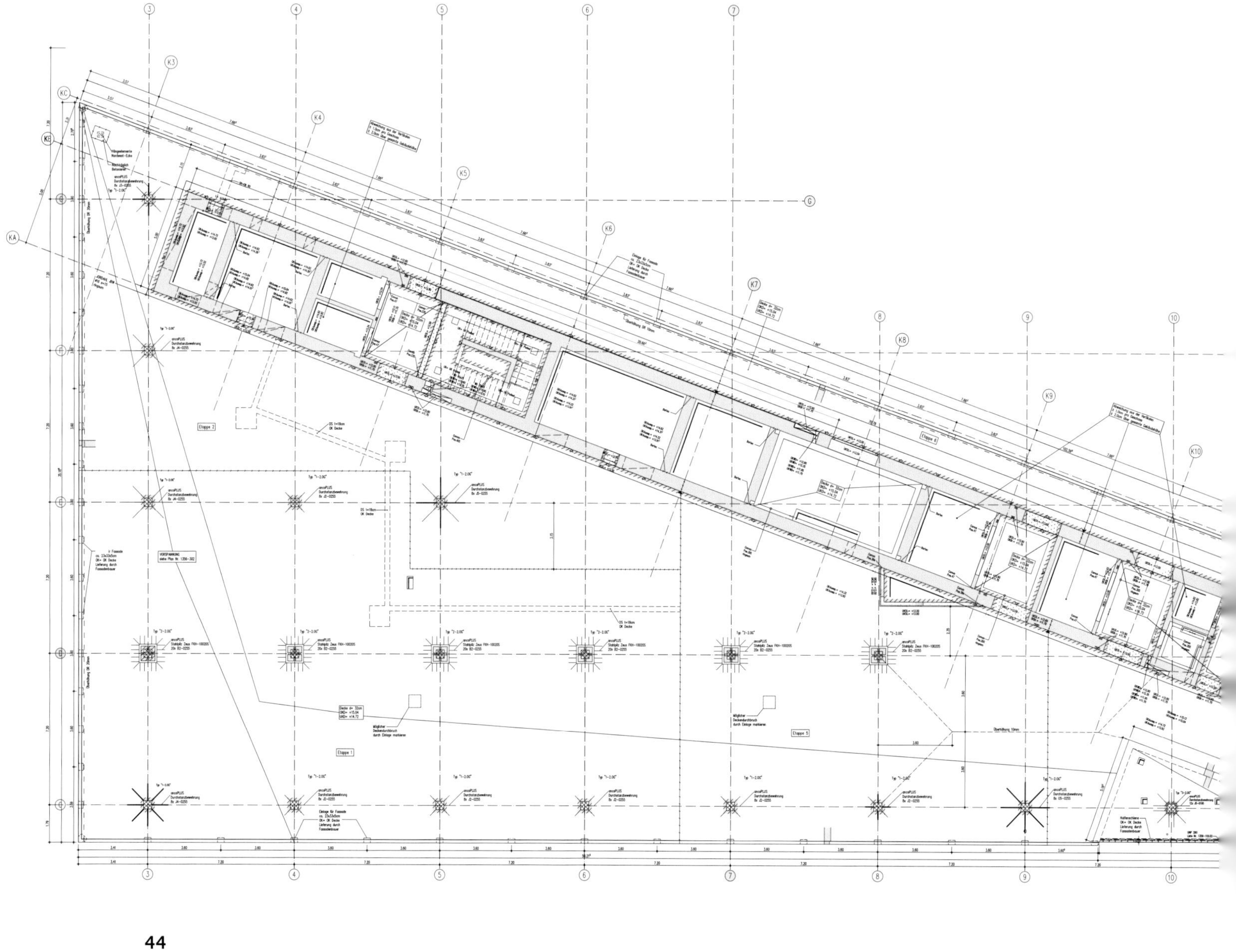

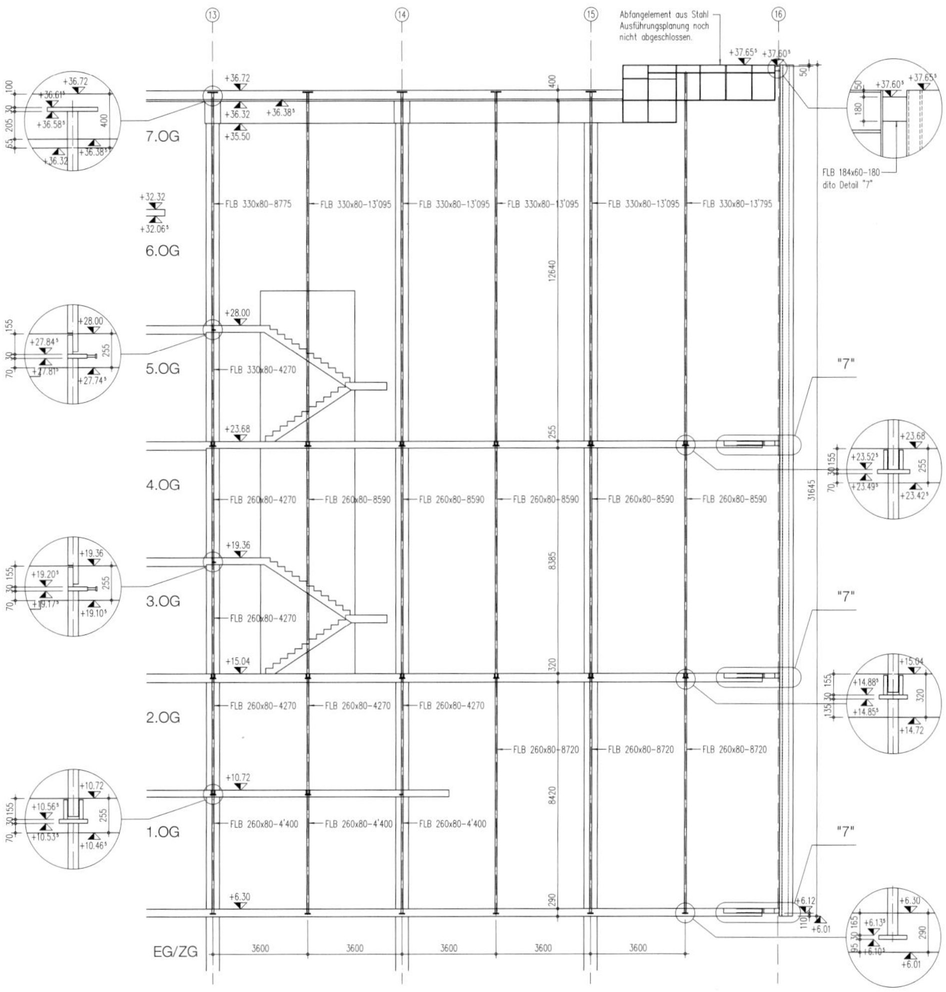

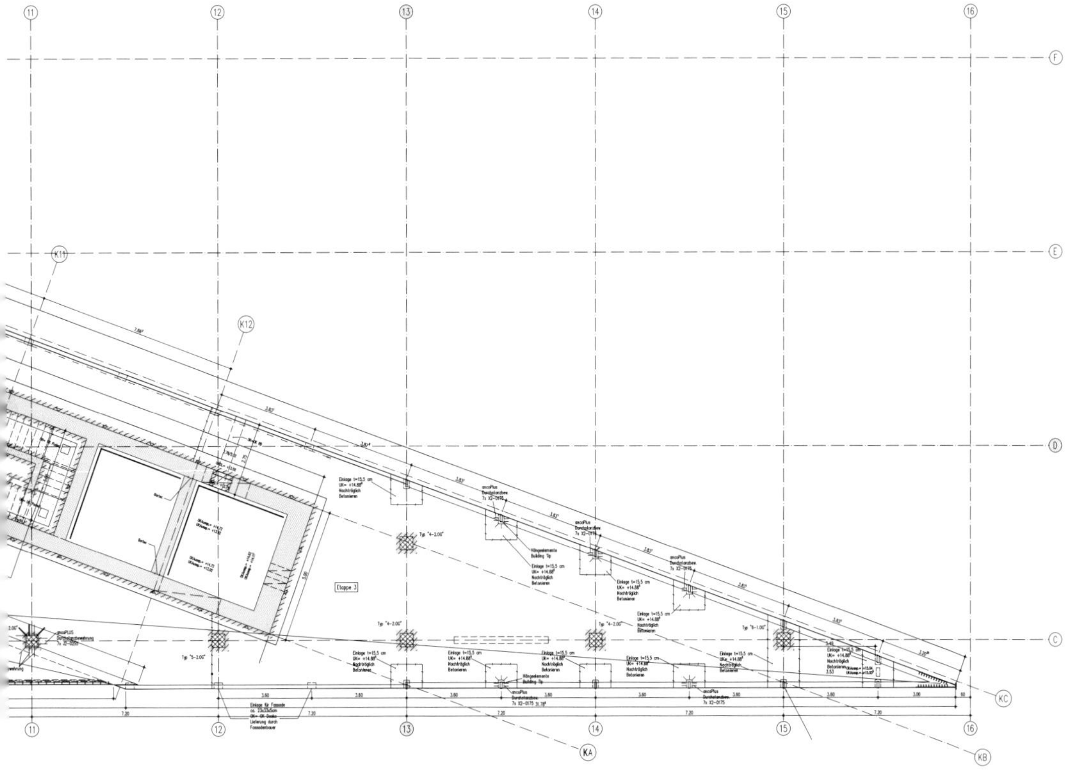

↖ Detail drawing of the steel building tip
← Plan of the top floor slab
↑ Suspended structure of the building tip
→ Interior view of the building tip
→→ Exterior view of the building tip

1 On Collaboration of Engineers and Architects
Laboratory Building WSJ 352, Novartis Campus

Hertenstein House
Weggis, Switzerland, 2012

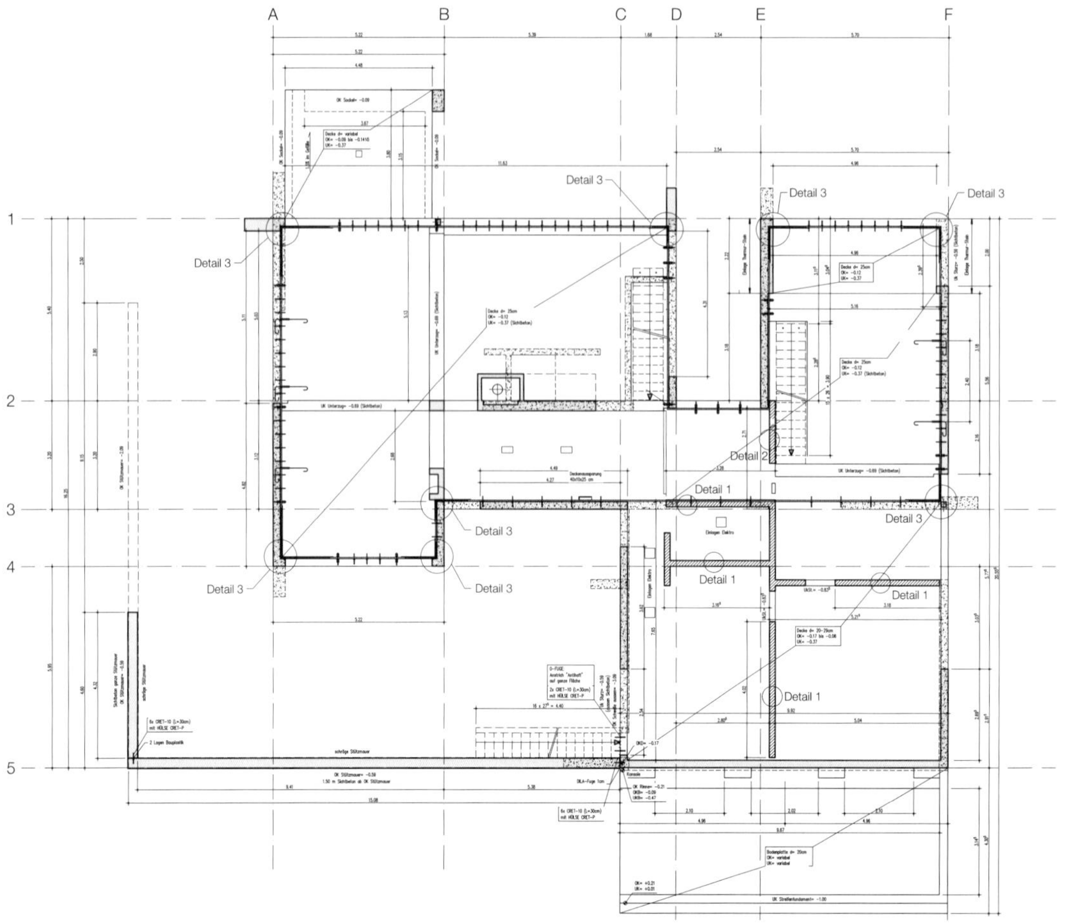

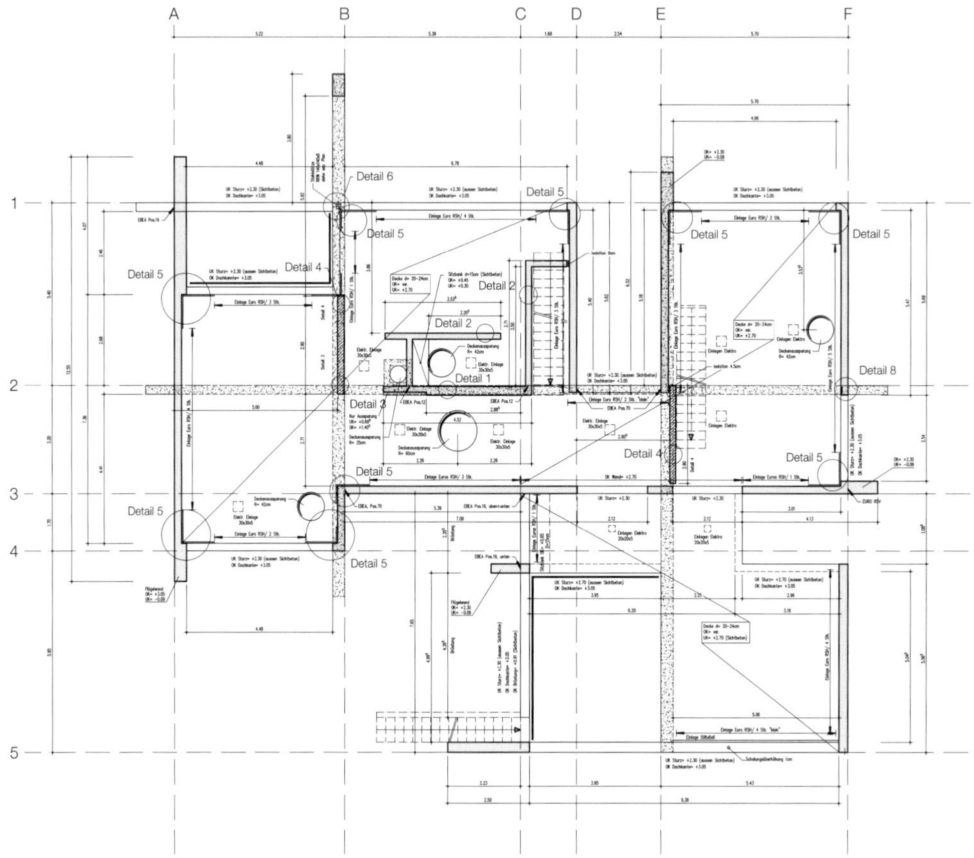

← Atmospheric image of the construction site
↗ Formwork drawing – ground floor
→ Formwork drawing – first floor
↗↗ Construction details of junctures between wall and slab
→→ Exterior view of the single-shell, load-bearing exposed concrete structure

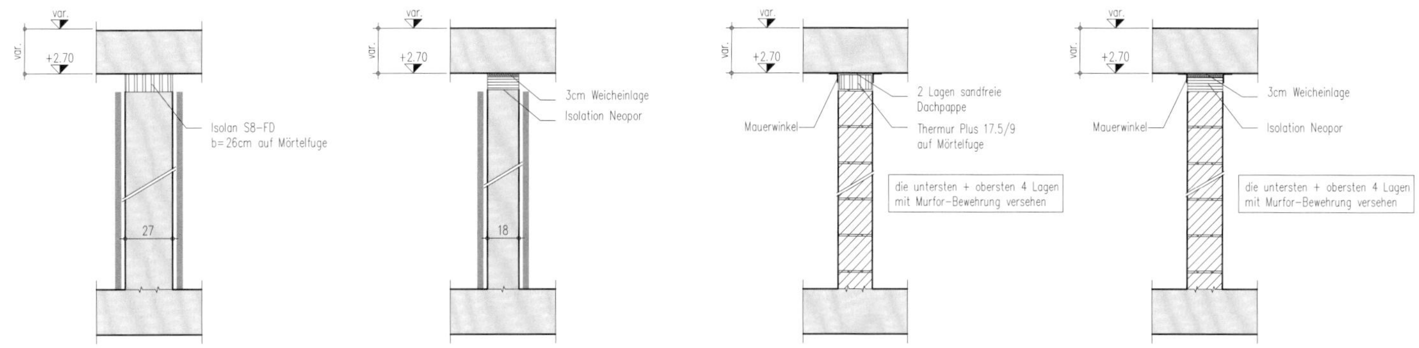
var.
var.
+2.70
Isolan S8-FD
b=26cm auf Mörtelfuge
27
var.
var.
+2.70
3cm Weicheinlage
Isolation Neopor
18
var.
var.
+2.70
2 Lagen sandfreie
Dachpappe
Mauerwinkel
Thermur Plus 17.5/9
auf Mörtelfuge
die untersten + obersten 4 Lagen
mit Murfor-Bewehrung versehen
var.
var.
+2.70
3cm Weicheinlage
Mauerwinkel
Isolation Neopor
die untersten + obersten 4 Lagen
mit Murfor-Bewehrung versehen

Expansion of the Swiss National Museum
Zurich 2013–2016

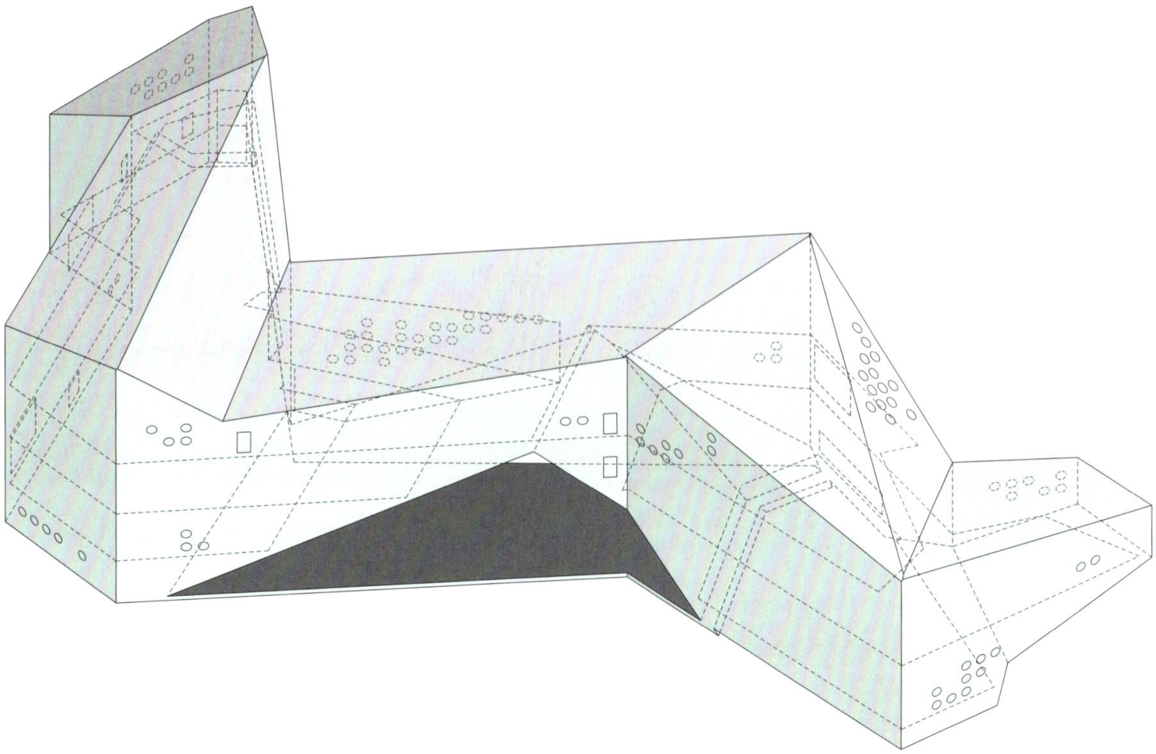

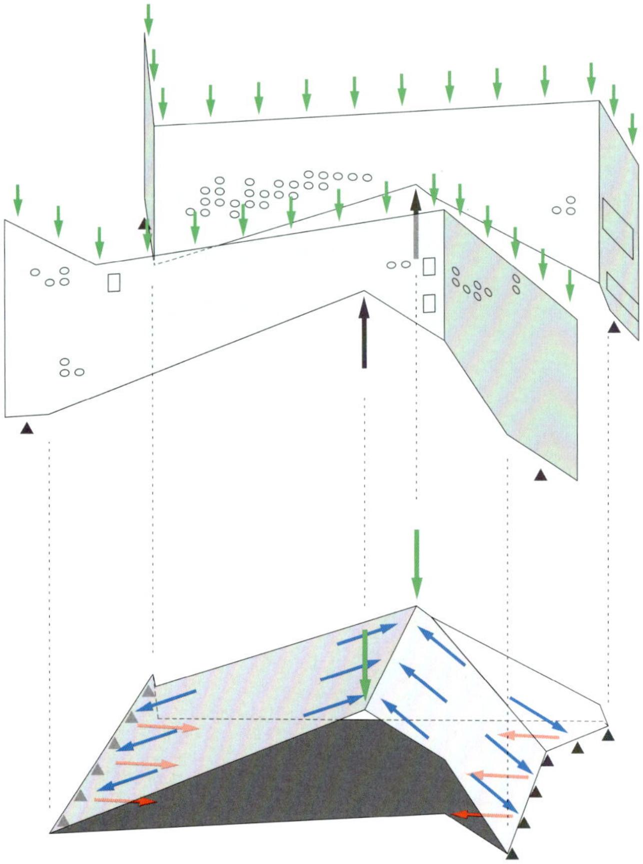

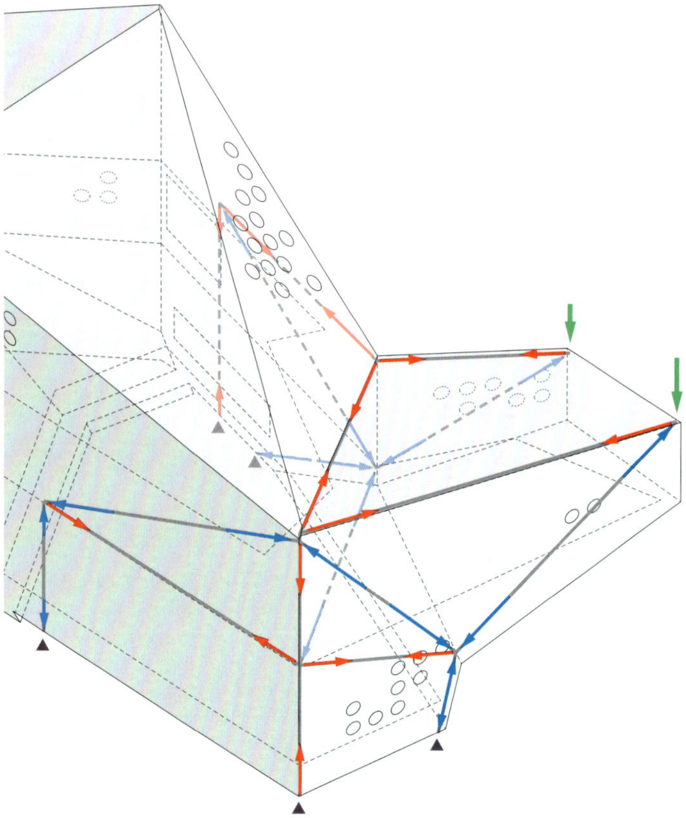

←← Full-scale model of the exposed concrete facade
↖ Portal-like built form as a vector-active three-hinged girder
← Docking site at the existing building, with torsion-resistant tube effect
↑ Model of the Swiss National Museum with the U-shaped annex that is varied as needed in plan and section to fit to the existing building
→ Concrete surface with volcanic tufa as aggregate

2 Development of Structural Concepts

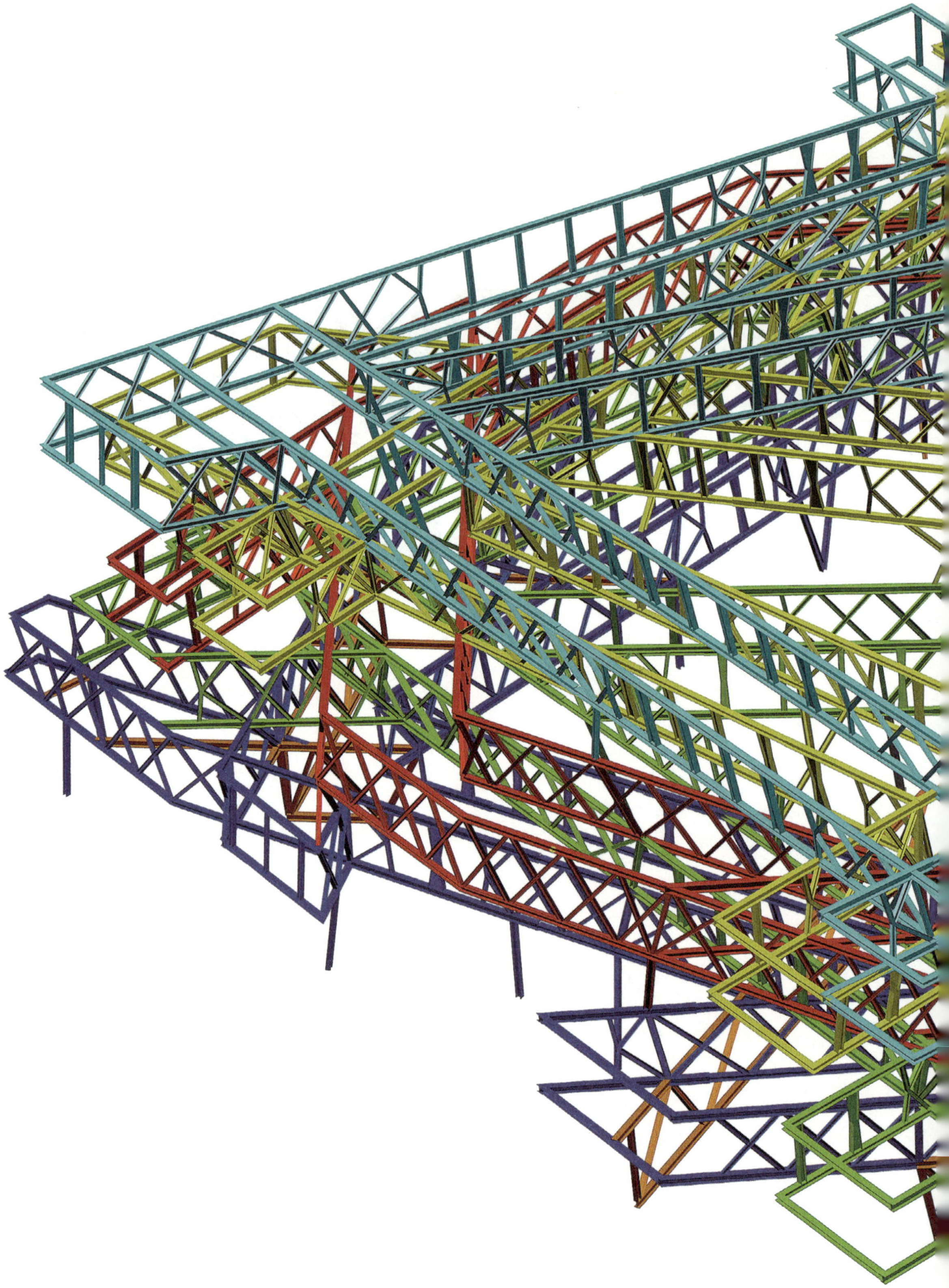

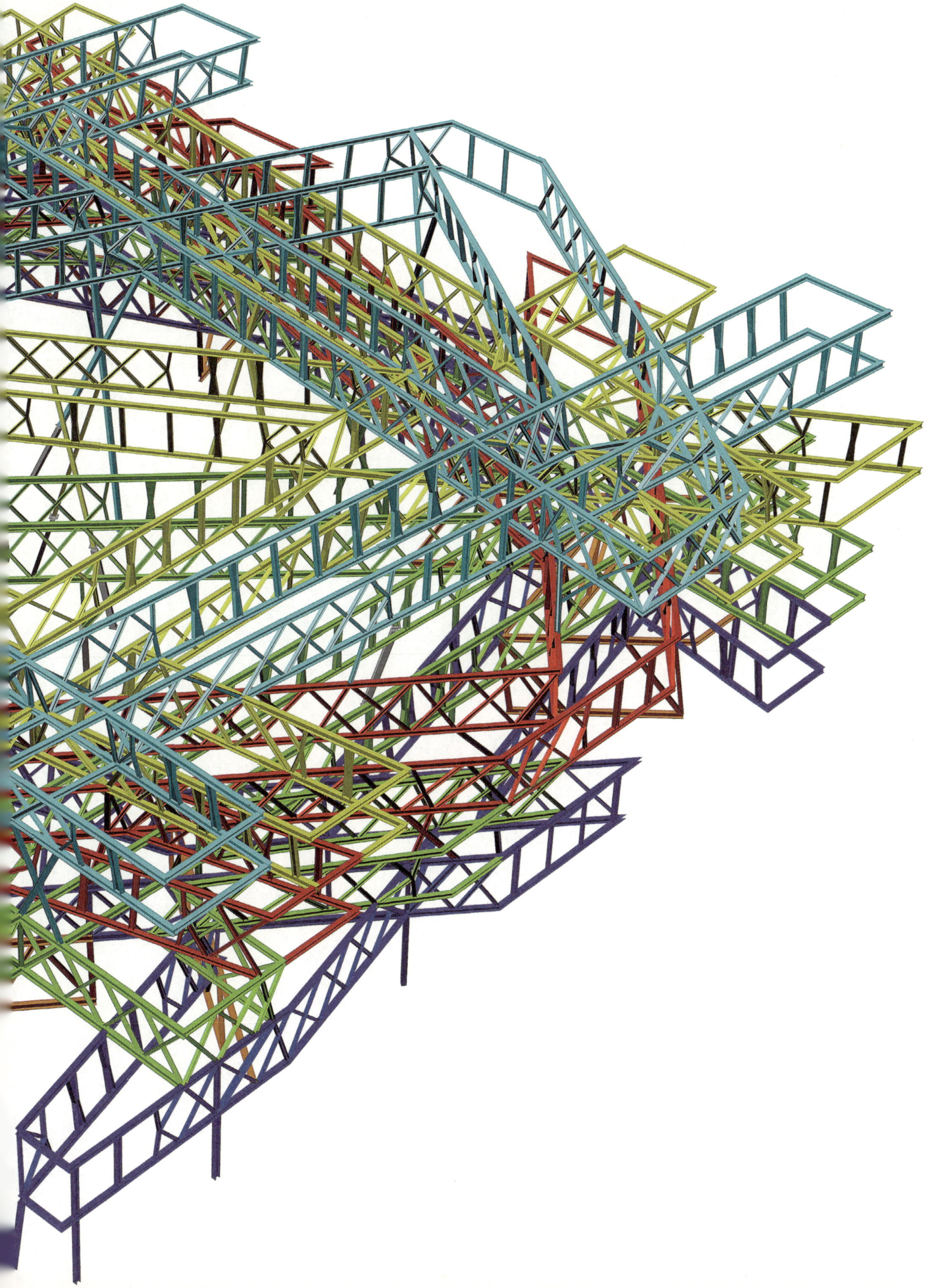

← Steel structure of the Actelion Business Center in Allschwil (computer model)

Conversations II

Where Do the Forms Come From?

AF As an architect, what interests me most of all is of course the form-finding of load-bearing structures – *where do the forms come from* – and how do your designs come about? In our informal discussions on this subject, Heinrich, you have often mentioned the notion of the "basic condition." Can that aspect really spark the productive act?

HS Architects are interested in spaces and surfaces, which they derive in a creative act from the client's requirements, from the spatial program. As an engineer, I think less in terms of spaces and surfaces, and more in terms of structures, or rather, structural elements. In so doing, what interests me are the force path and how to manage it as economically as possible. So thinking about it that way, spaces are simply structural constraints.

AF Tivadar, in our meetings you have often spoken about visual considerations that guide you when designing. Your statements had a tinge of associative architectural thought.

TP For me, design is an intuitive process – influenced, of course, by the knowledge and experience we have acquired. We have basic tools at our disposal, inasmuch as we know how to dimension a beam, a shear wall, etc. Building upon that, there are combinations of systems and principles like stacking that we can envision with the help of those basic skills. Novel spaces can also be the result – and the formation of space in general can be supported that way. In some of the projects we have done together with Herzog & de Meuron, for example, the space is actually formed by the structure itself, as demonstrated most clearly perhaps by Prada Aoyama in Tokyo.

AF So visual notions don't play a role for you?

TP Of course there are reference images – for instance, I can draw the bearing point from the University of São Paulo from memory. But what you are presumably addressing is contextual thinking – an analysis of the surroundings that, for instance, leads to a certain choice of materials. An industrial context might, for example, lead to designing an industrial bridge made of Corten steel.

SB My approach is to first lay out the parameters, such as requirements for sound insulation or fire protection, desired span lengths, possible construction heights and the like. The thinking that Tivadar has just outlined pertains more to special jobs, such as a stadium roof. In that case, no dynamic aspects must be considered or fire protection requirements met, meaning that, as opposed to situations with technically oriented or user-specific requirements, the design approach can have more importance.

TP The roof of our competition entry for the Hardturm Stadium in Zurich is a good example for outlining our design approach: The potential of a fifth facade inspired us to fold the roof, which is then continued on the facade. The folded surface cleverly incorporated solar panels at the ideal angle and, at the same time, was laid out with regard to the roof drainage. In that sense, it's the typical kind of integrally designed structure that we as engineers appreciate.

AF In the discussions I had with engineers on occasion of the "Constructors' Dialogue," I repeatedly ask the same question. Specifically, I'm interested in why engineer-architects – masters of space and structure in one and the same person, as exemplified by such twentieth-century luminaries as Nervi, Isler, Tedesko, Candela, etc. – no longer exist today. Why is it an anachronistic notion to believe that engineer-architects, who are guided by their own notions of form and effect, can generate form from the principles of design and construction in a shifting balance between the roles of contructor and entrepreneur?

HS The notion still exists, but the road is filled with hurdles. First you have to make a name for yourself and show that you have what it takes, like Calatrava has demonstrated. Looking back to Nervi goes too far – that was another era.

AF What was so fundamentally different?

HS Back then, engineers were viewed as having spatial competence, and today they are not seen this way – due to their education.

AF Doesn't it primarily have to do with the fact that these constructors also worked as contractors, who perfected the structures they were building in order to be economically competitive at all?

TP That was crucial. As entrepreneurs, the production process for Nervi, Isler, etc., was key to their search for form. But today that no longer has the same significance.

HS The production process still plays a key role today, but it's somewhat different: Earlier, a production process could be quite labor intensive, but it was always important to aim for economy of material. Today, a production process can be wasteful of material, but it's necessary for it to employ labor efficiently. The reason for this change is that lightweight, fine structures are no longer popular; in demand are "blocks" that can be piled on top of one another.

SB Added to that are, of course, also increased normative requirements and immensely greater interdisciplinary complexity (such as MEP systems and building physics), which make it impossible for an individual to acquire all the knowledge in sufficient depth.

AF Do you work with structural hypotheses and if so, how do these emerge? What do you draw upon? How does the idea relate to the concept? Methodology

TP From an initial idea we try to develop a concept from it, and structural hypotheses are the tools needed to do so. They can, for example, also include statements about the implications of different materials.

HS But the structural concept and the material always go together, they are not simply transferable.

TP Let's take the concept of a "cantilever" for a balcony slab – a routine task, so to speak. We can prefabricate it in wood, steel or reinforced concrete, or construct it on site. But the concept remains the "cantilever."

HS My statement pertains to overall structural concepts. These are always tailored to a specific building material.

SB The choice of materials usually crystallizes relatively early as a product of the basic conditions and the defined task.

HS In bridge construction, for instance, the selection of materials based on structural hypotheses is perhaps more common. I can, for example, think about how it would be to construct a steel arch as opposed to a concrete arch – often a complex thought process.

TP That applies exactly the same to buildings.

AF So you simultaneously investigate different materials that seem appropriate to you.

TP Yes, that's true at least through the design development phase. When sending the project to bid, at the very latest, you should know what material you want to use.

AF How would you describe the relationship between preliminary estimation and final calculation of a structure? What role does the processing power of the computer play?

HS Apropos working methods, my training gave me some interesting experiences: At the technical college, I had learned to comprehend structures and their flow of forces intuitively. At the ETH, that became somewhat muddier – there we were trained to rely on analytical calculations and assessment. It was not until I worked as an assistant for Professor Menn that I understood how the two approaches can be combined with one another. My approach today is generally very intuitive. I start by making a few small calculations by hand, which are then checked by computer. They're usually pretty close ...

SB Experience plays a big role in structural design. After ten or fifteen years, you have calculated every combination of loads with all kinds of geometries and dimensions – and developed a good feel for the forces along the way. Thus, an experienced design engineer also has a feel for dimensions and proportions. I only consult the computer in exceptional cases, when I'm not sure if something works. The processing power of the computer primarily comes into play for the construction calculations, which are usually made by the younger engineers. But if, for example, completely different reinforcement cross sections are suddenly calculated, we naturally question these again on the basis of our experience and intuition.

HS The possibilities for performing calculations on the computer these days are astonishingly diverse and effective. We should keep in mind: Heinz Hossdorf [the founder of WGG] could still approach a complex problem with just a model that was loaded and then used to measure the deformation. One of our major challenges today is the fact that the assumptions made for the computer calculations involve a great amount of uncertainty. So it is crucial for young engineers to understand that the computer does not per se make correct calculations – because the initial assumptions are tenuous. The validity of a result always depends on whether the initial assumptions are apparent. The engineer must therefore be capable of interpreting the result.

TP The assumptions are such things as methods of storage or material properties. What is actually important is to not rely solely on the seemingly accurate four-digit computer results, but to always also trust your own thoughts.

HS I just recently had an interesting experience with the large roof over the auditorium at the Elbphilharmonie Hamburg: Subsequent measurements of the building have shown that our calculations were correct – except for one support, where our load assumption was off by thirty percent. On the whole, this was not a problem, but due to the production process, higher loads were incurred in this one support. After the first cracks appeared and the concrete structure was in the end lowered completely, it fits again perfectly. In other words: workability and the whole always have priority with structures.

AF Similar to the way Nervi established a democratic distribution of loads and forces in his buildings.

HS Exactly.

AF How many stages does a project go through from a competition to the constructed building and how much do the structures change in the process?

TP Seen chronologically, there are five stages: competition, preliminary design, design development, bidding and construction. The financial feasibility is analyzed from the beginning and the analysis is continually refined in each planning phase. We use all available means to hold onto the guiding ideas developed during the competition, because they are central to any design. Core ideas need to evolve and be strengthened from phase to phase.

HS For a competition, the goal is generally to design very pure, statically beautiful structures. As the planning process continues, it's a constant struggle to keep these structures as consistent as possible through to their realization.

AF Don't you ever amend fundamental decisions?

SB On occasion. Ultimately, we have an obligation to the client to achieve the most ideal and most efficient solution for his needs. So we must also study variants and alternatives.

AF Can you give us an example of where, viewed over the course of the entire process, crucial decisions were overturned?

HS The Roche Tower is one such case. The "Two Spirals" were rejected just before things got serious. Structural considerations unquestionably played a role – the geometry and the means needed to build it were very challenging and thus would have been costly. But the crux of the matter was certainly the fact that, as of a certain point, Roche could not identify with a tower of that sort.

Tradition
Innovation

AF To what extent are there lines of development in your projects, in other words, correlations between the previous building and the next one? Do you aim for *continuità* in your work?

HS We have various "works-in-progress". One of them, for example, is the slab and plate system that we continue to develop in different buildings. For steel construction, too, we are continuously working on conceptual methods. In addition, we focus on lines of development for high-rise buildings and monolithic, load-bearing facades.

TP To discuss developmental correlations in our work, it is definitely necessary to think in the categories that Heinrich mentioned, that is: structural systems, construction materials and building types. But we are especially interested in structural innovations. On an industrial building for Ricola, for example, we are currently exploring the assembly of prefabricated, prestressed building elements.
And in current projects for high-rise buildings, we are investigating the possibilities presented by various typologies, such as layouts with eccentric cores. Of course, we are also interested in new building materials – but glass or carbon-fiber buildings tend to be the exception, and should be assessed more in terms of material innovation.

HS Our work is not focused on *one* theme. On the contrary, we have a firm with fifty employees who work on various themes. Moreover, we also work on buildings with no specific theme, and ones that have no developmental ambitions.

AF So continuities have arisen, but you are also guided by an enthusiasm for experimentation. Are your experiments within the established realm, or do you see yourselves working in the pioneering tradition of Heinz Hossdorf?

HS Our approach is very different from that of Hossdorf. Elasticity and preventing fissures were always his main concerns. Today, we have a "better" understanding of reinforced concrete, and, therefore, we can also design different buildings. The experimental themes cannot be ascribed to us as a firm in general. The experiments are all very different in nature and are related to the focus and interests of the respective lead engineer. But, of course, we influence one another through our mutual exchange of information.

TP The experiments that Hossdorf was still making on physical models are the same ones we make with digital models. Today's trials are not comparable with earlier experimental arrangements. We make smaller material analyses or examine the top floor of a high-rise building with regard to the comfort characteristics contingent upon oscillation. But nowadays, these are common analyses, not pioneering experiments.

SB Of personal interest to me are old, sometimes extremely delicate, ribbed concrete structures with slender structural members. For buildings with no fire or sound insulation requirements, it's appealing to approach those slim dimensions.

HS To illuminate the relationship between lines of development and experiments, I would like to outline an experience from the Forum 2004: The roof is a triangle, with sides measuring approximately two hundred meters each, that stands on eighteen points – a composite structure about four meters high. While we were designing it, the question arose as to whether the area between the points is like a beam, or if the points should be thought of as mushroom-head columns. In terms of the concept, I decided to view them as beams. But in retrospect, after going through the process of building it, I discovered that it would have been better to consider it as a mushroom structure. This large-scale experiment eventually led to an umbrella slab – that is, an inverted, point-supported slab – for the Caixa project in Madrid. Lessons learned from experiments affect the next project and influence lines of development.

AF In his introductory text "Constellations,"[1] Guy Nordenson writes that "engineering is necessarily typological." Is designing typological for you? Typologies

TP Typologies have been part of the art of engineering since the beginning. To stabilize a building, for example, we need three wall planes. That is a basic typology that exists and can be developed further.

HS There are developments in the way elements are put together and how these mutually function together. Slab and plate systems, for example, are much more prevalent than fifteen years ago. This is due to the fact that in Switzerland today, prestressing is easily employed in building construction. But in Germany, we have already encountered great resistance to a prestressing proposal. People are apprehensive of the unfamiliar. As far as construction is concerned, Switzerland is less conservative – even the builders here are quite willing to enter new territory.

AF An over-regulated environment prevents the implementation of new ideas and, thus, prevents innovation.

HS That is currently a highly observable development, especially in the USA and in Germany.

TP Returning to typologies: with bridges, you can certainly speak of types. This is also reflected in the approach to design, inasmuch as these bridge types are interpreted and you consider which one, such as an arched bridge or a cable-stayed suspension bridge, makes the most sense. With buildings, the typological decisions pertain to categories like frame or solid construction. Of course, there are also hybrid structures, but engineers enjoy the so-called "pure" structures most of all.

AF So are all the possible typologies already invented?

SB Certainly the basic typologies. But, of course, there are many opportunities for combining them – there are countless hybrid structures. Monolithic facades are further developments and combinations in which, for example, the inner shell carries the loads in one case and the facade does so in another.

HS Whether all typologies have been invented is an interesting question. With bridges, new construction methods and new building materials have led to new types. Thinking about production and building materials can, in other words, lead to new typologies.

1 See Guy Nordenson, *Seven Structural Engineers,* New York 2008.

The Engineer as Co-Designer

Daniel Meyer

The spectacular new business center for Actelion in Allschwil is composed of thirty-four prismatic forms that are set on top of each other to reach a height of up to twenty-two meters and cover an area of approximately eighty by eighty meters. The building is iconic and reflects the requirements for self-representation. Although the volumes are stacked casually and evidently without system, they nevertheless appear to be set precisely and thus convey the impression of sitting securely, stably and without forced effort. Upon realizing this, at the very latest, one is compelled to look closer at what Schnetzer Puskas Engineers have accomplished here.

The idea of stacking office units that span up to 40 meters, which was refined together with the architects, requires a supporting structure with a logic specific to the system. This is where the collaboration between the structural engineers and the architects begins to intensify: The design becomes a shared challenge. But this is about much more than merely making an architectural idea possible. Through their understanding of the spatial intentions and their tangible interest in architecture, Schnetzer Puskas Engineers are able to identify with the task, enabling them to meet the many challenges, some of which are truly exciting, with enthusiasm and innovative prowess. Issues of forming space, the placement of openings and how the structure is perceived are examined and discussed on equal footing with the many questions of technical and economic requirements placed on the structure. Thus, the design gains its structural clarity gradually and through a clear process – and this open and process-oriented attitude toward design problems characterizes the work of Schnetzer Puskas.

In this project, the structural challenges are of a kind that is mainly found today on the international stage. Digital technologies are gradually establishing themselves among architects as useful design tools, too, and by now already significantly influence the form and appearance of buildings. Highly complicated geometries for volumes and surfaces are becoming manageable and, thanks to parameterization, remain malleable and adaptable. In response to ever-malleable designs, structural engineers need adequate problem-solving approaches. These are, however, not found exclusively in the computerized structural tools that promise to solve everything, but rather in the strategy of disentangling the design to reveal a determinative and generating core that is based on simple and clear structures. These can then be described and explained comprehensibly with the fundamental laws of mechanics. This strategy of systematization is recognizable in the structural design solution for the Actelion building. Upon closer examination of its supporting structure, the following can also be noted:

The support system is ingenious, efficient and economical.[1]

Due to requirements of use and economy, a support system consisting of hybrid beam-like elements stacked on top of each other was designed. These "beams" are a skillful union of a Vierendeel truss and a conventional truss. In the areas of high shear forces – primarily at the bearing seats, where the individual beams are both structurally and functionally interconnected – cross-shaped panel divisions analogous to components of a truss are utilized. Toward the middle of the beam – where the bending moments increase gradually, the shear forces decrease and, in addition, the intended use calls for the greatest possible transparency – the struts used are all vertical, and in combination with the flanges of the beams they comprise a structural frame. The structural design was further optimized so that, by varying the spacing of the panel divisions, the stress in the chords is kept constant over their entire length, and hence the cross-sections are used to full advantage.

The overall structural concept is compelling for its clear-cut vertical and horizontal load transfer.

The stacked, beam-like steel elements generally direct the vertical gravity loads through their points of contact with each other along the most direct path, that is, from top to bottom with no complicated diversions of the forces. This has been attained by optimizing the local geometry of the panel divisions and the global location of the beam elements. At those few points where this direct mode of load transfer was not possible, the room-high trusses act as efficient transfer beams. The building is not braced in the ordinary way, using the rigid walls of the building's vertical circulation cores, but mainly through the panel divisions of the stacked beam elements. The horizontal earthquake and wind forces, together with the horizontally acting deflection forces of the inclined column elements that contribute to the vertical load transfer, are taken up by the trusses in concert with the rigid roof and floor construction. The stacked structural tubes, which each consist of two vertical, infilled trusses and horizontal roof and floor slabs, act together as a stable folded structure that is capable of carrying the entire load range with no additional bracing elements. The vertical circulation cores, which are usually always rigid, can thus be positioned freely in the building, thereby increasing flexibility and variation in the interest of sustainability.

The steel structure is clad, but it remains perceptible.

The way steel is used as a construction material is also interesting. The recurring discussion about revealing the steel in its manifold forms of legibility and in its tectonics is countered in the Actelion building with the opposing and thoroughly pragmatic approach of uncompromisingly wrapping the steel. By doing so, fire protection is assured in the simplest way possible and, in addition, the construction elements can be joined together economically. Regardless, the construction material steel remains perceptible: above all, the building's bold cantilevers, long spans and high percentage of openings still point to the steel construction.

In-depth examination of the structure clearly shows that Schnetzer Puskas Engineers have dealt with the challenge intensively, addressing all the facets of the structural design – and that profound and sweeping analyses of the structural and static modes of action were carried out. As an active participant within a large interdisciplinary planning team, they have set distinct but restrained traces and accents. Thus the simple principle of stacking spatial elements, together with the structure that was developed, ultimately comes across as a multi-faceted, complex and functionally differentiated spatial entity that, as mentioned at the outset, is perceived as balanced, self-evident and serene.

With regard to the design of buildings, Schnetzer Puskas Engineers perceive themselves primarily as equal partners of the architects and co-authors who are, on the one hand, capable of reading and understanding an architectural design and, on the other hand, can also actively participate in designing it to create a successful final concept. The approach used by Schnetzer Puskas Engineers is characterized by great openness. Strict dogmatic commitments to construction principles are alien to them. But an unambiguous means of load transfer, construction appropriate to the materials and an adherence to economic constraints are characteristics that can always be identified in their projects, even those that appear somewhat unprecedented, like the Actelion building. With their work, Schnetzer Puskas epitomize the modern structural engineer, who is open to new ideas and who, as an author, can show restraint or be demonstrative, depending on the task at hand. Thus, the working and doing of Schnetzer Puskas can unquestionably be seen in the tradition of exciting collaborations between structural engineers and architects as it was practiced by Pier Luigi Nervi and Gio Ponti, for example, or Dr. August E. Komendant and Louis I. Kahn.

1 For details on the subheadings, see: Heinrich Schnetzer, "Wirkungsvoll kombiniert" [Effectively combined], in: *Tec21* 3–4 (2010), pp. 2–6.

Development of Structural Concepts

Heinrich Schnetzer

Structural concepts are always developed contextually. For the engineer, the flow of forces, the material stress induced within the chosen building material, the distribution of that material, and the production process are thereby of great significance. These properties of a load-bearing structure, which are fundamental for the structural engineer, are closely interrelated with the use and the form of the building structure being designed.

For buildings, the use is defined by the architect, who translates the needs of the client into a spatial program and an enclosure. As part of the design process, the architect conveys these ideas to the entire planning team, from which a rudimentary building form emerges. For the architect, the structure is a possible means for defining or generating spaces. Sometimes, however, the structure disturbs the design process because it is supposedly not reconcilable with the design concept, the sequence of spaces, or the formation of space. In such cases, the spatial structure "collides" with the load-bearing structure – a fundamental problem arises that can only be solved through an in-depth examination and through analysis of the design concept. A design cannot evolve if the architect or structural engineer refuses to partake in the necessary dialogue, if there is no willingness to fundamentally reflect upon problem-solving approaches. Then the structure does not become part of the overall design concept and the structural engineer becomes a pure "problem solver" who can barely develop a coherent structural system for the overall architectural concept. This is usually the case when structural engineers are brought in too late in the design process, when decisions that have already been taken and which are important to the design are only reluctantly questioned.

Unlike the architect, who perceives the load-bearing structure primarily as a space-defining element, engineers understand it more as a machine: The formation of space is not so much the main focus, but rather the interaction of the structural elements, such as columns, slabs and shear walls – of the total complex of structural elements as a functioning mechanism or machine. The structural engineer always strives for internal order of the structure in terms of a harmonious whole. It must function simply and efficiently: the material needs to fulfill the requirements and the load-bearing structure needs to be relatively balanced and economically producible. In this sense, engineers do indeed have design aspirations. They focus less on the spaces or the surfaces of the structure and more on its inner qualities: functionality, efficiency, and proportions.

Geometric constraints are often the basis for the design of load-bearing structures. Geometric relationships with existing buildings, geometric restrictions resulting from the building form, or structural correlations can thereby significantly affect or determine the structural design. This is illustrated by the following three structural designs.

Actelion Business Center, Allschwil, Switzerland, 2010

The structural system for Actelion's new office building in Allschwil was developed in close collaboration with the architects Herzog & de Meuron. The building structure consists of thirty-four interlocking, beam-like forms stacked on top of each other, each of which are thirty to one hundred meters long and, depending on the office use, either five or seven meters wide. These forms are stacked twenty-two meters high on an area with overall dimensions of about eighty by eighty meters. Each form essentially consists of the floor and ceiling slabs plus two room-high trusses that are glazed to form the facades. Five to seven of these prismatic forms constitute each office floor in plan. Each floor level has its own beam layout, which does not lie directly over the story below, but

1

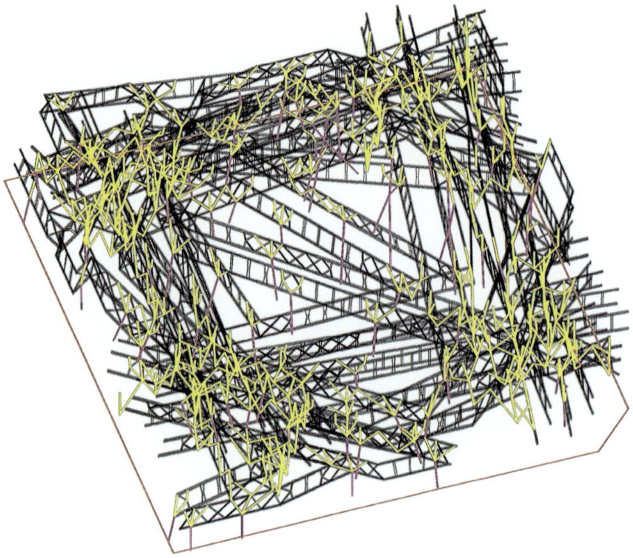

2

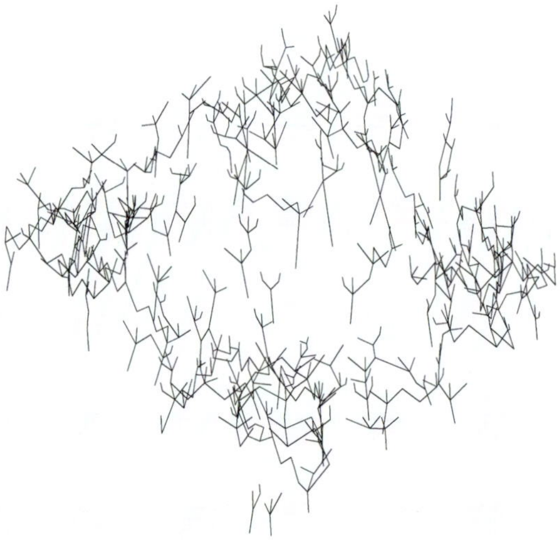

1 Model of the individual steel members comprising the trusses and columns of Actelion Business Center's load-bearing structure. Members participating directly in the load transfer are shown in yellow.

2 Depiction of only those structural members participating directly in load transfer

3 Stuctural diagrams of a hybrid truss, depicting loads and reactions, axial forces, bending moments, and deformations

transfers its loads to the level beneath via a few individual points.

To make large window openings with no disruptive diagonals for the office spaces, we developed a load-bearing system that combines the properties of conventional triangulated trusses with those of Vierendeel trusses. This enabled us to combine the advantages of both structural systems: truss members are only subjected to tension and compression; the cross-sections can therefore be designed to save material. Because the nodes are also considered to act as hinges, meaning they experience no bending stresses, they can be easily constructed. However, floor-to-ceiling trusses are generally not desirable in office buildings because the diagonals impair the view through the windows. By contrast, Vierendeel trusses, which consist solely of chords and vertical members, or posts, permit rectangular window openings, but are not very economical: The internal forces of the posts and chords consist not only of axial forces, but also shear force and bending. Furthermore, only the moment arm of the individual member's height has structural relevance – the static height of the entire truss cannot be used efficiently. This leads to a soft structural system with relatively high production costs for the materials and their assembly.

To lower production costs, a construction is needed that is as simple as possible. Therefore, despite the building's complexity, the structural system for Actelion should a have certain simplicity and feature a systematic organization. Economical

3

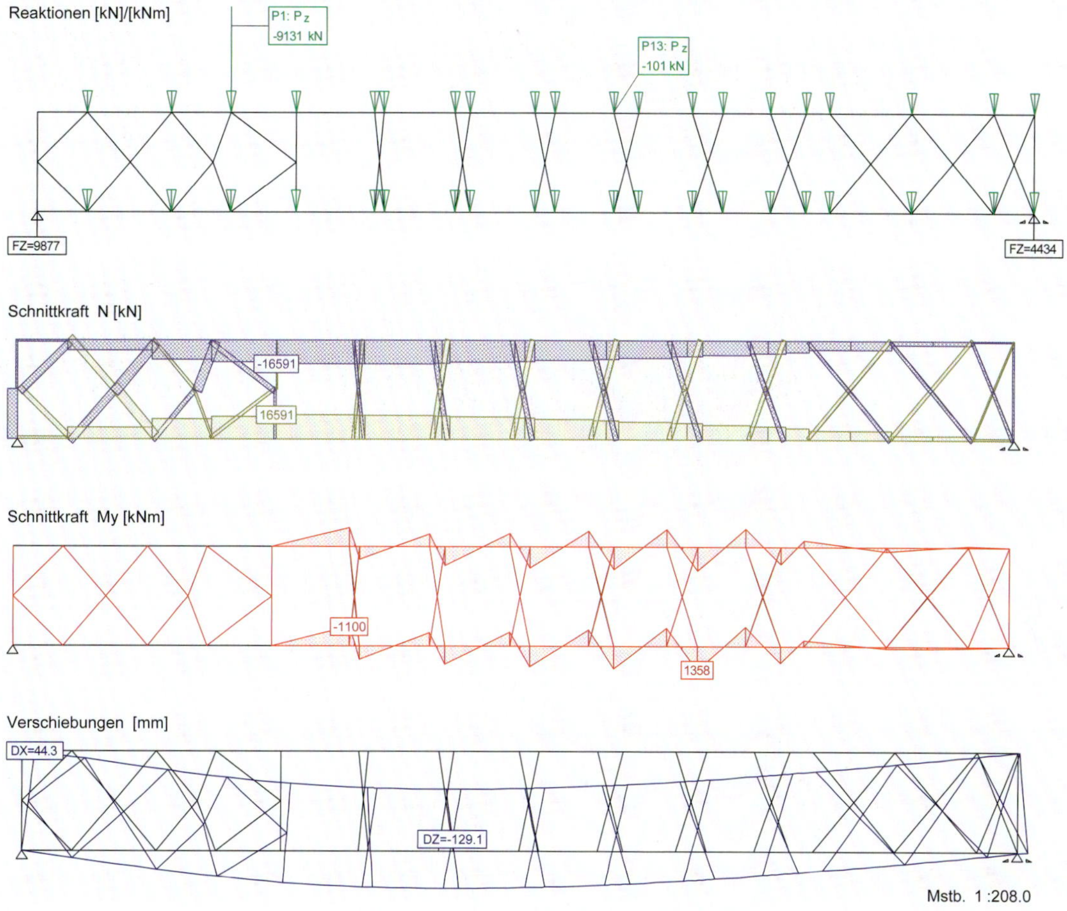

load-bearing systems require that the employed cross-sections are consistently utilized to a high degree and that the costs of production are kept to a minimum. With the optimized combination of conventional and Vierendeel trusses, the internal forces within the chords can now be regulated in such a way that the cross-sections of the members are fully utilized throughout their whole length: In areas with high amounts of shear force, especially at the supports, the stress – and thus also the deformation – of Vierendeel trusses is very high. With conventional trusses, on the other hand, the stress and the deformation are high where the moment is the greatest – especially mid-span –, since the tensile and compressive forces of the chords follow the moment diagram of the beam. In order to counteract these undesired stresses in the beam, we varied the panel division. Instead of attaching the struts of the cross-bracing to the chords at regular intervals, the points where the struts are connected are literally "pulled apart": at the bearings, where the shear forces are highest, a pure diagonal member truss is formed; toward the middle of the span, the angles of the crossed struts change, closing up to form a single vertical post, so that a pure Vierendeel truss with the desired rectangular openings results there.

Forum 2004 Building and Plaza, Barcelona 2004

The shape and orientation of the roof of the Forum 2004 in Barcelona are based on urban axes: The architectural and urban relationships yield a nearly equilateral triangular roof with sides that are each roughly one hundred eighty meters long. The sheer size is hard to comprehend: The roof is supported on just seventeen points and has spans of sixty meters. The corners of the triangular roof cantilever out as much as twenty-five meters. The exhibition level is suspended from the roof structure, which covers the conference center and the plaza, which is grouped around the forum. The

4

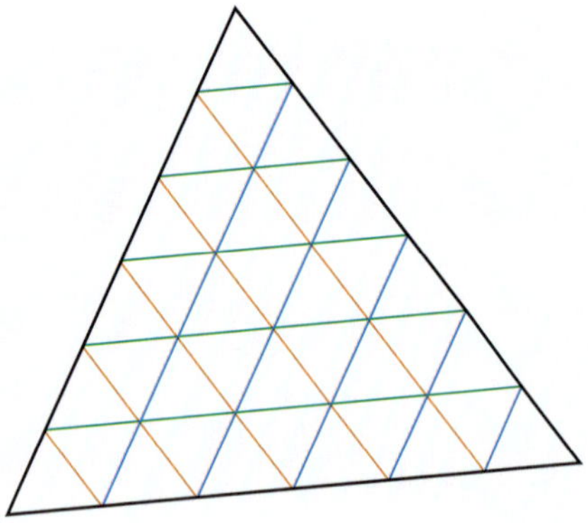

5

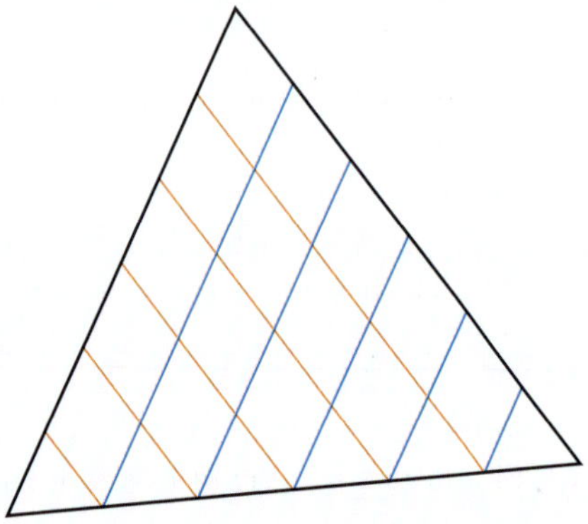

6

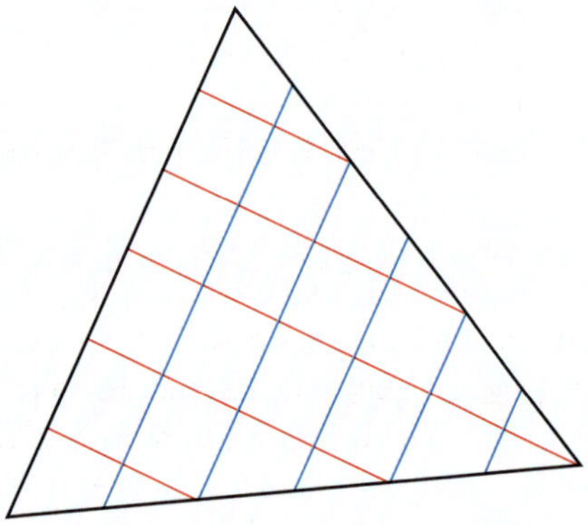

4–6 Alternative basic geometric arrangements of trusses needed to achieve biaxial structural behavior for the composite construction, which consists of the trusses sandwiched between two concrete slabs.
4 Three sets of trusses at angles of 60 and 120 degrees
5 Two sets of trusses at an angle of 60 degrees.
6 Two sets of trusses at an angle of 90 degrees.
7 Calculated volumetric form of the roof structure for an elliptical plan, under the assumption that the tensile and compressive forces are constant at every point when uniformly loaded.

steel and concrete composite construction has a static height of four meters and resembles a sandwich: Between the thirty-centimeter-thick upper and lower concrete slabs, which function as compression and tension zones, are steel trusses that are substantially responsible for the shear connection between the two slabs.

When designing such large, biaxial load-bearing slab structures, there is the question of the geometric arrangement of the trusses in relation to the location of their points of support. While circulation cores and use requirements usually precisely dictate the location of the points of support to within a few meters, the layout of the trusses is fundamentally undefined. But in order for the biaxial load-bearing behavior to actually work, at least two sets of parallel trusses are needed, and these are best arranged orthogonal to each other. Three sets of trusses would be even better for the load-bearing behavior to develop and for more flexibility in positioning the columns, but that is practically impossible to construct, since the fabri-

7

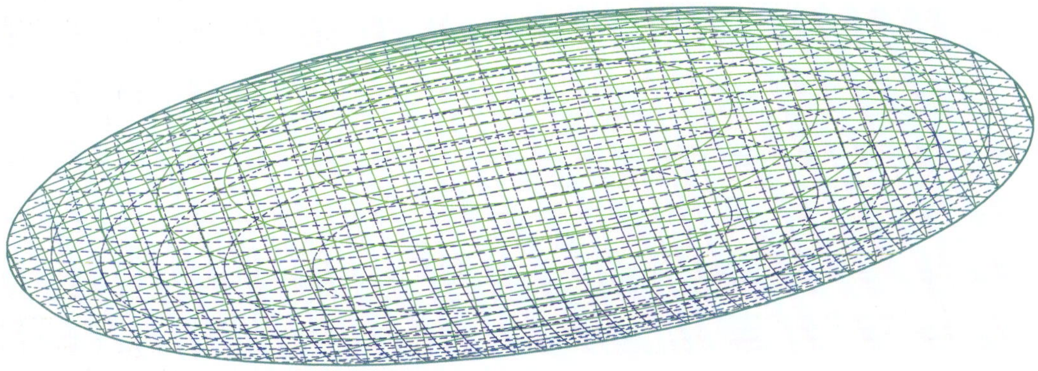

cation of nodes with ten truss members each – six attached sections in the chord plane plus one post and three diagonals – would be much too complex. The design with two sets of trusses is still complex, because one node still unites seven truss members – four chord members, one post and two diagonals – that must all be joined together structurally.

The two sets of trusses in the Forum's roof structure are arranged orthogonally to each other and their precise positions are aligned with one side of the triangular roof. Places where points of support are intended are simultaneously the intersections of two trusses with a corresponding post. With such a relatively large number of trusses resting on just a few points of support, most of the trusses are naturally supported indirectly.

The structure's design also raises a second fundamental question: How should the sandwich slab carry the loads? The load-bearing structure can be viewed as a "mushroom slab," which mainly carries the loads via negative moments directly to the column area, or it works as a "girder floor," which transmits the loads to the columns via field moments. A combination of both load-bearing structures is also conceivable. In this case, the roof structure is designed more as a "girder floor." This concept is reinforced by additional prestressing. In order to reduce the deformations, unbonded prestressing is specified in the interstitial space between the two concrete slabs, arranged parallel to the main trusses. Due to the chosen construction method with large, prefabricated steel space frames, however, in retrospect, the structural concept of a "mushroom slab" would have better suited the flow of forces and the construction process.

St. Jakob Indoor Ice Rink, Basel 2002

The development of the roof geometry for the St. Jakob indoor ice rink in Basel is based on structural and geometric interrelationships. The formulated architectural intention was to have a corporeal roof that floats above the ice as a separate, tangible entity. With this in mind, we designed a structure that, on the one hand, ingeniously and efficiently forms the roof volume and, on the other hand, is clearly separated from the grandstand area by a circumferential band of windows. The shape of the ice pad itself and the grandstand that completely surrounds it, resulted in a roof form that is roughly elliptical in plan.

The structure of the roof rests on twenty-four support points and consists of twelve lenticular steel trusses – so-called fish-belly trusses – that span crosswise to the ice pad at irregular intervals of 7 to 8.4 meters. They bear at their ends on reinforced concrete columns that stand on the concrete substructure behind the grandstand. The lenticular trusses were developed from the structural concept that the tensile forces in all the chords remain unchanged over their entire length. This is the case when the truss geometry runs affinely to the moment line. Thus, all the lower chords could be fabricated with the same slender cross-section of double angles. Since the upper chord is subjected to compression, an H section was selected for use there and, together with the roof sheeting, formed a roof plate or roof shell. Furthermore, due to the geometric relationship, the upward-curving upper chords prevent the lower chords, which are subjected to tension, from bowing out of the truss plane. Hence, the system is stabilized solely in the upper plane of the roof. In the lower plane, additional longitudinal connections and diagonals can be largely omitted.

The structural condition of consistent tensile and compressive forces along the entire chord length is, of course, only determinable for a uniform load. This means, however, that the diagonal and vertical forces are very small, because the truss made from these verticals and diagonals is essentially only subjected to asymmetric loads. Hence, these truss members can be very slender; like the lower chords, they were also constructed with double angles.

Actelion Business Center
Allschwil, Switzerland, 2010

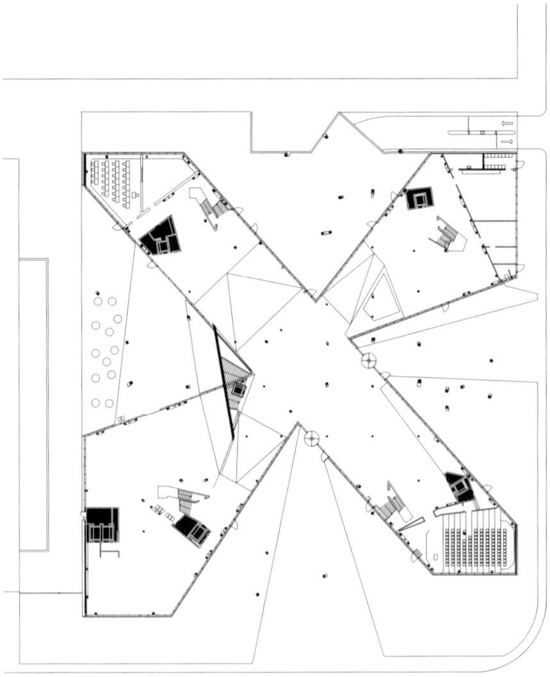
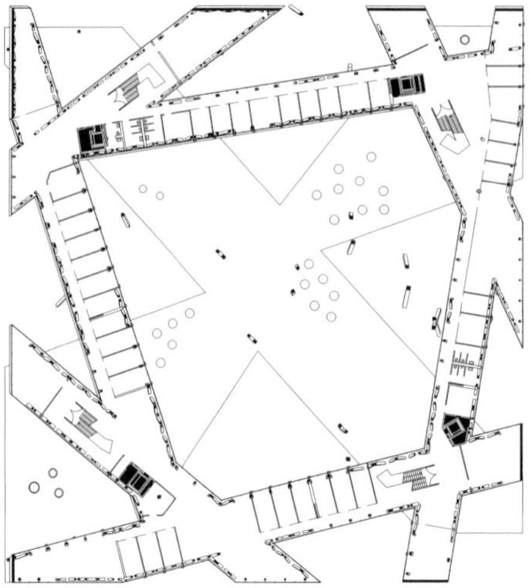
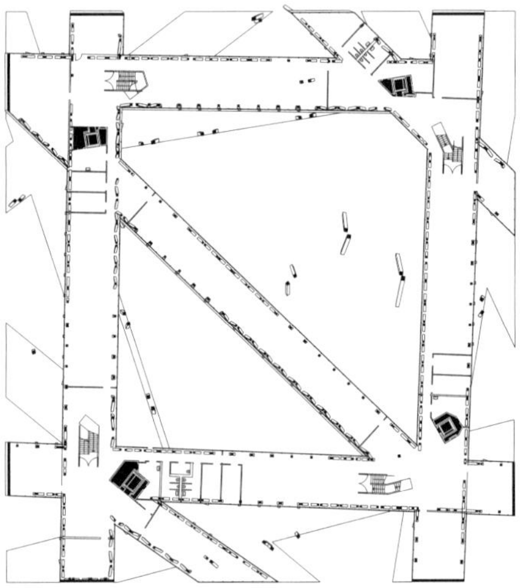
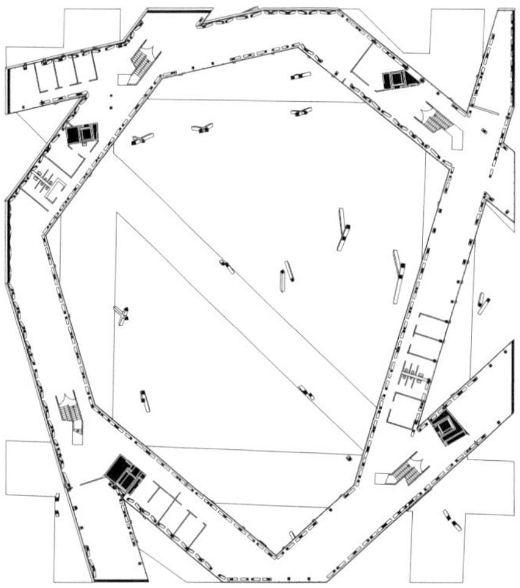
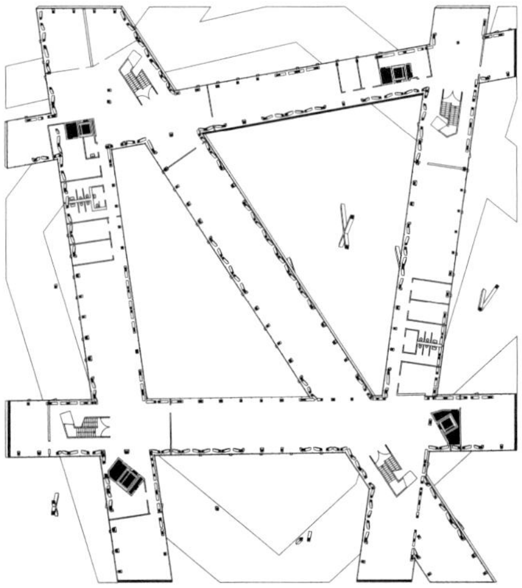
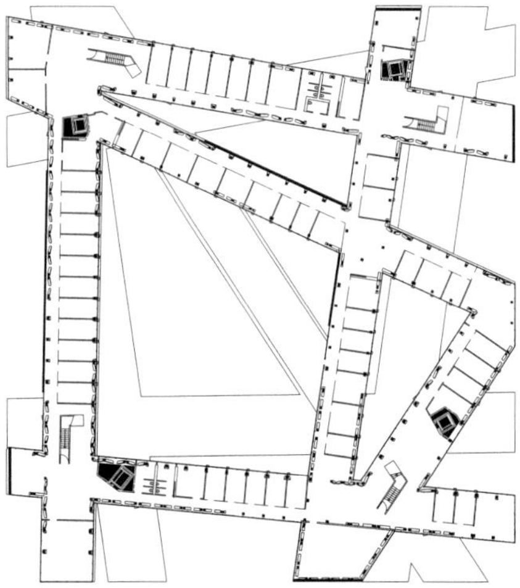

←← Installation of the second floor trusses (construction site photo)
↑ Ground floor plan (left); first floor plan (right)
↖ Second floor plan (left); third floor plan (right)
← Fourth floor plan (left); fifth floor plan (right)
→ Fifth floor plan, showing the floor beams of the fifth floor superimposed on the ceiling beams of the fourth floor

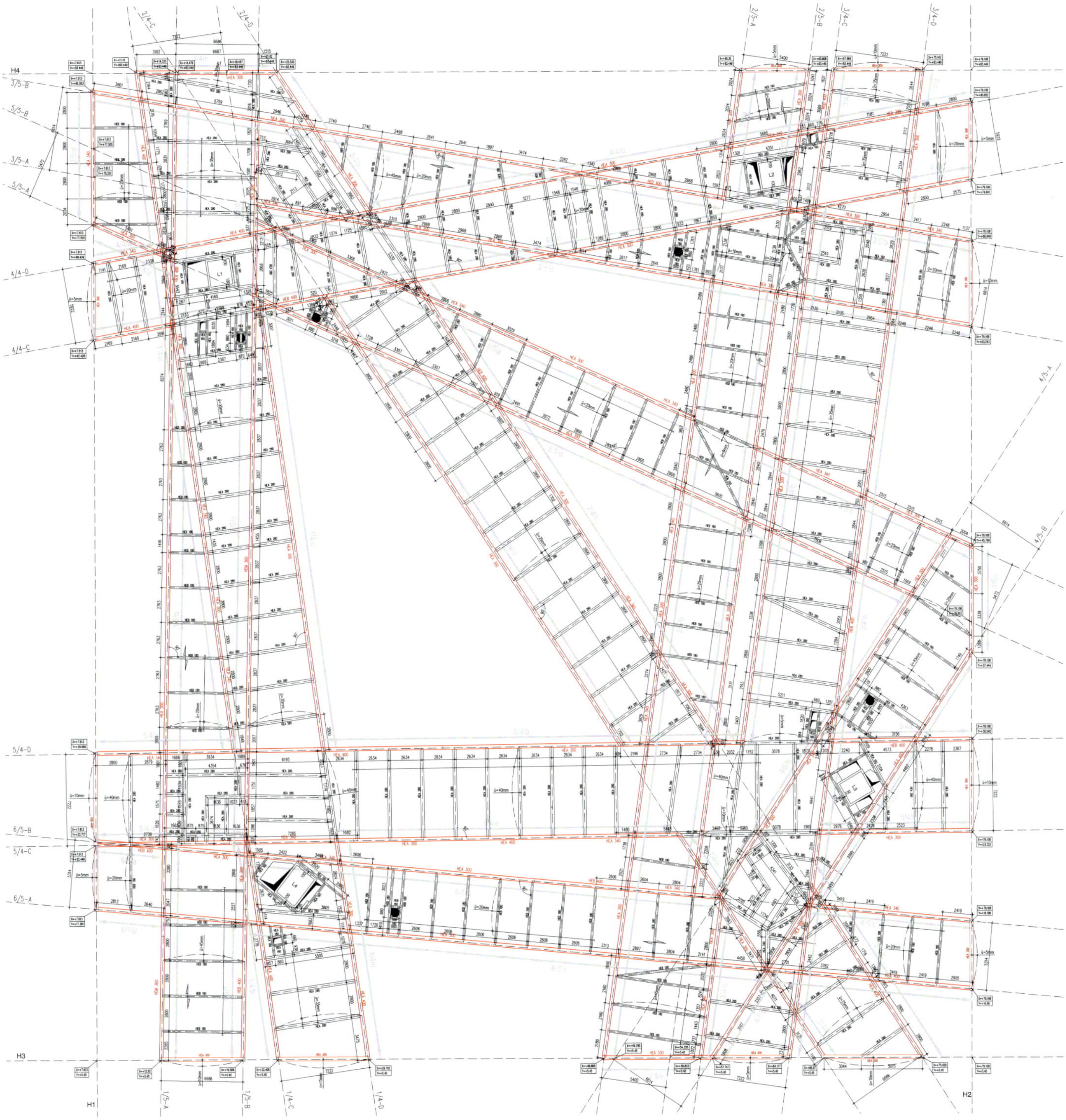
H4
H3
H1
H2
3/5-B
5/5-B
3/5-A
5/5-A
4/4-D
4/4-C
5/4-D
6/5-B
5/4-C
6/5-A
1/5-A
1/5-B
1/4-C
1/4-D
2/4-C
2/4-D
2/5-A
2/5-B
3/4-C
3/4-D
4/5-A
4/5-B

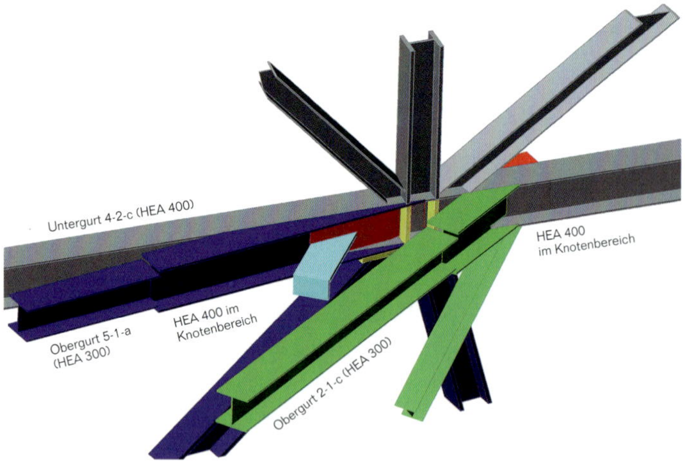

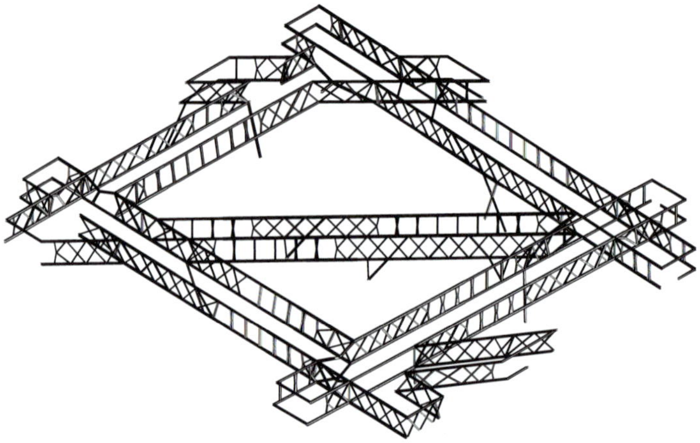

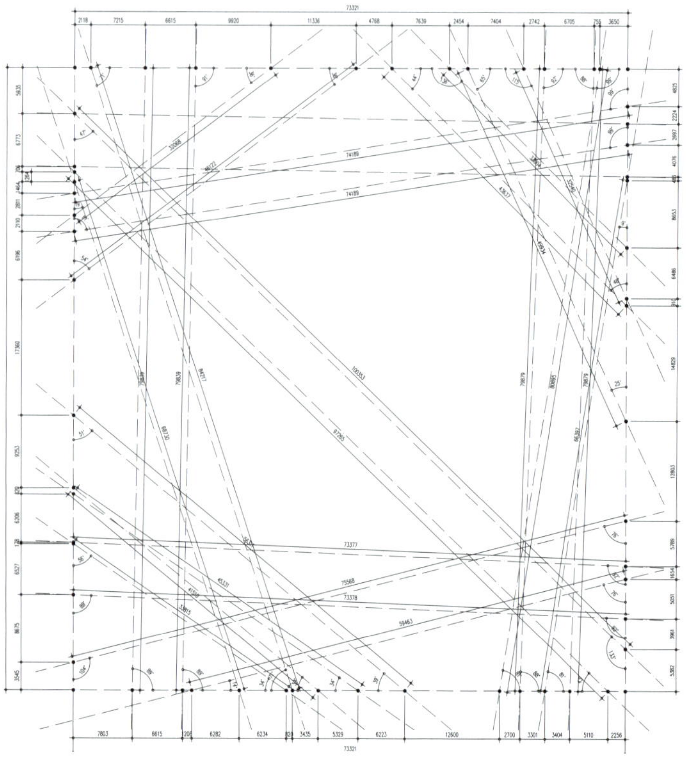

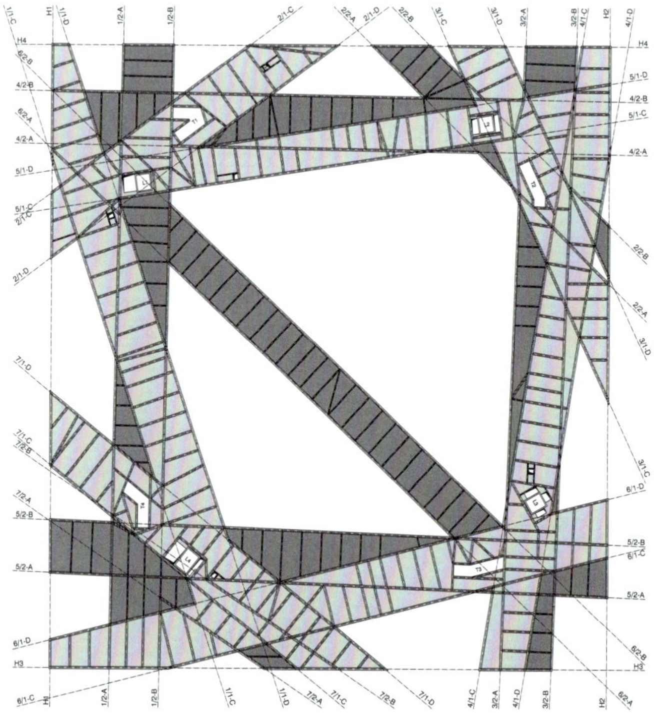

↖↖ Complex truss node with elements differentiated by color in chronological order
↑↑ Isometric view of the second floor
↖ Geometry of the ceiling of the first floor and the floor of the second floor
↑ Beam layout of the ceiling of the first floor and the floor of the second floor

← Detail of flattened nodes (construction site photo). The nodes were flattened at the suggestion of a contractor.
↑ Facade
→ View of the courtyard

Forum 2004 Building and Plaza
Barcelona 2004

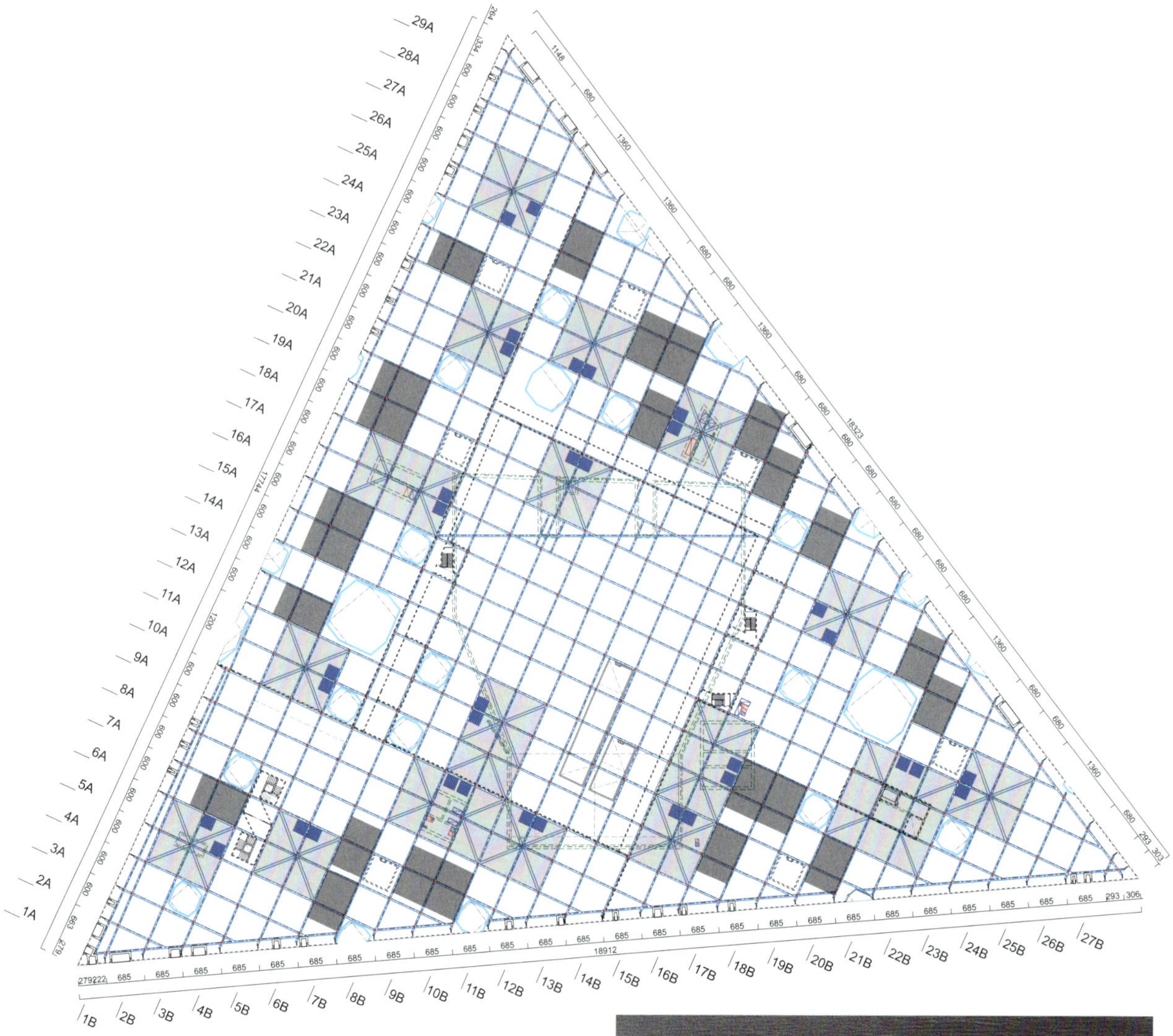

← Assembly of the steel structure. Large steel elements are welded together into a space frame supported by temporary columns, which subsequently also serve as scaffolding for constructing the upper and lower composite slabs.
↑ Plan of the structural plane with the column capitals and voids (holes)
→ The Forum shortly before completion

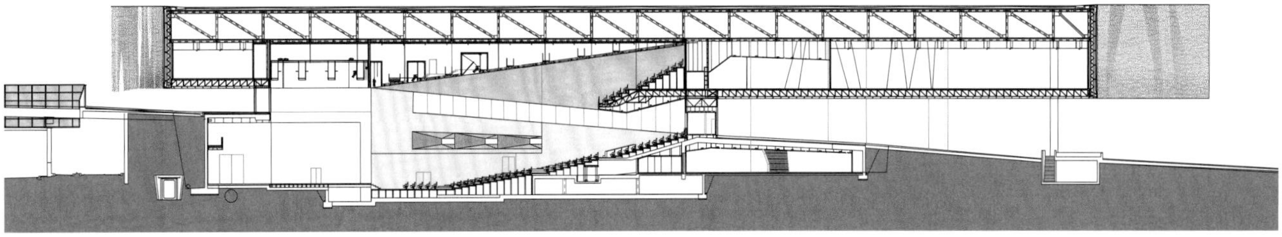

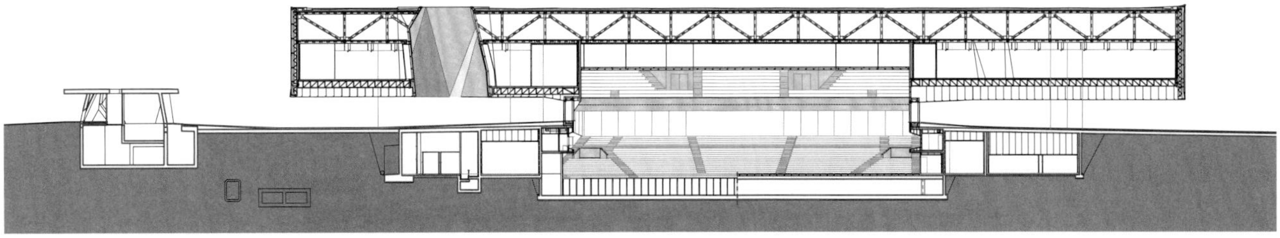

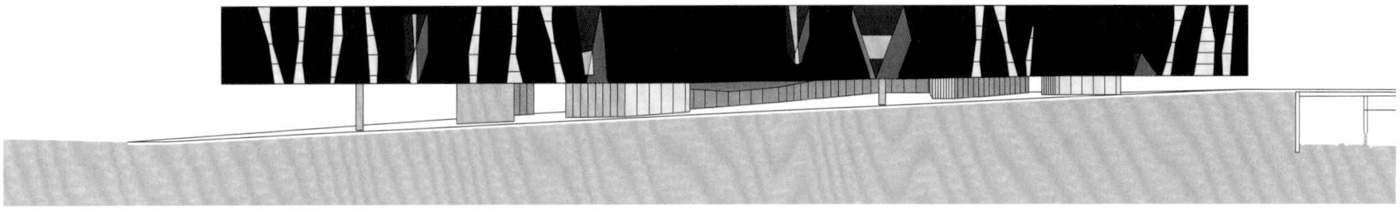

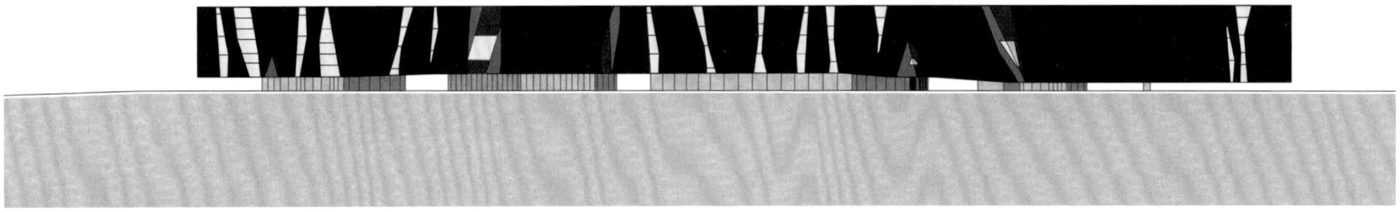

↑↑ Longitudinal section 1-1 through the roof structure, suspended exhibition level, and auditorium. The auditorium is roofed over by the structure (above). Longitudinal section 2-2 through the roof structure, suspended exhibition level, and auditorium (below)
↑ Southwest facade (above). North facade (center). Southeast facade (below)

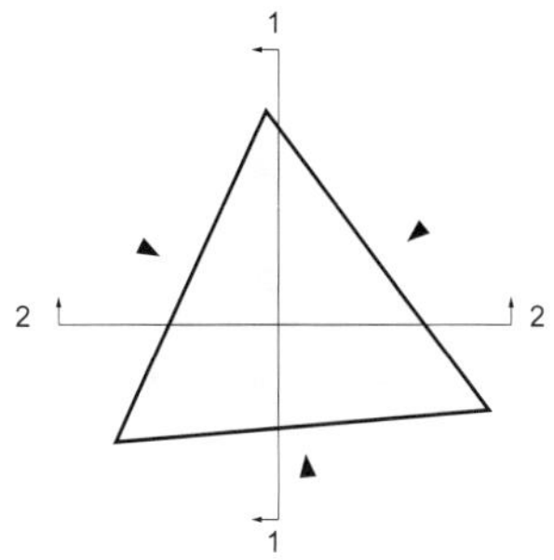

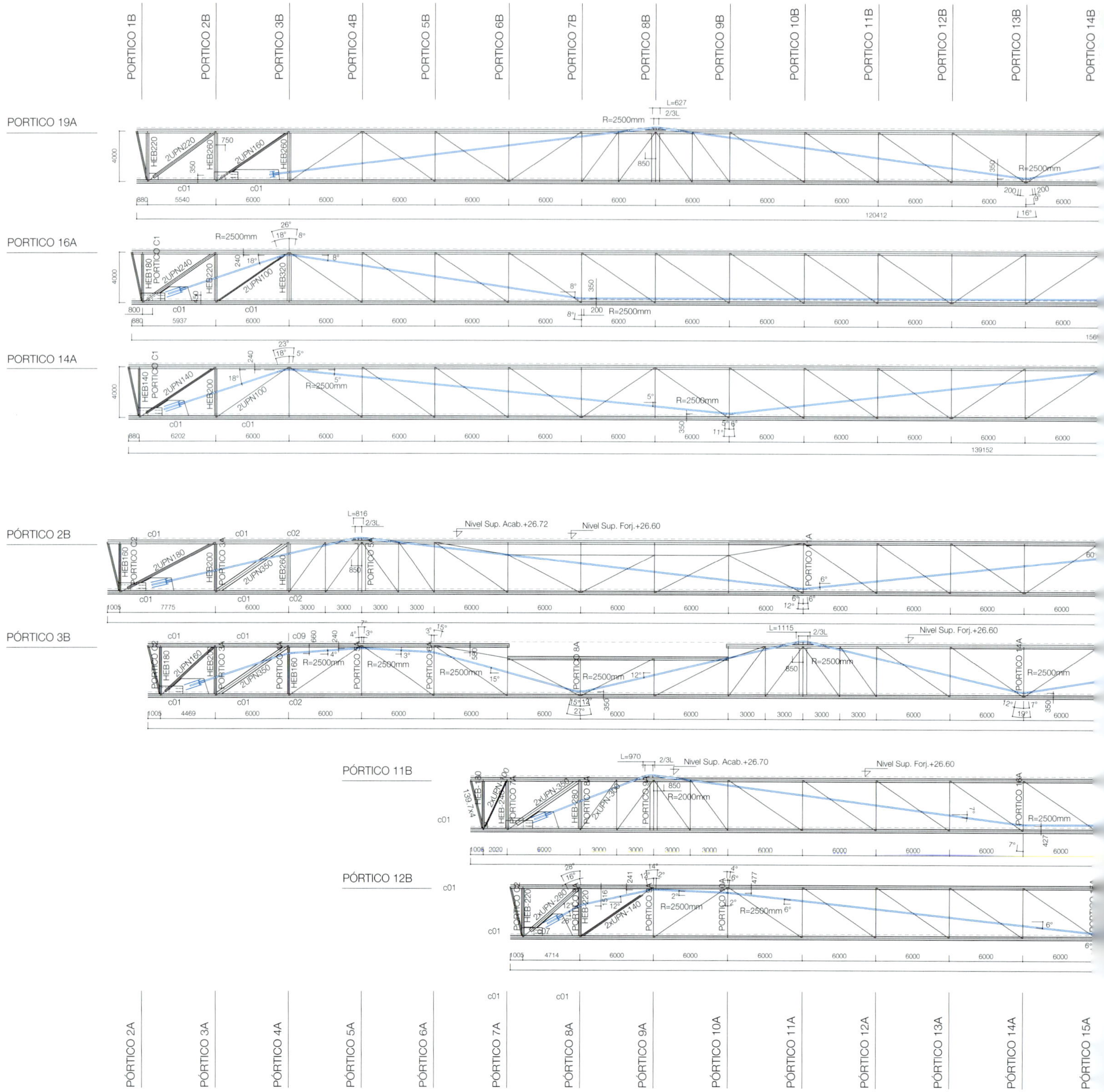

↑ Tendon layout for unbonded prestressing of the roof structure

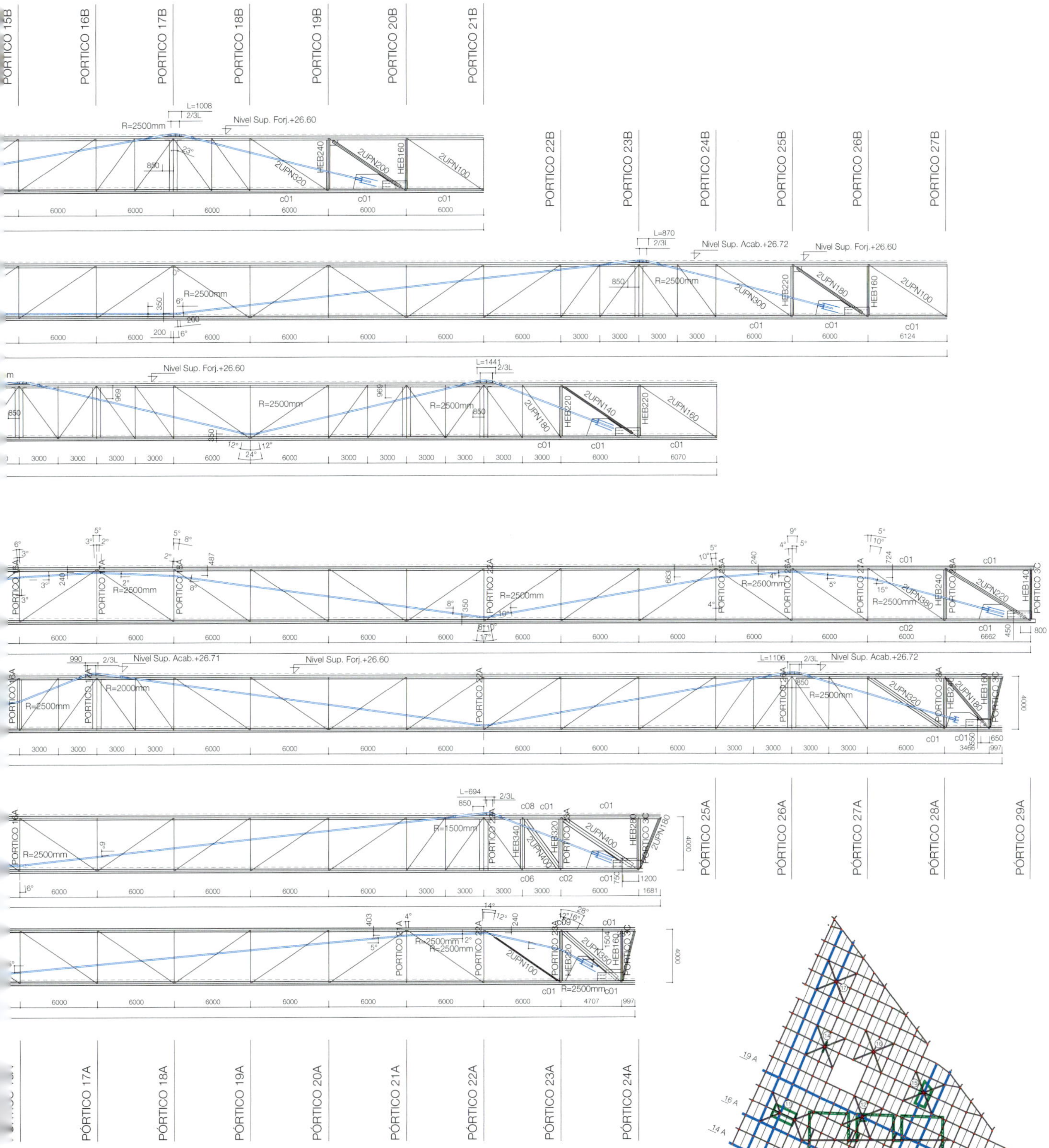

PORTICO 15B
PORTICO 16B
PORTICO 17B
PORTICO 18B
PORTICO 19B
PORTICO 20B
PORTICO 21B
PORTICO 22B
PORTICO 23B
PORTICO 24B
PORTICO 25B
PORTICO 26B
PORTICO 27B
Nivel Sup. Forj.+26.60
Nivel Sup. Acab.+26.72
Nivel Sup. Acab.+26.71
R=2500mm
R=2000mm
R=1500mm
PÓRTICO 25A
PÓRTICO 26A
PÓRTICO 27A
PÓRTICO 28A
PÓRTICO 29A
PÓRTICO 17A
PÓRTICO 18A
PÓRTICO 19A
PÓRTICO 20A
PÓRTICO 21A
PÓRTICO 22A
PÓRTICO 23A
PÓRTICO 24A

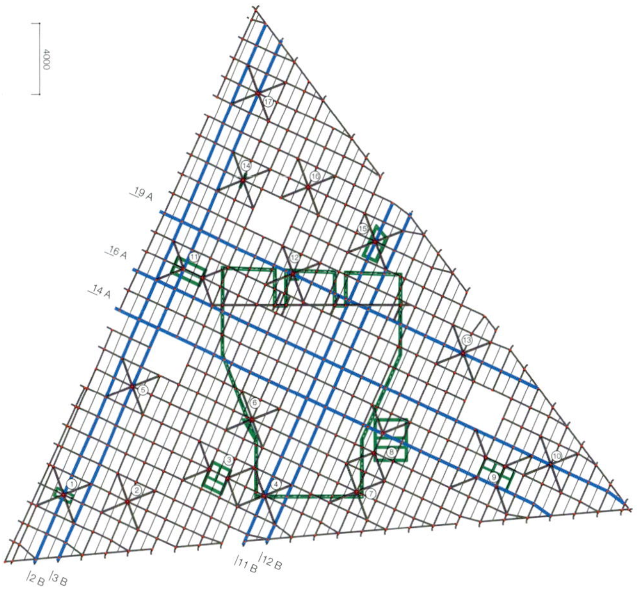

19 A
16 A
14 A
2 B
3 B
11 B
12 B

↘ Computer-generated surface texture for the ceilings above the outdoor space and the auditorium. The water pattern was generated with mathematical equations and the use of random variables.
→ Detail view of the generated surface texture
↓ Ceiling of outdoor space (construction site photo)
→→ The building's northern corner, which cantilevers out twenty-seven meters, seen here with the clad cores and ceiling of the outdoor space

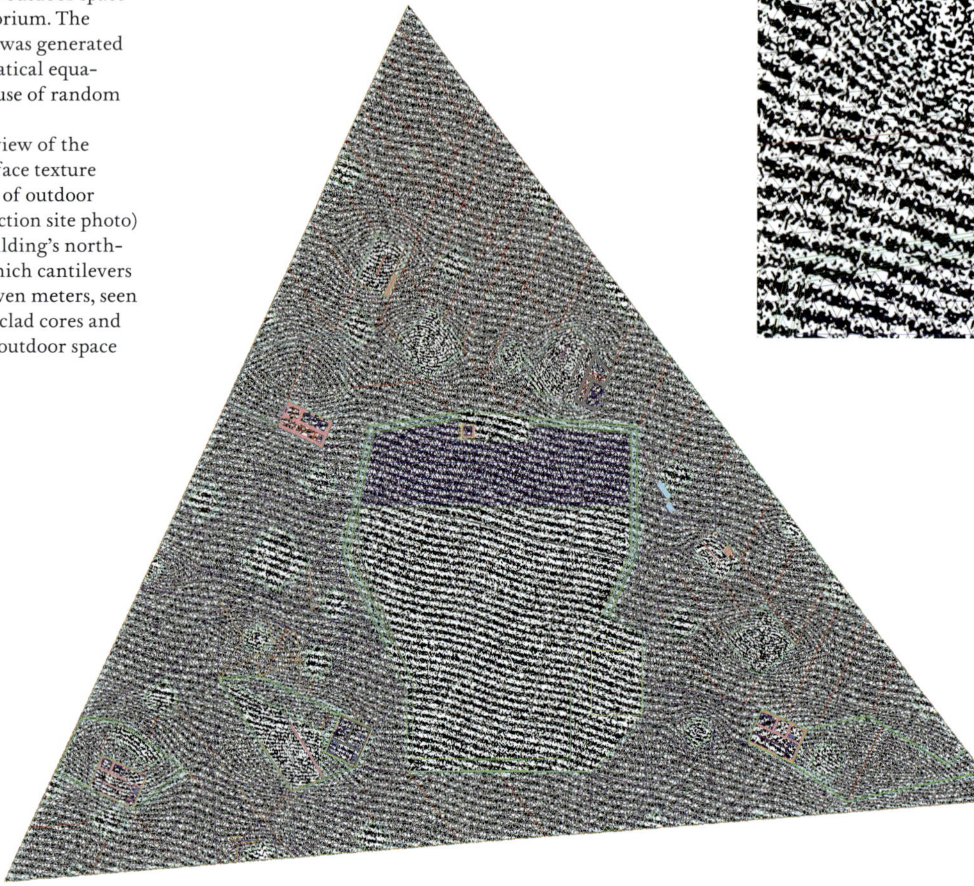

St. Jakob Indoor Ice Rink
Basel 2002

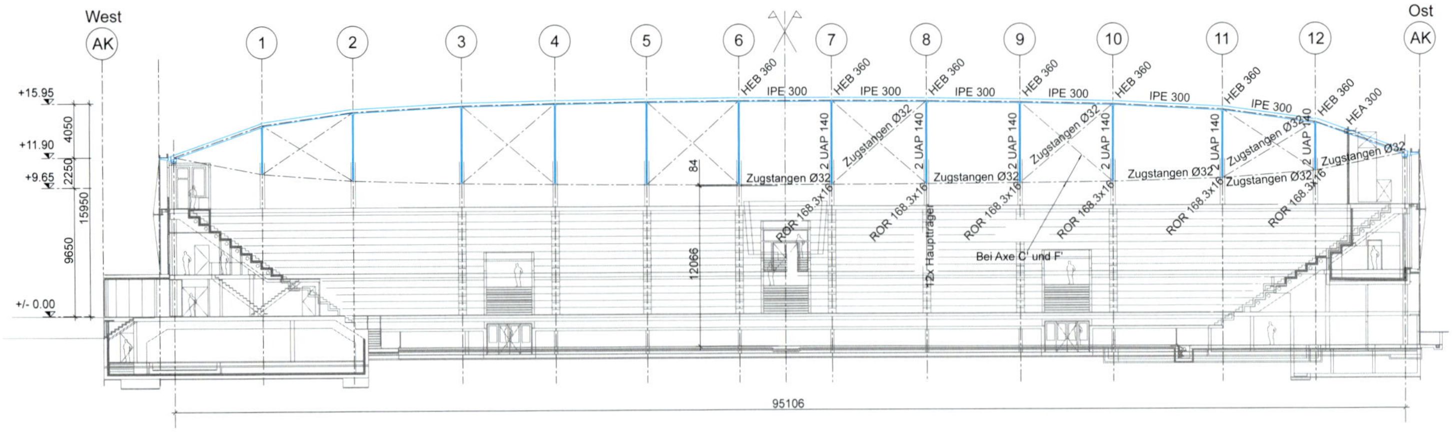
West
Ost
AK
HEB 360
IPE 300
2 UAP 140
Zugstangen Ø32
ROR 168.3x16
12x Hauptträger
Bei Axe C' und F'
HEA 300
+15.95
+11.90
+9.65
+/- 0.00
4050
2250
15950
9650
12066
95106

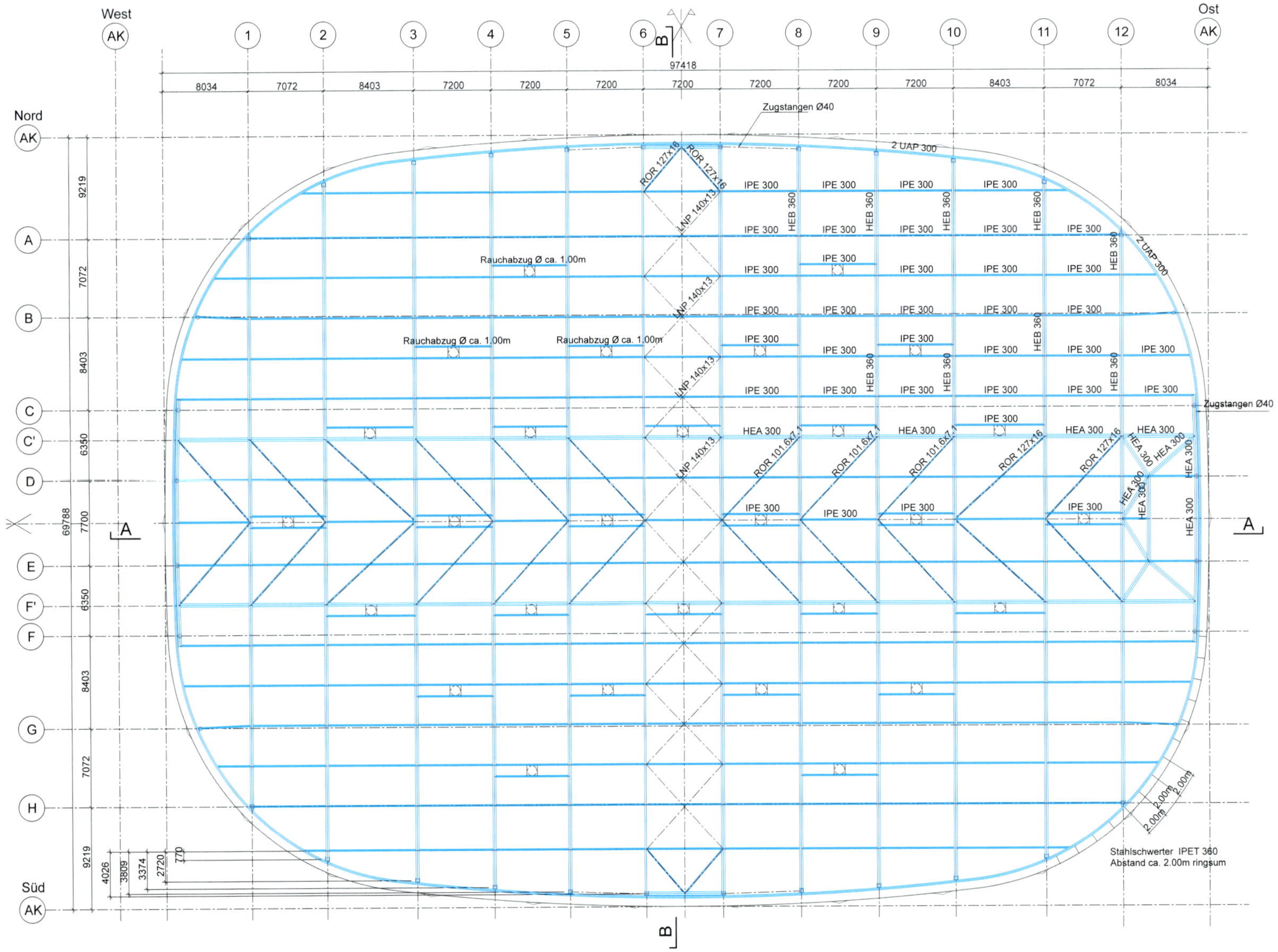
West
Ost
Nord
Süd
AK
97418
8034
7072
8403
7200
Zugstangen Ø40
2 UAP 300
ROR 127x16
LNP 140x13
IPE 300
HEB 360
HEA 300
ROR 101.6x7.1
Rauchabzug Ø ca. 1.00m
9219
7072
8403
6350
7700
69788
4026
3809
3374
2720
770
2.00m
Stahlschwerter IPET 360
Abstand ca. 2.00m ringsum

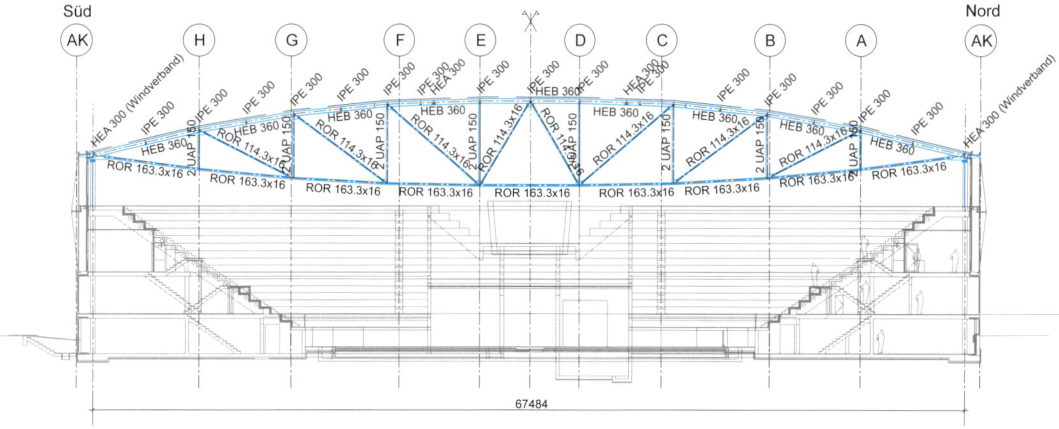

←← Assembly of the roof structure (construction site photo)
↖ Longitudinal section through the indoor ice rink, with roof structure and grandstands
← Plan of the roof structure
↑ Cross-section through the indoor ice rink, with roof trusses and grandstands
↗ Crosswise view at the height of the trusses' lower chords
→ Longitudinal view with wrap-around band of windows

3 Structure and Space

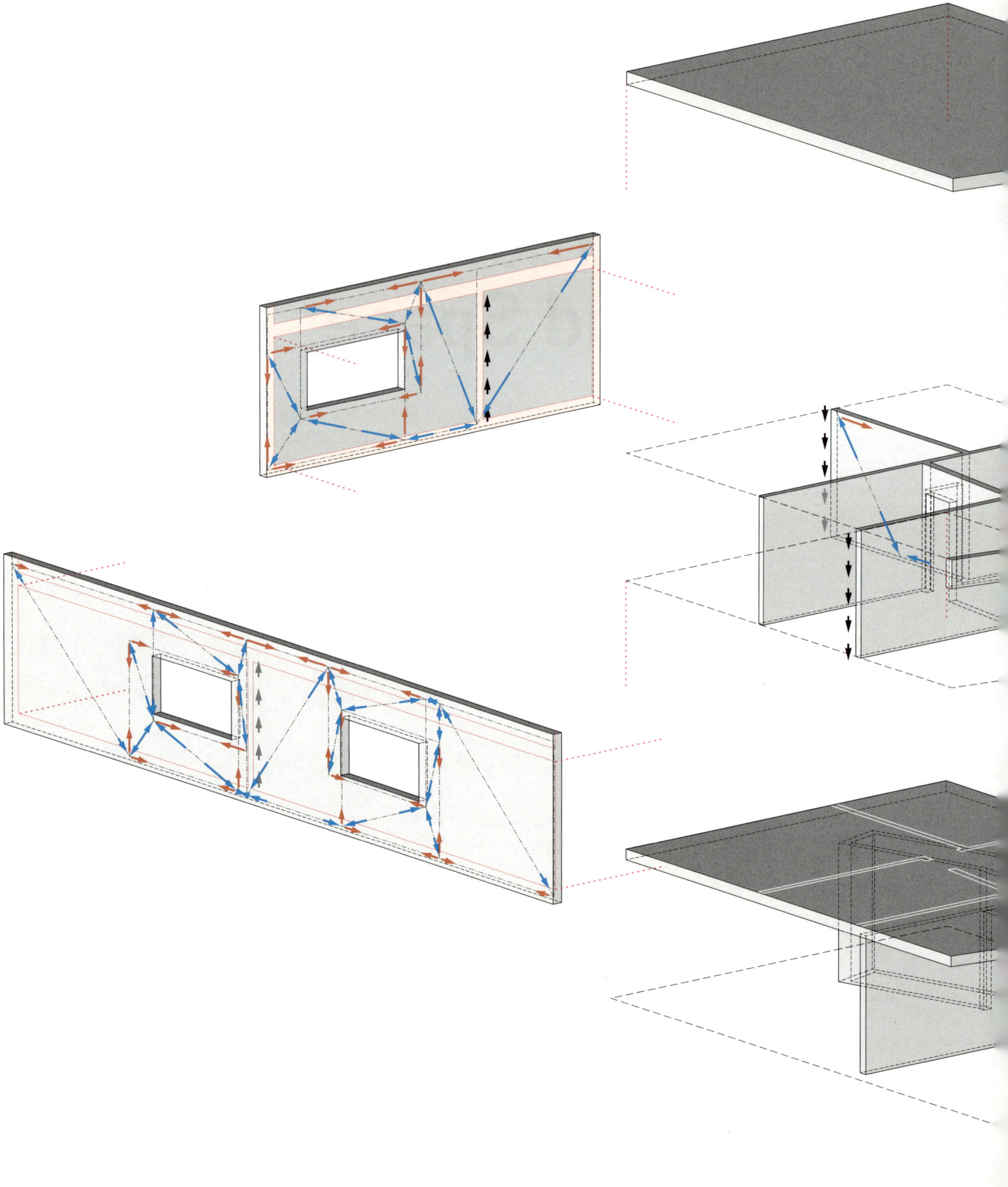

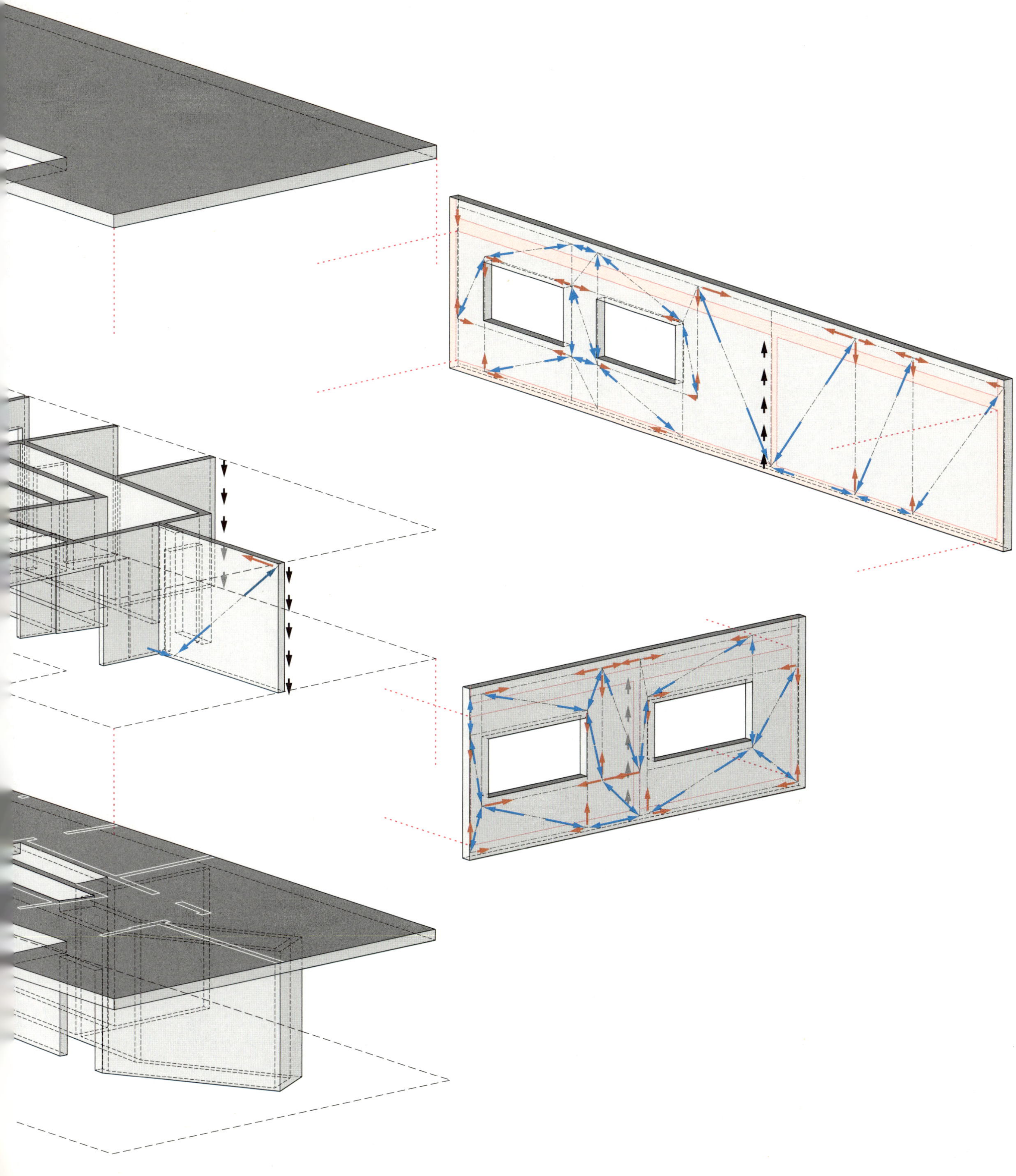

← Assembly of walls and slabs into a spatial structure

Conversations III

AF Let's talk about the relationship between architectural projects and structural projects. What is the relationship between space and structure?

Architectural Project / Structural Project

TP On a good project, the architectural part and the structural part come together as a whole. One of our interests is to comprehend space and structure as a unity.

HS Our goal is to find the ideal structure for a particular space – one that is neither exchangeable nor capable of being further optimized.

AF In other words, you aspire to space-defining structural engineering.

TP Yes. The merging of structure and space is an important proposition.

AF Let me formulate it in a simplistic and provocative way: All of your projects should then be shear-wall/slab designs, as that's where the intersection between structure and space is the greatest. With skeletal frame constructions, the structure has less of a space-defining function; the frame is rather separable.

HS A shear-wall/slab system is not always optimal. Maybe a skeletal frame that allows non-load-bearing walls to be inserted in certain places defines space just as well and, at the same time, is optimized in terms of the manufacturing and building process.

TP Above all, a skeletal frame allows flexibility in using the spaces. On some projects, we certainly do consider space-defining structures. But with industrial buildings, for example, flexibility of use has priority, so frames are more usually employed for those.

AF So space and structure should establish a dialogic relationship to each other. Sometimes the structure is more present, but depending on the use, it can also be made less apparent.

TP Many programs – such as retirement centers, hospitals and industrial buildings – simply require a more flexible framework for use, which can usually be achieved better with skeletal frames and cores.

AF Nevertheless, a new architectural trend toward three-dimensional structural systems can be recognized, toward spatial concepts where everything does not lie neatly on top of each other. Such three-dimensional solutions challenge the interfaces between engineer and architect – and they surely interest you as well.

TP As soon as you stack load-bearing structures, very close collaboration is required. In these concepts, the forces don't always take the direct route down; they make detours and that can be very appealing, as in life itself. The shortest route is not always the most exciting and that's why, yes, we are naturally interested in such structures, which also amplify the interfaces with the architect.

HS The indirect flow of forces – as brought about by the modern trend of stacking – interests us, but less in the sense of trying to optimize it and more as a search for the flow of forces, within the structure, that once again has a certain directness, being economical and relative.

Program

AF Your work is distinguished by a very broad range of different projects. Do you have any preferences or specializations for particular programs? How does the use, the program, generally affect your structural and design approach? Are you involved in analyzing the spatial demands of the program?

TP We are always required to develop the proper structure for the program.

HS Of course, we also analyze the programs in terms of their purposefulness. When discussing where which rooms should be situated, the structural concept is, of course, a crucial part: are the spaces with the longest spans at the very top or at the very bottom? This joint analysis is important in setting the course, for both the architectural and the structural project.

TP In school buildings, for example, the location of the gymnasium is important. In high-rise buildings, the same question applies to the auditoriums and meeting rooms. There, we ask ourselves whether all the people can be brought up in elevators, etc.

SB Sometimes there are great contradictions between the spatial program and the load-bearing structure: Operational needs call for locating public-use spaces with the longest spans as low down near the entrance as possible. In terms of the flow of forces, however, it would be easier to put the long spans at the top and ramifying the forces as they move down. These are issues that we encounter very often. There are certainly exceptions, like the Leutschenbach school, where certain compromises may indeed have been made in operational terms, but which has a 'proper' arrangement of spaces from a structural point of view. Because such solutions are not always possible, we often face the challenge of using transfer beams, shear walls or the like, so that a gym can be located *beneath* the classrooms.

TP These, too, are interesting problems and we can equally well tweak the layout of spaces, for example, by having the partition walls between classrooms act as offset shear walls. A gym can be spanned over very elegantly by a structure situated above it, as familiar examples by us and by other engineers beautifully demonstrate.

AF Which aspects of building interest you the most?

SB I am especially fascinated by the interdisciplinary aspect: taking into account all the factors and dealing with other areas of expertise.

TP Collaborating with architects in designing a structural system.

HS What I find exciting is teasing out a solution that is space-defining and easy to build as well as cost-effective.

AF For architects, as we all know, space is primarily sensory space, emotive space and the space of appearance. Do you see your work as a sensual undertaking – or how much of a role do spatial effects play in your structural designs? To what extent do you want to address perception by means of the structure? Space

HS First, we want to understand the architect's perception and goal. Sometimes structures are built that involve considerable expense and effort, even though simpler ways to transfer the loads would be possible. That's why it is important to understand the architect's sensory perception.

TP Being sensible is, after all, part of the engineer's nature, but that it should also be sensuous is another matter ...

HS Sensuousness has to be comprehended if the work is to make sense!

SB What's crucial is whether the load-bearing structure is visible in the end. If it is clad and thus has a purely supporting function, then I'm unemotional. If the structure can be shown or even highlighted, it has a hugely different meaning.

AF So you fight for attention to be paid to the structure?

SB It always depends on the task at hand.

AF With sensory perception, I actually meant that a design must address the psychological needs of its users: by ensuring stability and security, for example. Is visual stability not an issue for you?

TP That's a difficult question. We engineers, too, build for people – for the users – and we share common interests with the architects.

AF Would you ever structurally over-dimension something, just so it looks optically stable?

SB That's not at all necessary – it's in the nature of things that the proportions agree if the dimensioning is right.

HS I would describe it as an interaction: there is an upward limit and a downward one – in other words, a limit to slenderness and a limit to massiveness. On that, the engineer and the architect need to find a common position.

Structure and Space

Tivadar Puskas

"[...] for all adornment hides what it adorns."[1]
Friedrich Nietzsche

In his discourse on the subject of art and technology, Peter Behrens wrote in 1910: "It is unlikely that a special profession which could be called engineer-architect will develop. Instead, I believe that the future will make a close coexistence of artist and engineer necessary. Neither the architect nor the engineer should be subordinate to the other."[2]

In a contemporary dialogue between architect and structural engineer where the planning partners operate on equal footing, buildings are created with spatial and supporting structures that are mutually dependent; an intensive exchange for specific architectural and structural priorities and intentions to crystallize.

The desire for flexible use and a light appearance, for example, can give rise to skeleton structures that only configure space in combination with the inner, non-bearing walls and the outer, non-structural skin. The desire for a building shell in which the structural and spatial elements coincide can, in another case, give rise to shear-wall/slab systems.

An ingenious interpretation of a three-dimensional structural system that is based on a principle of stacking, for instance, or is developed from bridge-like elements such as triangulated or Vierendeel trusses, represents a particularly demanding challenge for the engineer.

In each of the following three examples, which represent completed and ongoing projects of considerably different sizes and functions, structure and space are bound together in a unique way.

House, St. Gallen, Switzerland, 2012

The house on Rosenberg in St. Gallen displays a strong interweaving of space and structure. The structure of the unpretentious building, developed together with the architect Daniele Marques, is surprising at first glance, because the inside is not directly legible from outside. The architectural and structural expression of the house is shaped by the exposed concrete facade, which features different hues in each story, as well as the interior walls – all of which are designed to carry loads. The outer skin was prestressed at the spandrel bands by means of an internal tensioning device inspired by technology typically used for silos. Since the outer shell functions as a supporting element to help carry the building loads, large, column-free spans are possible inside the house. This roomy, column-free spatial organization is made possible by a crosswise arrangement of shear-walls on the upper floor.

The insulation layer is consistently located between the outer skin and the inner wall or slab levels, which presented significant challenges in terms of building physics and construction detailing. Individual couplings link the internal structure to the load-bearing outer skin, which made it possible to design the building with just a supporting shell. The flow of forces to the exterior is ensured with shear dowels and debonded anchors as well as vertical frictional and sliding bearings. The interplay of the load-bearing planes from inside and outside with the corresponding flow of forces through the insulation layer was, in this case, a structural challenge in itself.

University of Applied Sciences and Arts Northwestern Switzerland, Muttenz, Switzerland (in planning)

With the planned new building for the University of Applied Sciences and Arts Northwestern Switzerland (FHNW), the schools of Architecture, Civil Engineering and Geomatics; Life Sciences; Social Work; and Teacher Education – which have until now been spread across thirty-six sites in the cantons of Basel-Landschaft and Basel-Stadt – shall be brought together in a single location in Mut-

1–5 Schematic sequence of stacking slabs and shear walls into a three-dimensional structural system (house in Riehen)

1

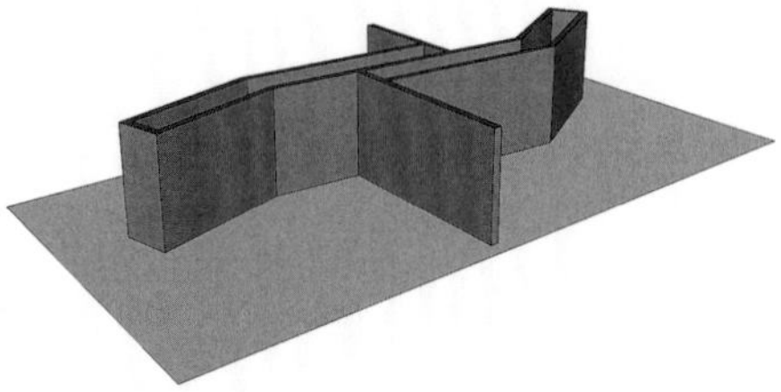

2

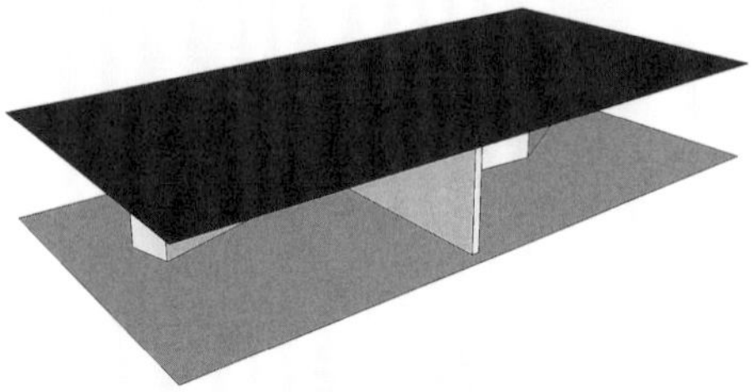

3

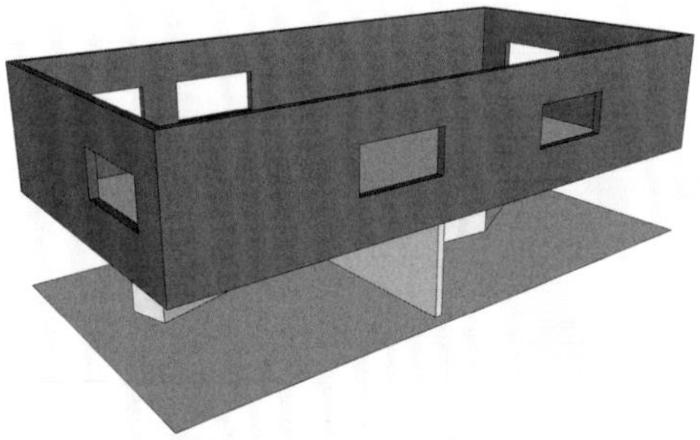

4

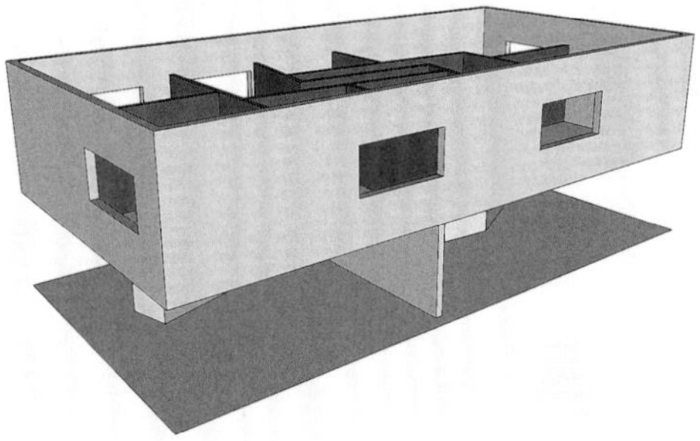

5

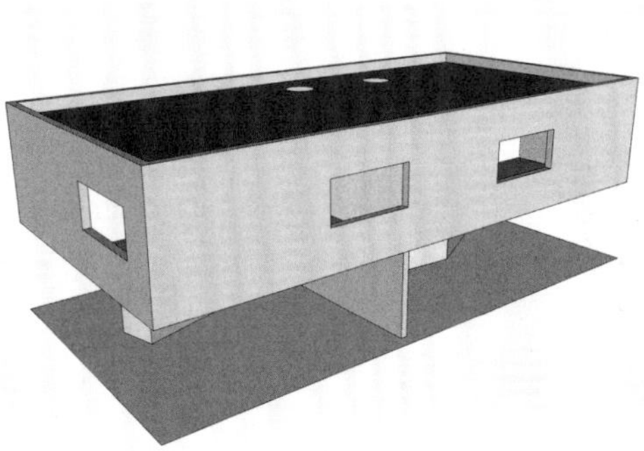

tenz. Pool Architekten won the competition with a distinctive, emblematic building.

The programmatic components are stacked vertically – the plinth, entrance level and teaching levels, as well as the library and faculty levels are clearly discernible from outside and give the building its spatial and structural identity.

A central, two-story circulation space constitutes a void that extends upwards through the highly concentrated uses, splitting to become two atria on the upper floors, around which the faculties are arranged. The building's base contains laboratories, gymnasiums, and mechanical equipment rooms. The entrance level contains a foyer, auditorium, cafeteria, and administrative offices. Above the classrooms is the open-plan library level, which also has additional work areas for student use. Above that are stories that can be flexibly subdivided. Along with a sublime panoramic view, the top floor provides a multipurpose sheltered roof terrace.

To keep the spatial division flexible, the upper stories are designed as a skeletal frame with a column spacing of fourteen meters. Along the slab edges, the floor spans are reduced by half, so that along one axis, prestressed flat or ribbed slabs with a thickness of forty-eight centimeters are possible. By virtue of their mass alone, no additional measures against structure-borne noise are necessary.

The fourth floor is comprised of peripheral, story-high shear walls that transfer loads to the circulation cores. On the library level beneath, the vertical load-bearing elements are reduced to the facade columns and the walls of the cores. The inner columns on the fourth floor and the library's ceiling slab are thereby intercepted and supported. The same principle makes it possible to keep the ground floor free to a large extent: the loads are intercepted and supported in the first and second floors by double-story, peripheral shear walls and two additional main columns connected by beams.

The prestressed slabs allow here spacious, cantilevered hallways around the stair court. Edge parapets ensure distribution of the loads at the connecting stairs, which boldly transverse the space of the entrance hall, thereby symbolically uniting the faculties and students.

Housing for the Elderly Im Büel, Cham, Switzerland (in planning)

Due to its recessed base, this apartment building for the elderly in Cham seems to float above the ground. The replacement building's floating form relates to the outlines of the nearby Kirchbüel school and the old hospital, and it accentuates the space along the hill range stretching between the Kirchbüel and Röhrliberg schools. To the north, the building is close to a proposed traffic corridor. The perforated facade, the walls between apartments, the floor slabs, and the core define both the load-bearing structure and the space, adding clarity and distinction to the building.

The commission arose from an architectural competition won by Schneider & Schneider Architekten. In order to obtain structural alternatives for the free-floating form above the transparent ground floor, an engineering competition was held subsequent to the architectural competition. We presented four structural concepts: The first option was to set the six upper stories on a supporting table of sorts, the second proposed hanging the floor slabs from a roof structure, and the third alternative represented the attempt to combine the first two concepts. The fourth and final structural concept, which was our favorite, was based on the idea of interpreting the surrounding facade as a load-bearing and space-defining structure. The perforations in the facade and the cantilevered loggias with their parapets evoke the image of a wraparound Vierendeel structure, whose loads can be transferred to the circulation core by four concrete shear walls inside the building. The shear walls are connected to the core in a radial, pinwheel-shaped arrangement and simultaneously constitute the party walls between apartments. Due to uneven normal load distribution in the reinforced concrete boxes, the core also transmits the differential horizontal loads to the reinforced concrete box in the basement.

1 Friedrich Nietzsche, *On the Advantage and Disadvantage of History for Life,* translated by Peter Preuss, p. 64.

2 Peter Behrens, Kunst und Technik, in: *Elektrotechnische Zeitschrift* 31 (1910), H. 22, pp. 552–5, here p. 554.

House
St. Gallen, Switzerland, 2012

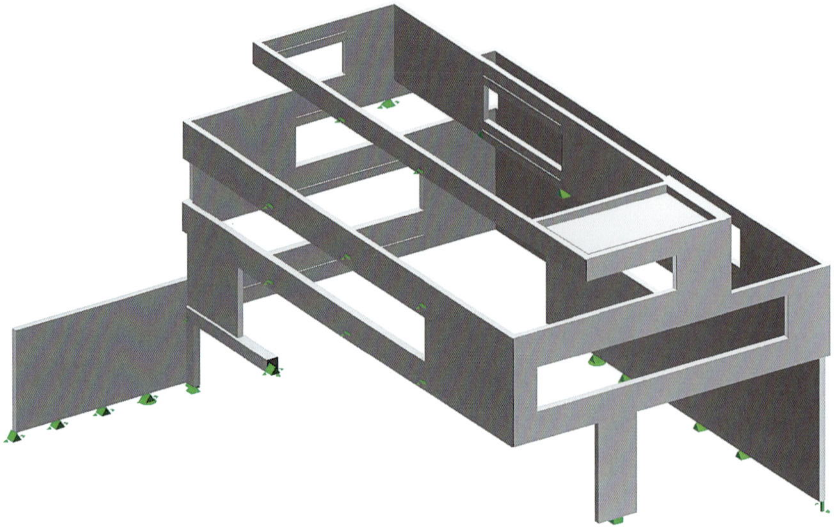

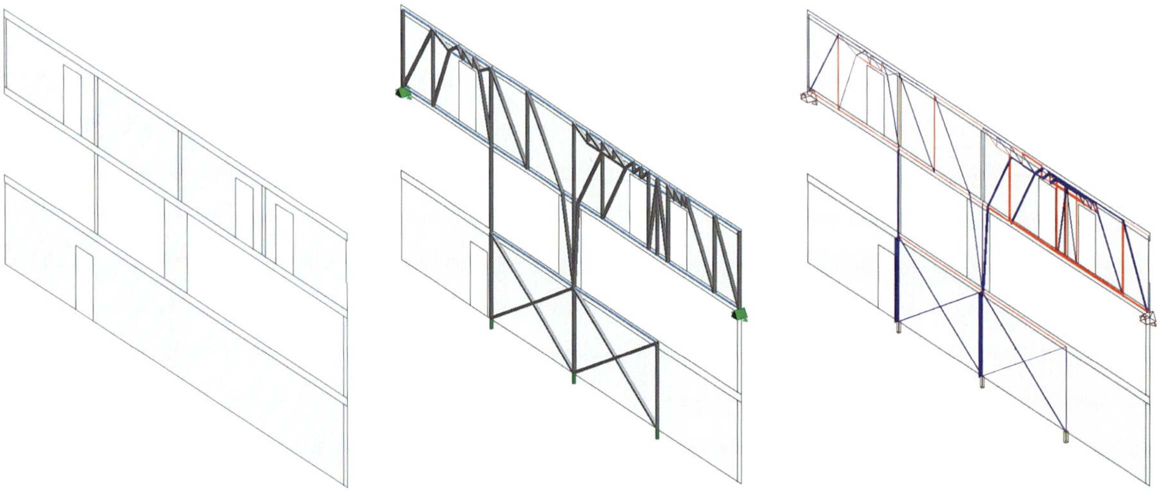

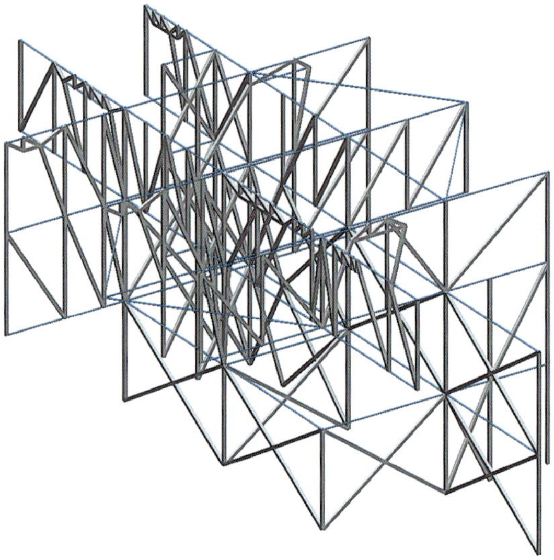

← South elevation
↑↑ Three-dimensional, interlocking plate-like structural system
↑ Parallel, three-story shear wall models
→ Three-dimensional truss analogy
→→ Column-free space at the corner of the top floor

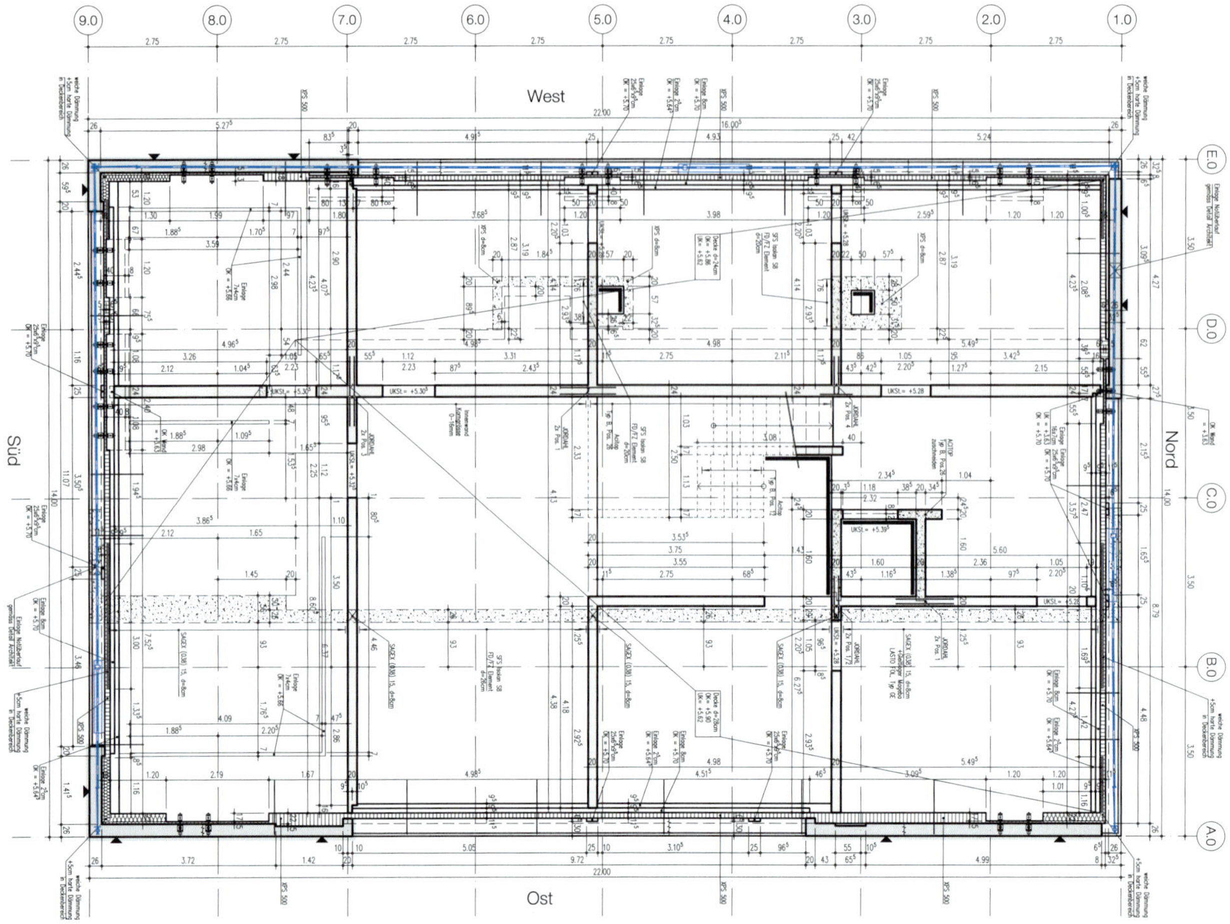
9.0
8.0
7.0
6.0
5.0
4.0
3.0
2.0
1.0
West
Ost
Süd
Nord
E.0
D.0
C.0
B.0
A.0

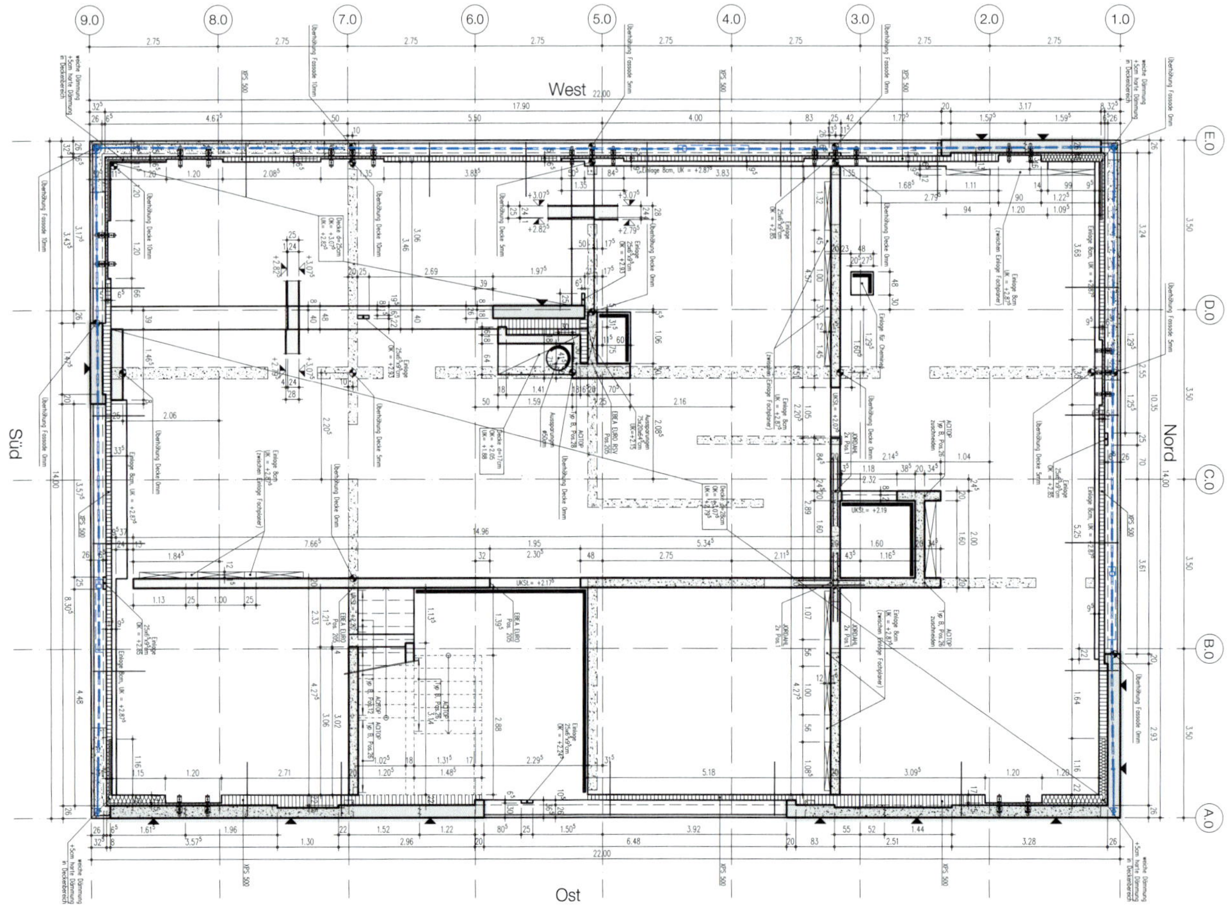
9.0
8.0
7.0
6.0
5.0
4.0
3.0
2.0
1.0
West
Ost
Süd
Nord
E.0
D.0
C.0
B.0
A.0

←← Formwork drawing, first floor (above), ground floor (below)
↙ Elevations of the prestressed spandrels in the first floor, with specifically located anchor niches

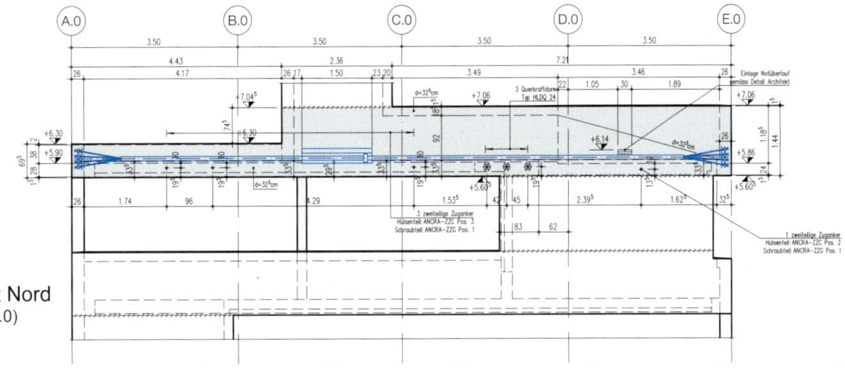

Ansicht Nord
(Achse 1.0)

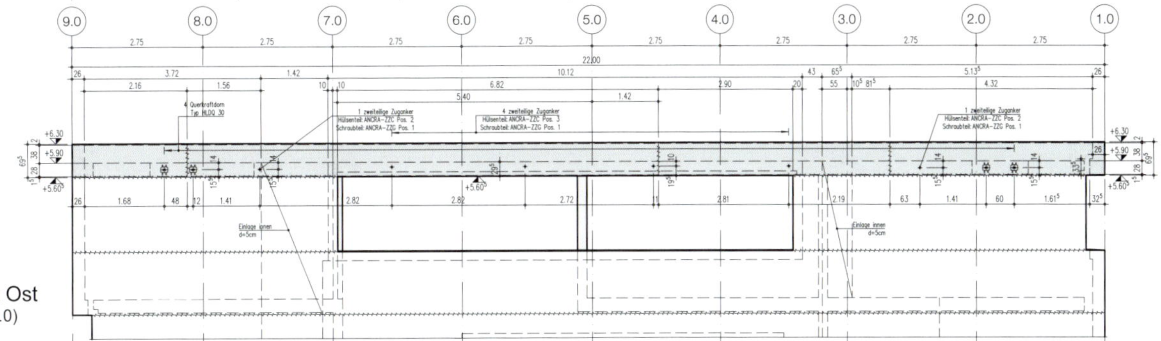

Ansicht Ost
(Achse A.0)

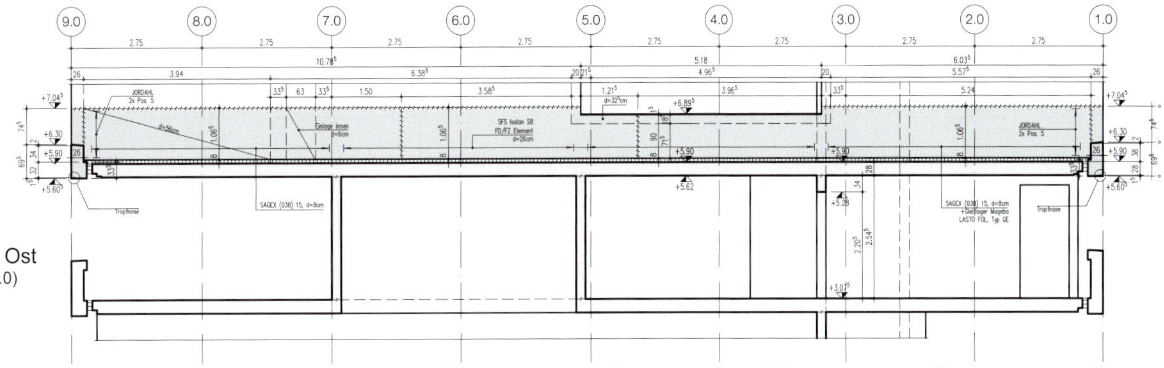

Ansicht Ost
(Achse B.0)

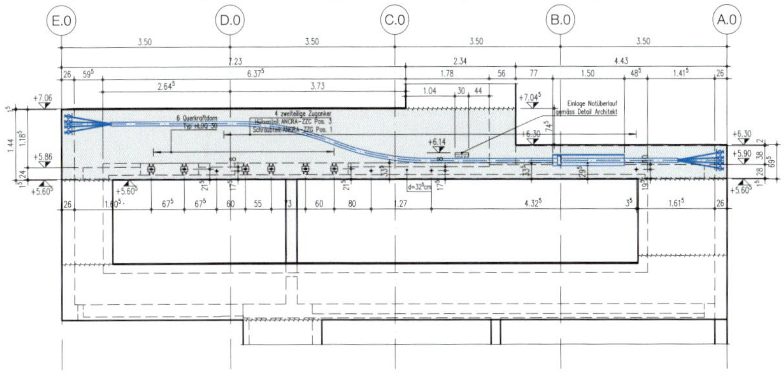

Ansicht Süd
(Achse 9.0)

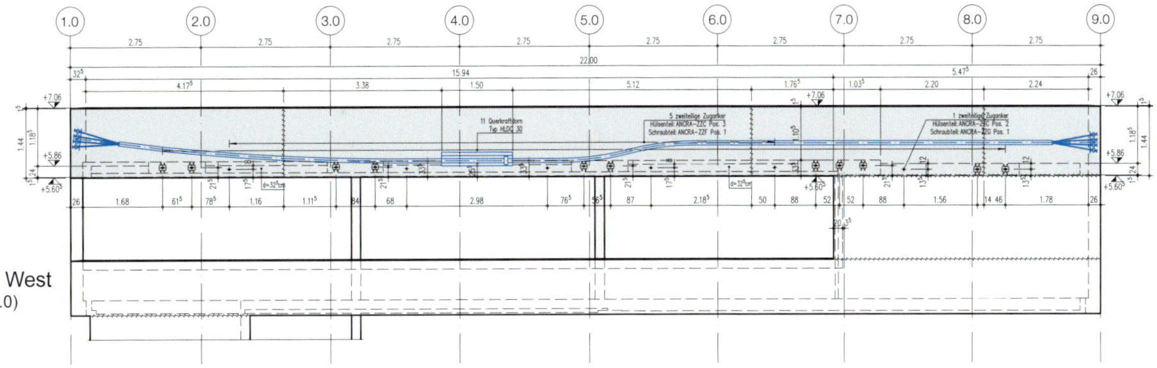

Ansicht West
(Achse E.0)

University of Applied Sciences and Arts Northwestern Switzerland Muttenz, Switzerland (in planning)

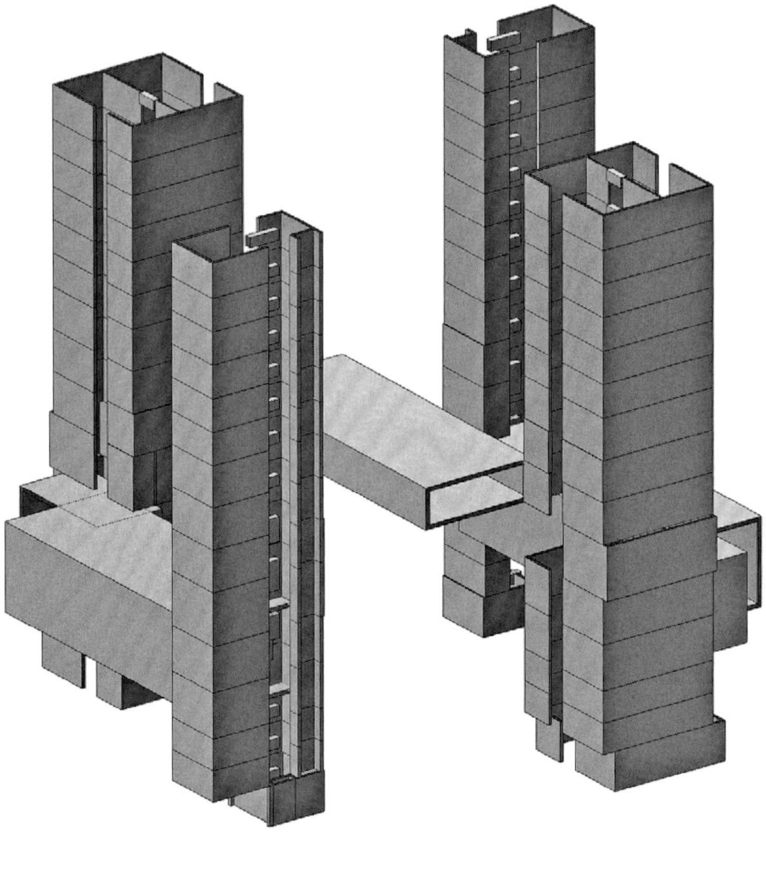

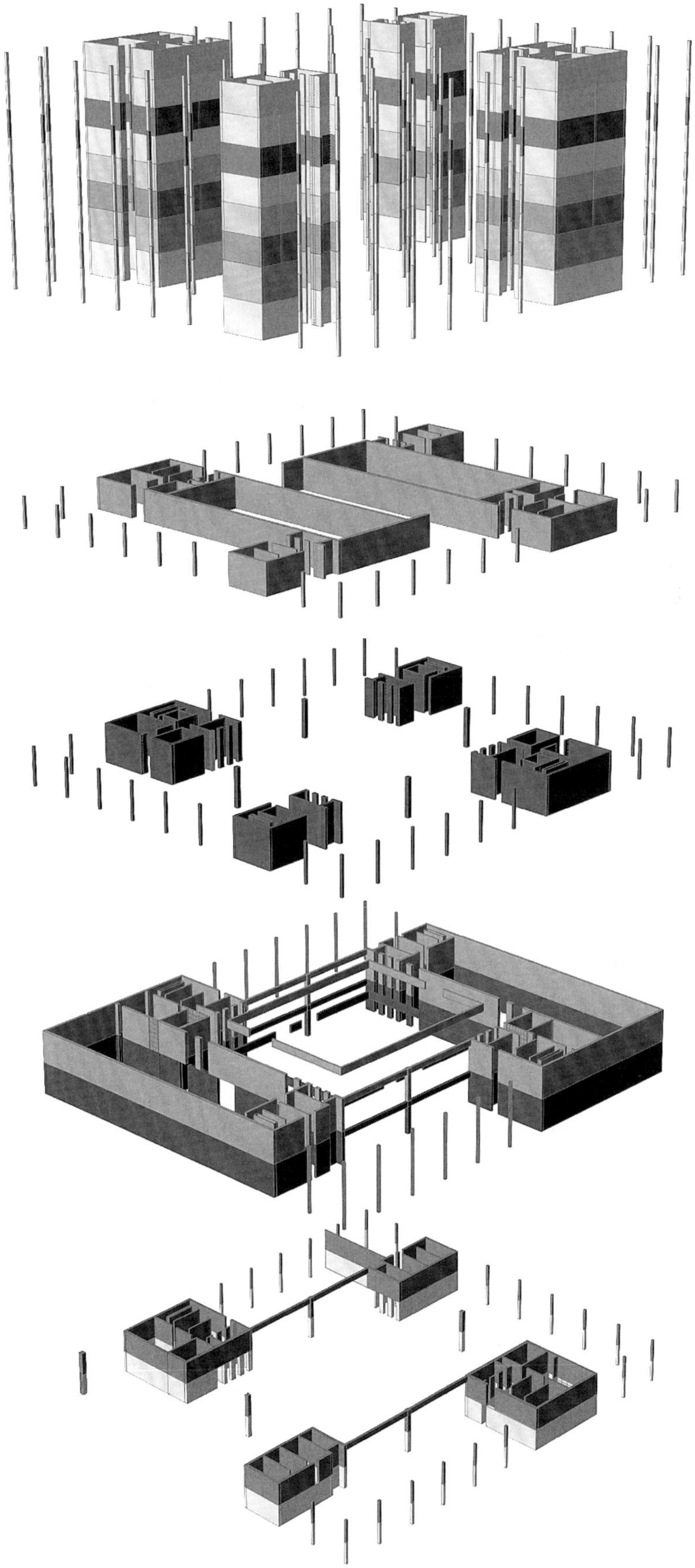

←← Interior view of foyer area (visualization)
↖ Interior view of circulation space (visualization)
← Four tube-like, stabilizing vertical circulation zones with connecting library level (computer model)
→ Isometric diagram of the vertically stacked load-bearing elements

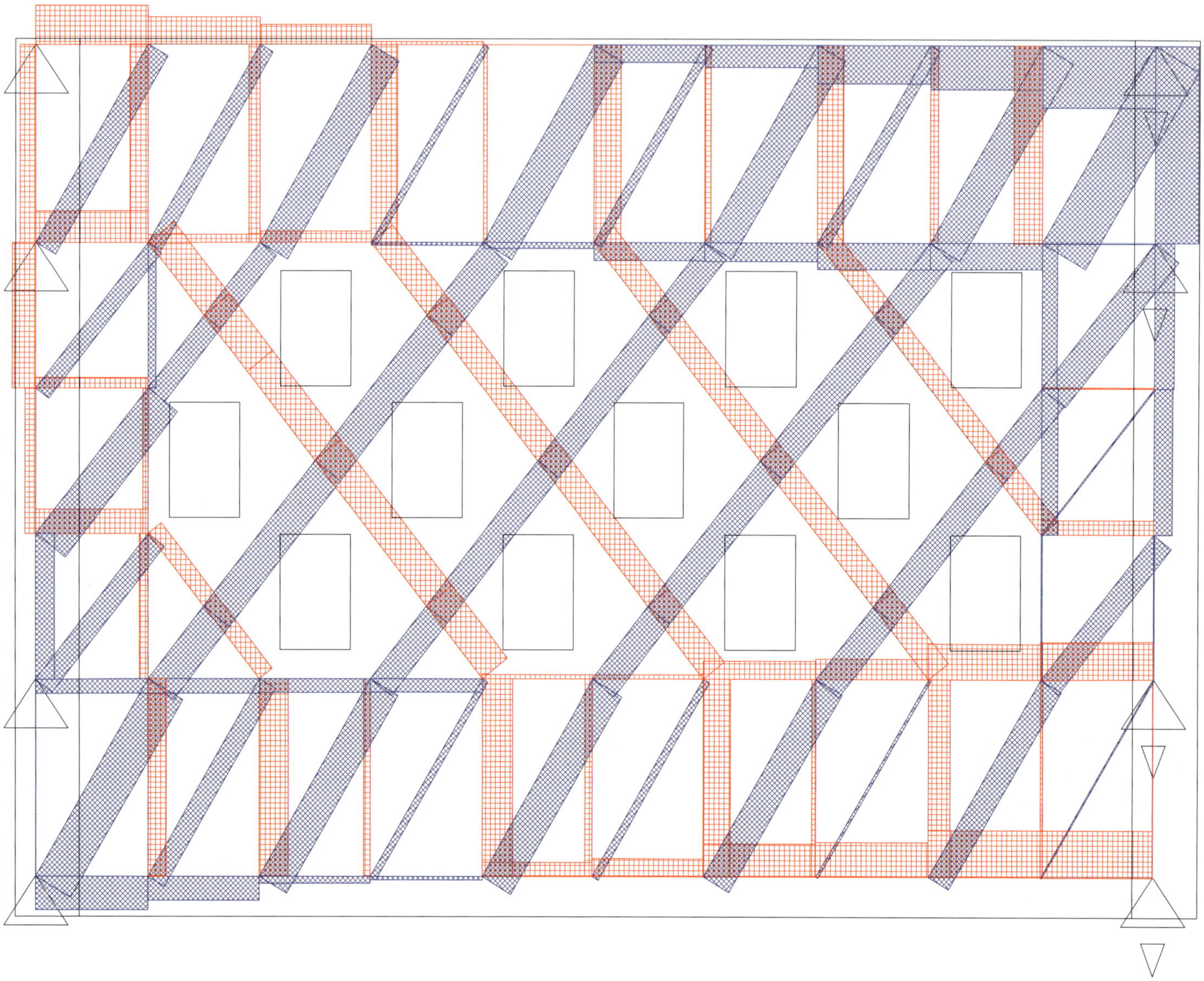

↑ Flow of forces in a cantilevered, perforated shear wall
→ Exterior view (visualization)

Housing for the Elderly Im Büel
Cham, Switzerland (in planning)

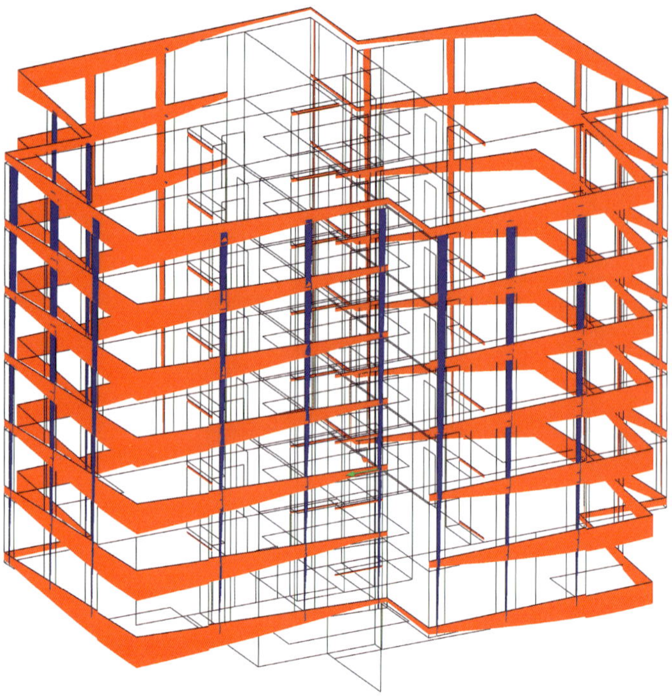

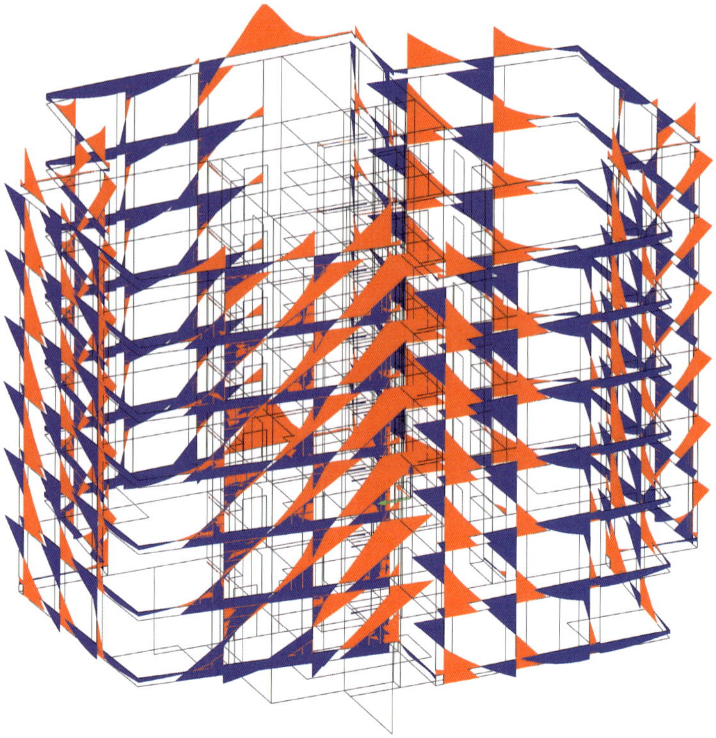

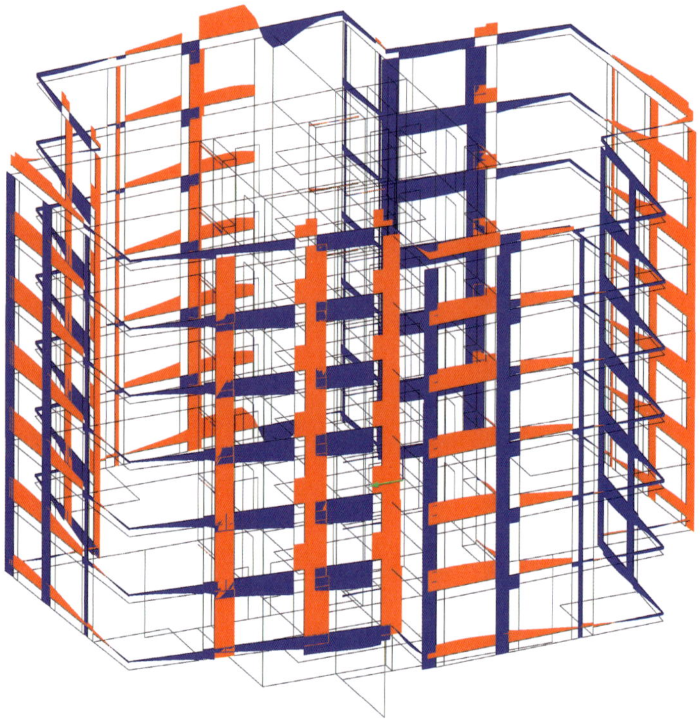

← Exterior view (visualization)
↑ Deformations in the load-bearing Vierendeel facade
↗ Bending moment diagram in the Vierendeel structure
→ Shear force diagram in the Vierendeel structure

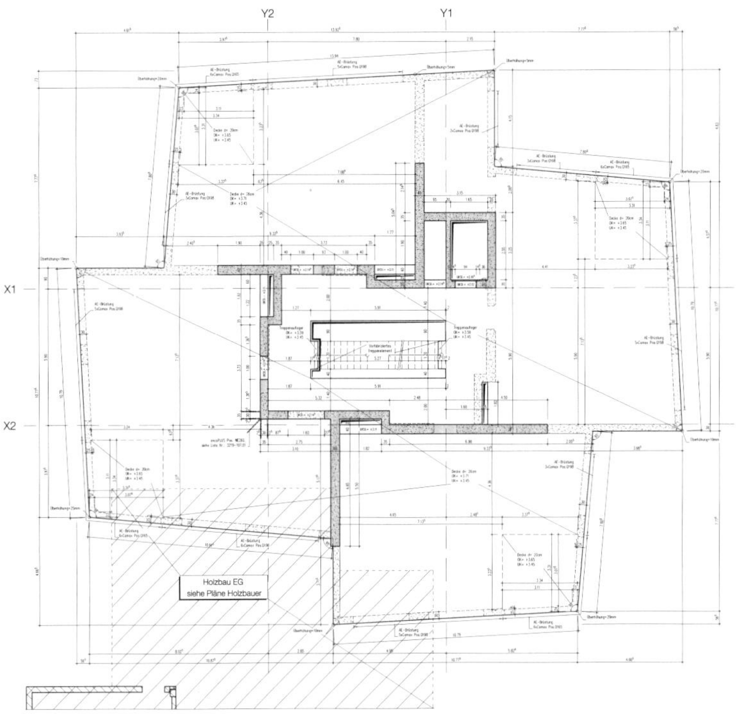

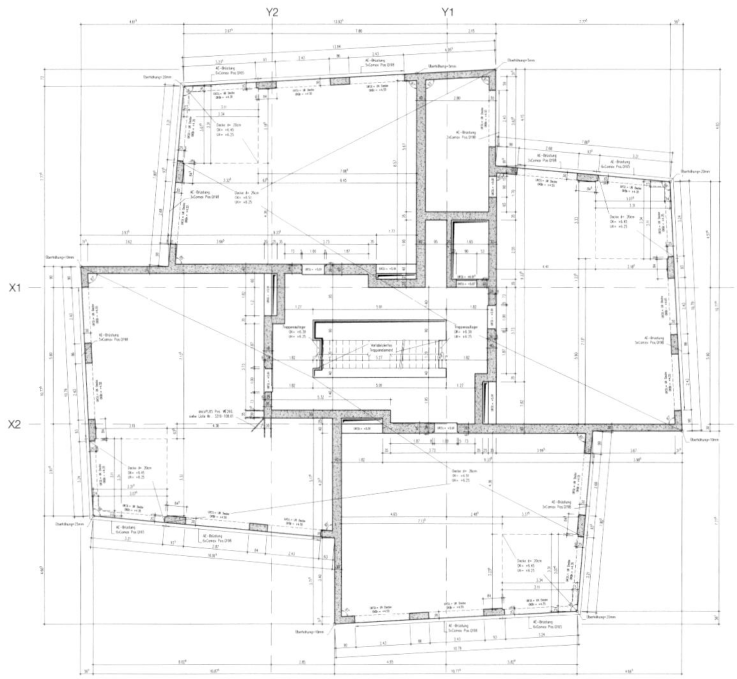

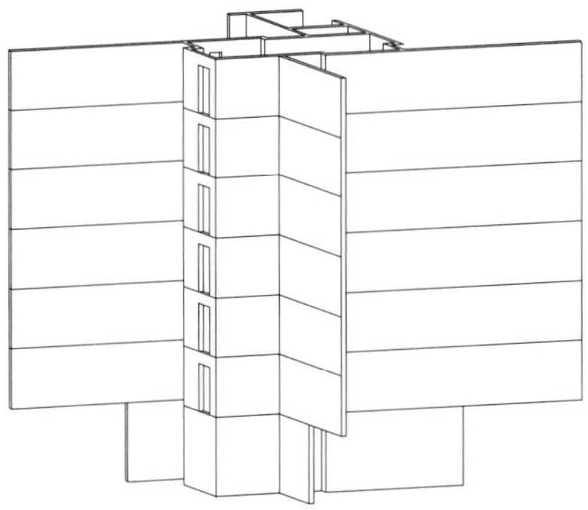

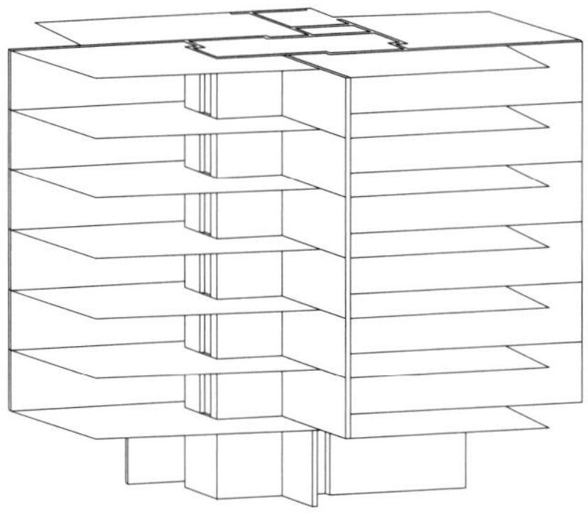

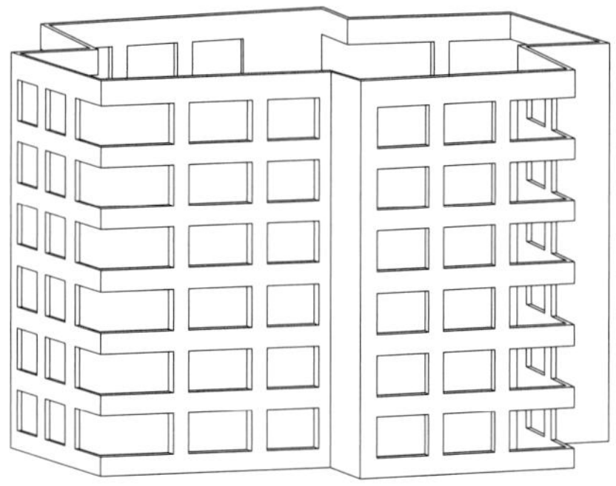

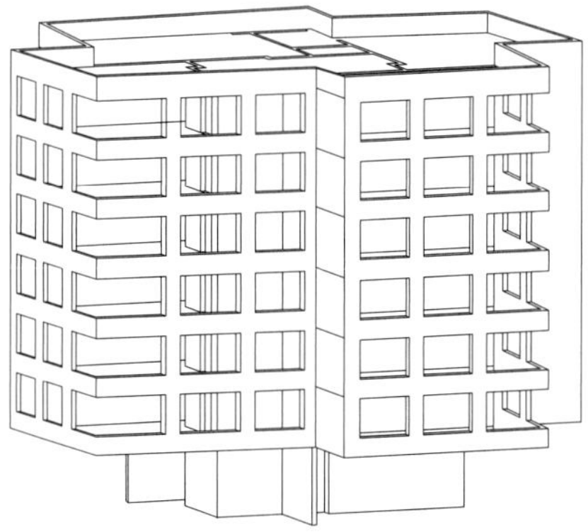

↑↑ Formwork drawing ground level (left); upper floor (right)
↑ Core with cantilevered party walls projecting on all four sides
↗ Core, party walls between apartments, and slabs
↗→ Load-bearing Vierendeel facade
→ Entire structural system with core, partition walls, and Vierendeel facade

4 Structure and Roof

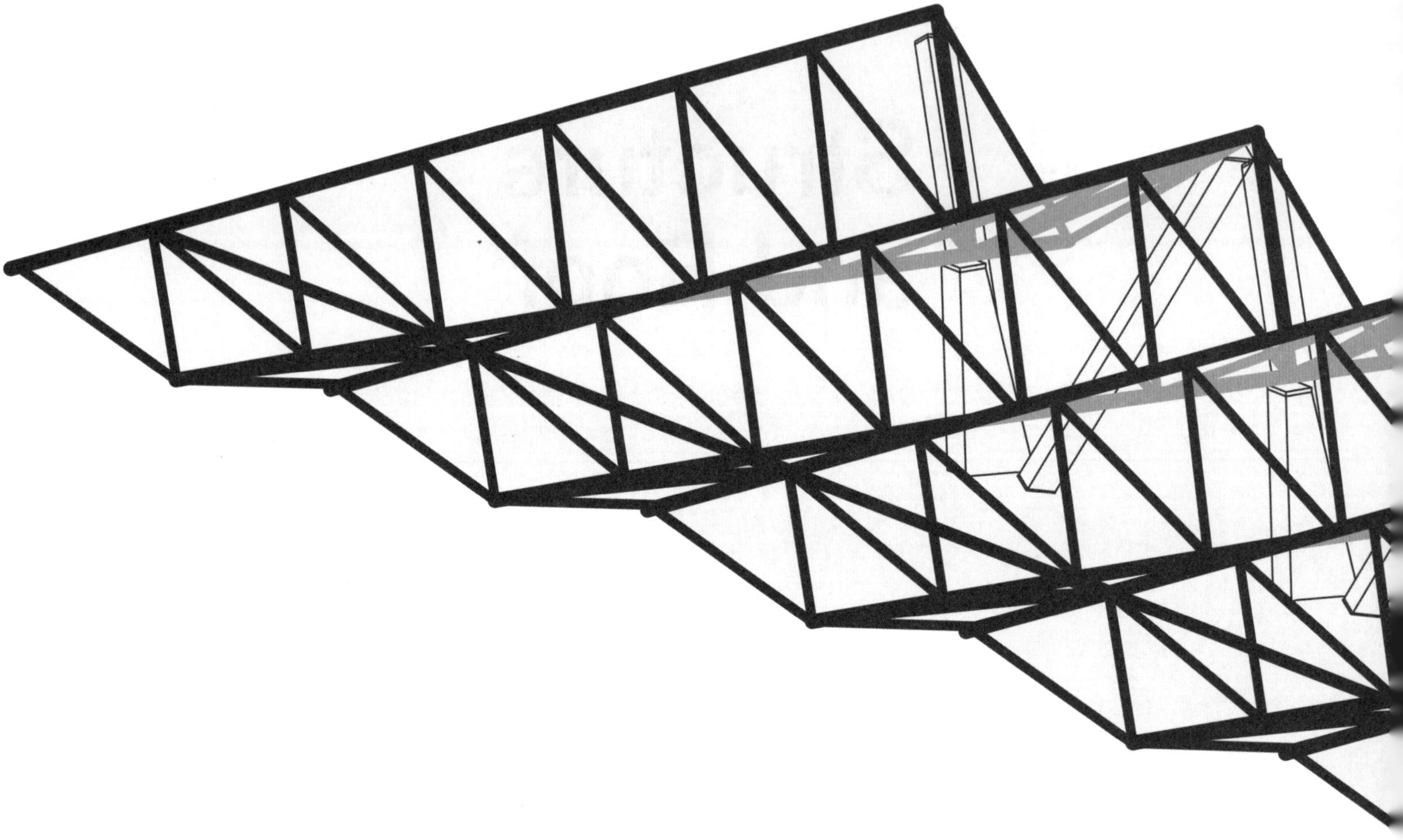

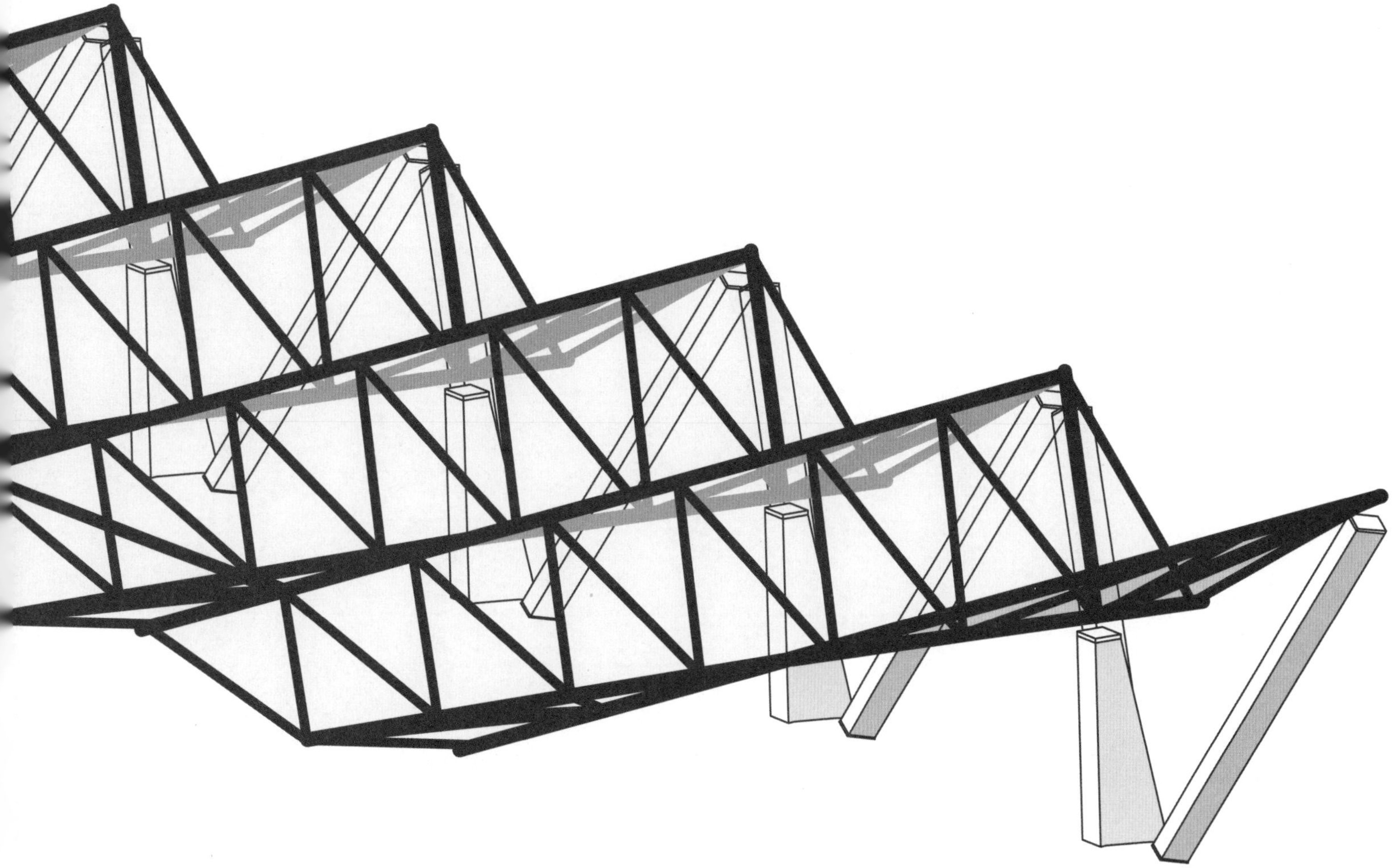

← Roof structure of Hardturm Stadium in Zurich (visualization)

Conversations IV

Strong Structures

AF What concepts do you associate with strong structures? What role is played by the purity of the material, the massiveness of the construction, or the load-bearing structure? Or are strong structures first and foremost efficient structures, which can also be comprised of hybrid, high-tech systems and clad?

HS Fundamentally, strong structures have nothing to do with clad structures. In the development of long spans that has taken place to date, however, hybrid structures are more efficient than pure structures. In the latter, the degree of purity becomes a question of efficiency. A strong structure is a structure that is efficient, has a clear geometry, is geometrically formed as a whole, has a strict inner logic, is suitable for the material and is readily buildable. In this respect, a strong structure may also be clad.

SB Strong structures excel over long spans. Indeed, they are only necessary and seen above a certain span length. A clear use of materials and a visible, legible flow of forces are further aspects of a strong structure.

AF So is a legible flow of forces a prerequisite for a strong structure?

SB Yes.

TP I would define it as both and. Of course, there are strong structures that are visible, whose force path is legible, and which are very pure. On the other hand, structures with smaller spans – which may even be clad – can be strong, too, if they constitute ingenious systems.

AF The opinions seem to diverge here – would you each like to give an example?

SB The Aare bridge is a prime example: an arched bridge, in which the forces are clearly visible. The arch, with its relatively powerful dimensions; its concrete abutments, in which the pressure is tangible; and also the delicate tension rods on which the roadway is suspended – a very clear system.

HS The way this bridge in particular works cannot be fully ascertained. Few people recognize the contribution made by the road deck to the torsional behavior of the bridge as a whole. Or take the example of Prada Tokyo – there you see the structure at first glance and think you understand it. But there, too, the floor slabs make a key contribution, which is probably not recognized right away. Both are structures that only seem at first glance to be understandable.

AF So is refined simplicity of the construction an issue?

HS Sure. Simple, refined structures fascinate me. A matchbox set atop two bearing lines is already a refined, simple structure.

AF Is there a crisis of scale – structures that have lost all sense of scale? Exaggerated somewhat: buildings that become objects, like swim rings, tires, or birds' nests. Can the scale of a structure be varied at will, back and forth?

SB There are purely scientific model factors that set limits.

AF That is a scientific, mathematical reason why things are not infinitely scalable. But it seems possible to give that impression, as various stadiums demonstrate, for example.

TP The need to accommodate huge crowds provoke stadiums into real jumps in scale, especially in relation to their context. And, of course, you can't derive a stadium from basketwork. Generally, for me this certainly raises the question of whether there is a confusion of scale, as with the Burj Khalifa Tower in Dubai.

HS But the question is, whether it is legitimate to take analogies of objects and scale them up to become a facade.

TP In my opinion, it's acceptable. If, for example, we recognize the possibility that a mesh can be stabilized by rods in a rhomboid arrangement, we can transfer this principle to a larger structure.

Dimensions
Spans

AF Inherent in roofs is always the challenge presented by large overall dimensions and long span lengths.

SB For the engineer, roof structures are the 'bridge construction' of building design. They have the fewest constraining parameters: no fire protection requirements and no building services connections. In an open stadium, there are neither thermal nor – at least not in the majority of cases – dynamic load requirements, if the roof is not used programmatically. This allows considerable design freedom. In this respect, long-span roof structures have a very special significance in our work.

AF Steel construction has a unique status for long spans. The delicateness of the load-bearing structure becomes the motif that characterizes space; this is manifest in your projects, too.

TP Owing to its strength, steel allows much more slender designs. As a result, a high degree of transparency can be achieved, for example in a ceiling. Moreover, steel can be prefabricated.

HS With long spans, steel's lighter dead load makes it superior to concrete structures, for which scaffolding must first be built. This gets very costly for spans of forty meters or more, because the structure is then very heavy. Other considerations can still make a concrete slab necessary, which is why one soon ends up using composite structures. With roofs, the geometry is what's most interesting – here, the geometry really lies in the hands of the engineer and depends upon his expertise!

Construction Site

AF Especially with large roofs, the fascination of the construction site is palpable. Communicating with the contractors, technical directors, and tradesmen on site is undeniably key. How much potential do you think there is in such close cooperation? To what extent can a design fail if this exchange isn't possible or if it's restricted?

SB As mentioned, the advantage of large roofs is that they are less dependent on other disciplines. The key aspects are feasibility, the element sizes, their transportability, and the assembly weights. These are the factors to be kept in mind. The construction process itself is often actually relatively easy.

TP With the extension to the soccer stadium in Basel, however, the construction process was crucial, because we had to build the new roof while the building remained in use. It was necessary to work in close cooperation with the steel fabricator and the general contractor. But that's true in general: with challenging load-bearing structures, there's usually a very intensive dialogue between the engineers and the contractors.

HS I would like to add two comments to that. The first is that we as a firm are reliant on building and implementing. Only sufficient experience in construction gives us the feedback needed for design. The processes of manufacturing and construction must always be present in an engineer's mind. The second point is that any design and any structure is ultimately subjected to the market. In order to gain a contract, a steel fabricator must be able to offer as low a price as possible and, accordingly, he offers constructive solutions that are as inexpensive as possible. If we want to retain our ideas, we need to give consideration early in the planning phase to how the details can be produced inexpensively – because otherwise, we won't achieve the desired result. This is very important, because we engineers are ultimately judged by what gets built.

AF You once remarked that the design side, i.e. the architects, attaches too little importance to carrying out the work.

HS Yes, this aspect plays a role. If the architect has no construction experience, the work is executed quite differently on a practical level and a certain competence is lost.

TP The contractor is usually selected according to the price.

HS But, of course, we try to manage things somewhat. When putting work out for bid, we select relatively solution-oriented contractors, similarly to how architects choose consultants with whom they can work well on developing designs.

Hidden Efforts

AF The hidden efforts of the construction process – do they interest and excite the engineer as much as the finished building does?

TP The engineer gets satisfaction from cleverly devising the construction process, that's clear. I'm thinking about the tip of Tadao Ando's building for Novartis, which weighs around six tons. The floor slabs were attached to the tip and then tied back with upstand beams – a very exciting process, which, of course, you cannot see now in the building. Naturally, I'm also interested in the finished spaces, but as an engineer, it is undoubtedly the shell that's primarily of interest – creating the building shell.

HS Or the Rhine bridge – built with the cantilever erection method, where we used presses to keep the sections apart until the final joint was closed – those are very impressive moments. The production process is, in any case, interesting as a part of the whole. Of course, we are especially pleased when the shell is so good that nobody wants to clad it anymore!

Structure and Roof

Stefan Bänziger

Designing and planning roof structures with large spans poses a particularly exciting task for the engineer, since many factors that restrict other structures are not relevant: usually, there are absolutely no noise or fire protection requirements, and if the roofs are not used programmatically, the deflection and oscillation behavior can often be dealt with easily from a structural engineering perspective. The vertical loads are relatively small when, apart from the dead load, no live loads but only snow and wind loads need to be considered; with very delicate structures, at most wind suction forces can be decisive for the dimensioning. This "free" starting point ensures a great amount of design leeway and offers optimization potential with regard to the use of materials and in planning the construction process, which can ultimately have an impact on the aesthetic appearance.

For span lengths up to approximately forty meters, the selection of supporting structures is limited for economic reasons. Usually, single-axis systems with glulam beams, prestressed concrete beams, precast girders or steel beams are of prime consideration. For span lengths over forty meters, the options expand to include spatial structures formed with trusses, post-tensioned beams, frames, or compressive arches.

Two soccer stadiums and a market hall follow as examples used to present different roof structures.

St. Jakob Park Stadium Expansion, Basel 2008

For the 2008 European Soccer Championship, we enlarged the St. Jakob Park soccer stadium in Basel together with Herzog & de Meuron. With a new, third tier of spectator seating on the long side to the north, around seven thousand additional seats were created. As all the seats were to be roofed over as before, the existing twenty-six-meter-deep grandstand roof first had to be dismantled and then replaced by a new structure that cantilevers out roughly thirty eight meters – and all this needed to be done without causing any disruptions whatsoever to the normally scheduled soccer games.

From both a structural and an architectural viewpoint, the new construction is well integrated into the existing structure. The added portion attaches to the existing structure like a chair with an umbrella, and with a deep cantilever, it roofs over the old and the new. At the sides of the new roof, sloping intermediate areas join it to the existing roof. The new facade is comprised of illuminable air cushions that uniformly enclose the entire volumetric form. Each of the cushions has a span of twenty-four meters and is fixed every eight meters to the rear column row. Both the form and the dimensions of the entire grandstand structure, including its foundation, are significantly impacted by loads from the roof structure – the loads imposed on the bleachers by the spectators only play a minor role.

The roof structure consists of triangulated main trusses laid out in the existing eight-meter-wide planning grid and purlins that span in between to support the metal roof decking. Along the front edge of the roof and roughly through the middle of the main trusses, longitudinal trusses add stabilization and a certain amount of redundancy. Whereas most of the purlin sections and metal decking from the old roof structure could be removed and reused, the existing trusses had to be completely discarded because of their geometry. The main trusses of the entirely newly constructed trussed framework each sit atop a slightly inclined steel column that rests in turn on a sawtooth-shaped grandstand girder. This girder formally constitutes the extension of the existing lower grandstand construction. A tension rod anchors the main truss on the short end at the back. Vertical tensioning cables in the new concrete structure conduct the tensile forces from the roof edge down into the pile foundation.

1

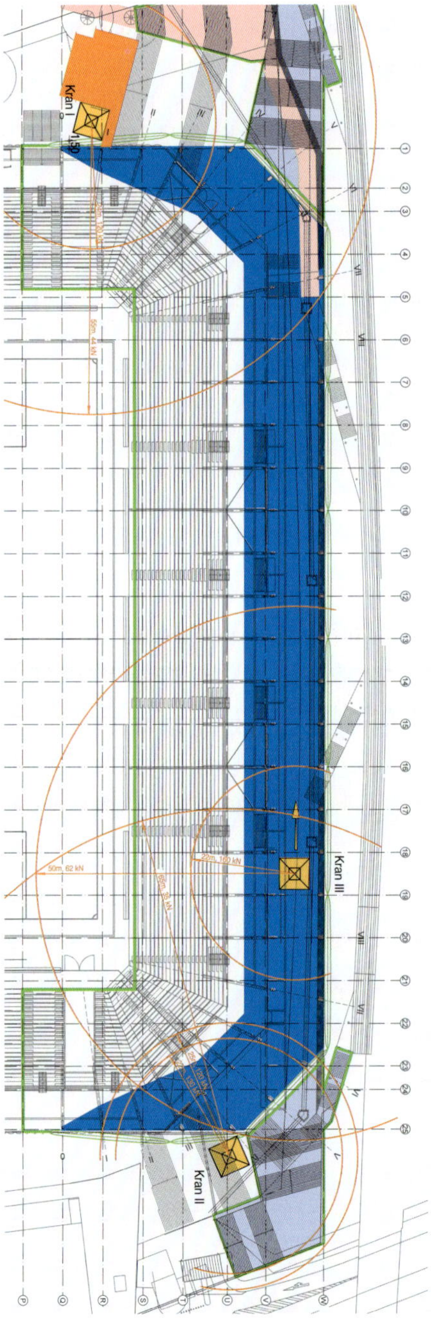

Functional, architectural and structural aspects do not always influence one another positively. In these cases, it is important to find sensible, refined, and effective compromises. The position of the roof support in this project represents one such case: The dimensioning of the entire structure is determined to a significant degree by the position and inclination of the support on the grandstand tiers. The lower it is positioned, the more economical the construction. Conversely, the obstruction of view increases and, as a result, the quality of the new seats decreases. The selected disposition is, in this respect, a sensible compromise between structural engineering and desired spatial effect, and moreover, it made it possible to keep within the prescribed tight budget.

Im Viadukt Market Hall, Change of Use of Viaduct Arches, Zurich 2010

In the triangular interstice between Wipkinger Viadukt and Lettenviadukt is Zurich's first enclosed market hall, which we were able to develop and realize together with the architects EM2N for the PWG Foundation. Whereas the rough cyclopean masonry from the railroad viaduct built at the end of the nineteenth century still appears on the walls, the space of the hall itself is spanned by a folded roof structure that mediates between the geometries of the viaduct arches.

The roof's hybrid structure, built of wood elements and an integrated steel truss skeleton, makes it possible to optimize the component dimensions and create elegant solutions for the bearing details at the areas of concentrated forces in the quarry stone masonry. The wood elements of the roof span between the trusses. They consist of glulam ribs, forty centimeters high, that are insulated with mineral wool and clad to resist shear with three-ply plywood on the top and sheets of OSB on the

2

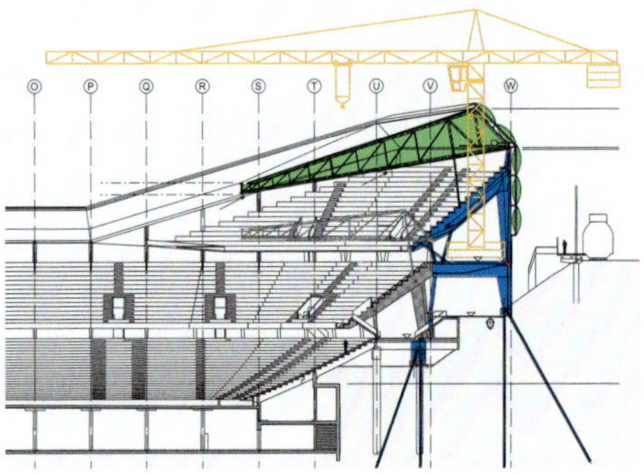

1 Construction site overview of St. Jakob Park Stadium, with three rotary tower cranes
2 Cross-section of the stadium, with mobile crane on tracks on the new peripheral gangway

bottom. In the finished state, these elements also seem like planes that stiffen the roof areas horizontally and stabilize the steel trusses laterally. The vertical surfaces between the trusses were insulated with mineral wool and clad in situ.

An important requirement was to eliminate the transfer of vibrations from the railroad operations above. In combination with the dissimilar thermal expansion of the two different viaducts and of the roof, which is comprised of different materials, the bearing concept therefore presented a special engineering challenge.

Fundamentally, only the steel structure is connected to the existing masonry at isolated points. Wipkinger Viadukt continues to serve railroad operations, so no vertical forces were permitted to be conveyed its piers. Because only slight horizontal forces were allowed for stabilizing the hall roof, the structure needed to be self-supporting between the viaduct piers. With the decommissioned Lettenviadukt, by contrast, all the trusses sit on grouted steel sleeves, whereby sliding bearing were used that only transmit vertical forces. On the inside of Wipkinger Viadukt, shock-absorbing tie rods, which establish the fixed point in the transverse direction of the hall, were drilled into the piers. In the longitudinal direction, a floating bearing is provided inasmuch as at Wipkinger Viadukt, the roof rests on V-shaped steel supports that form a rigid frame together with a continuous roof truss and the gallery's girder grid. The steel columns, in turn, are fitted with shock-absorbing bearings.

Competition for Hardturm Stadium, Zurich 2012

A soccer stadium lives from the game for 90 minutes each week and, for the remaining 9990 minutes, from architecture and thus also from the supporting and spatial structure. Since, in this case, the load-bearing structure has a significant impact on the architectural appearance, in our submission to the competition for Zurich's new stadium, it should be legible and also palpable for the spectators: that's a special task for the structural engineer. Thus, in addition to the structural parameters, design aspects also had significant influence on the structural design for this competition project.

The structure of the roof is folded and tilted toward the center of the stadium, thus creating a bowl-like effect that enhances the atmosphere. The load-carrying capability comes from three-dimensional steel trusses that are tapered and cantilever out to the roof edge. Each truss is restrained by a vertical extension of the grandstand girder and two lateral splayed columns. Both structural elements jointly take up the moment arm that is necessary for the restraint: the grandstand girders act as compression elements and the columns absorb the tensile forces. Slightly warped and aligned next to one another, the clad trusses form the folded roof plane that typifies the project. They give the stadium a distinctive silhouette and identity, and their attributes are thematically continued on the vertical surfaces of the facades.

The cantilever is typically just under nineteen meters with an effective height between one and three meters (at the front edge of the roof and at the restraint, respectively). The spacing of the trusses is determined by the 10.6-meter planning grid of the grandstand girders. By elevating the roof at the playing field side such that the lower chord has a slope of about three percent, the roof drainage is solved "naturally" with a gutter in each of the bottom folds. A circumferential horizontal stiffening girder at the second panel point near the tip of the truss compensates the horizontal forces that result from the folding. The peripheral catwalk for roof maintenance is also located here. In restraining the cantilevers, the splayed columns absorb the horizontal forces from the folds.

St. Jakob Park Stadium Expansion
Basel 2008

EINMAL ASAG – IMMER ASAG
EINMAL VW – IMMER VW
EINMAL BASEL – IMMER BASEL
EINMAL ASAG – IMMER ASAG
Panasonic
MANOR

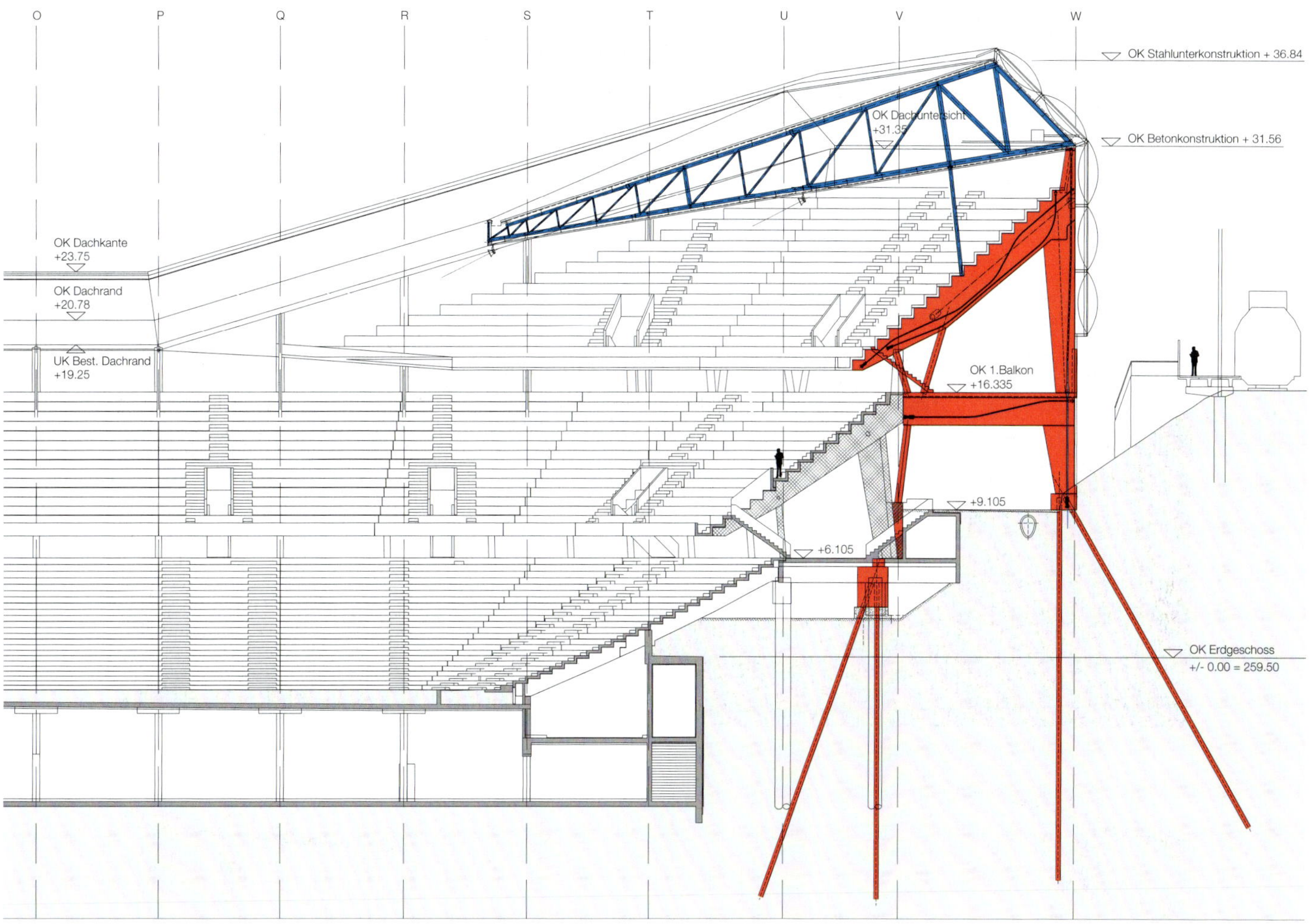

←← Staged construction in cross-section, with old roof as interim storage and work area
↖ Truss lying on ground, pre-assembled
← Ongoing construction during normal stadium operations
↑ Cross-section with new elements (red)
→ Truss being assembled with a crane

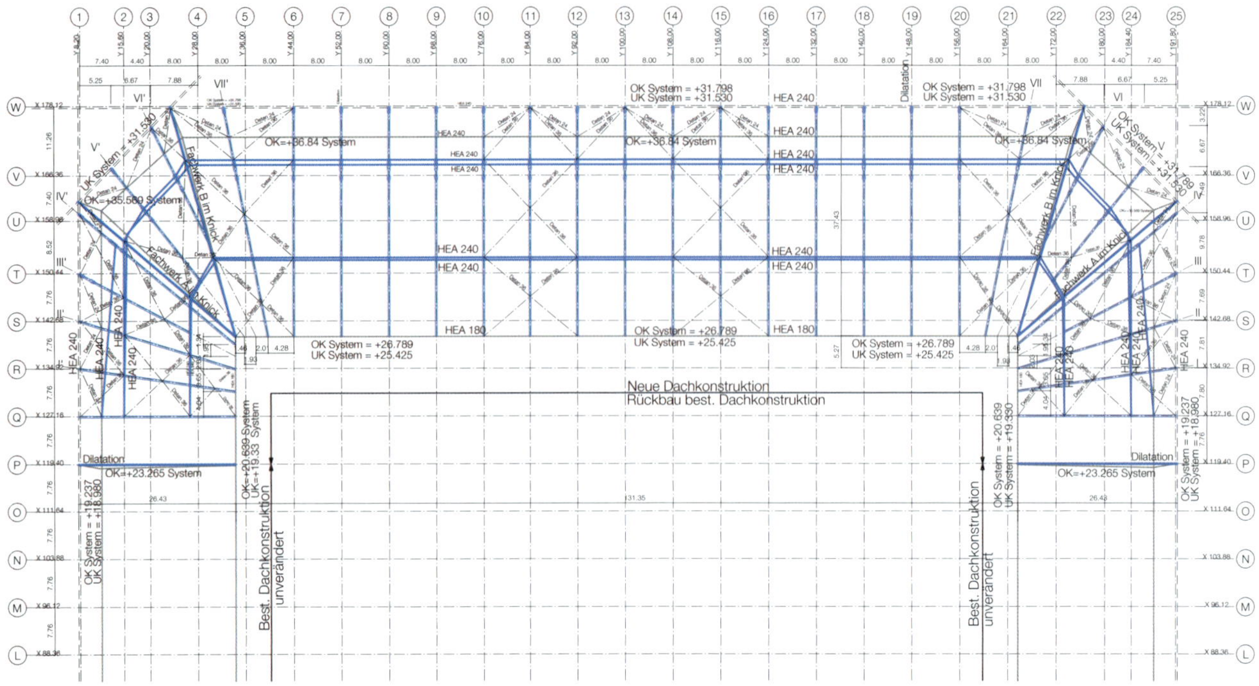
OK System = +31.798
UK System = +31.530
OK=+36.84 System
OK=+35.569 System
Fachwerk B im Knick
Fachwerk A im Knick
HEA 240
HEA 180
OK System = +26.789
UK System = +25.425
Neue Dachkonstruktion
Rückbau best. Dachkonstruktion
Dilatation
OK=+23.265 System
OK System = +19.237
UK System = +18.980
OK=+20.639 System
UK=+19.33 System
OK System = +20.639
UK System = +19.330
Best. Dachkonstruktion
unverändert

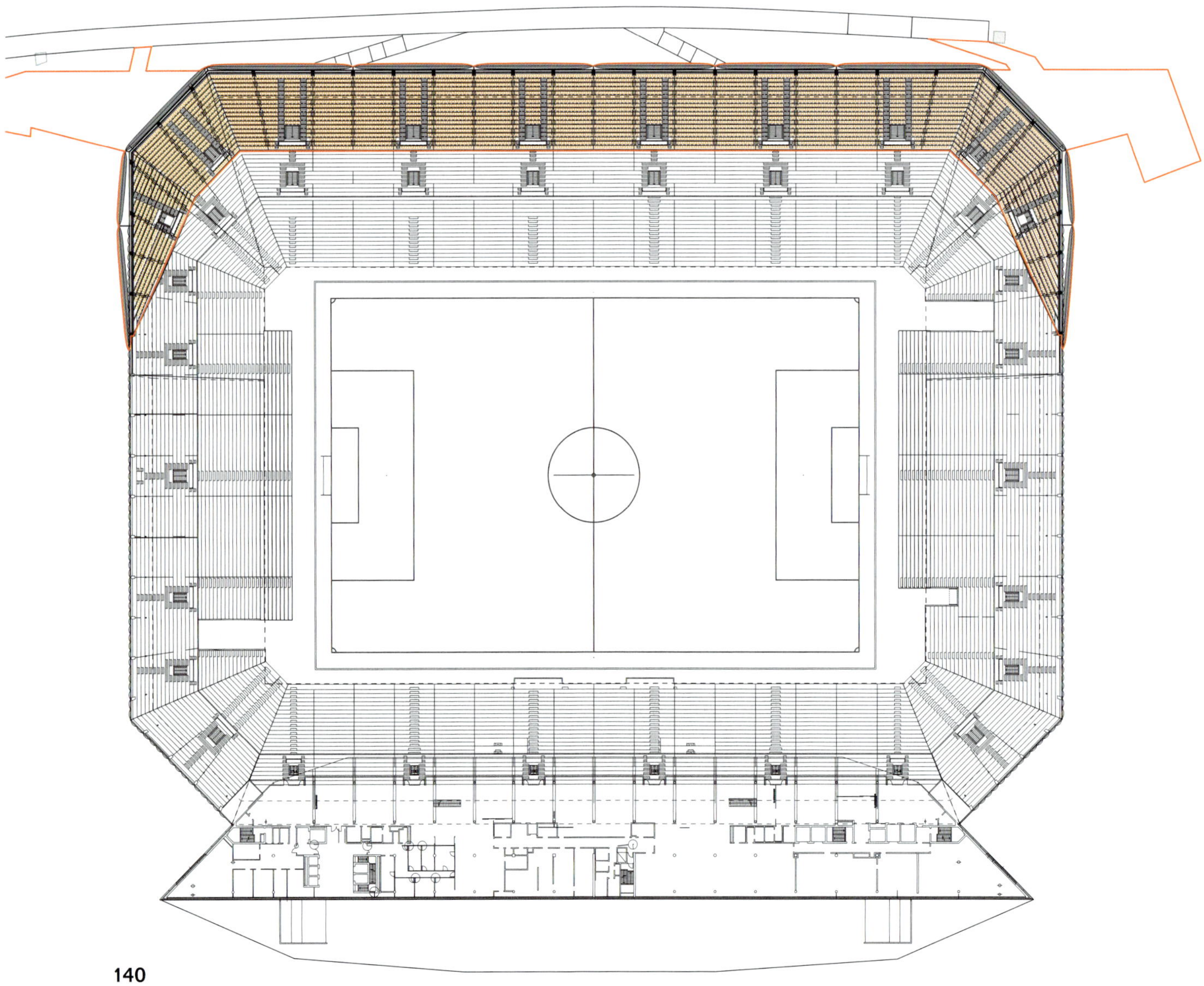

↖ Overview of steel structure for roof extension
← Plan of entire stadium at the level of the enlarged seating gallery
↑ Sloping transition from the existing roof to the new, higher roof. In the background: St. Jakob Tower
→ Exterior view at night with illuminated air cushion facade

Im Viadukt Market Hall, Change of Use of Viaduct Arches Zurich 2010

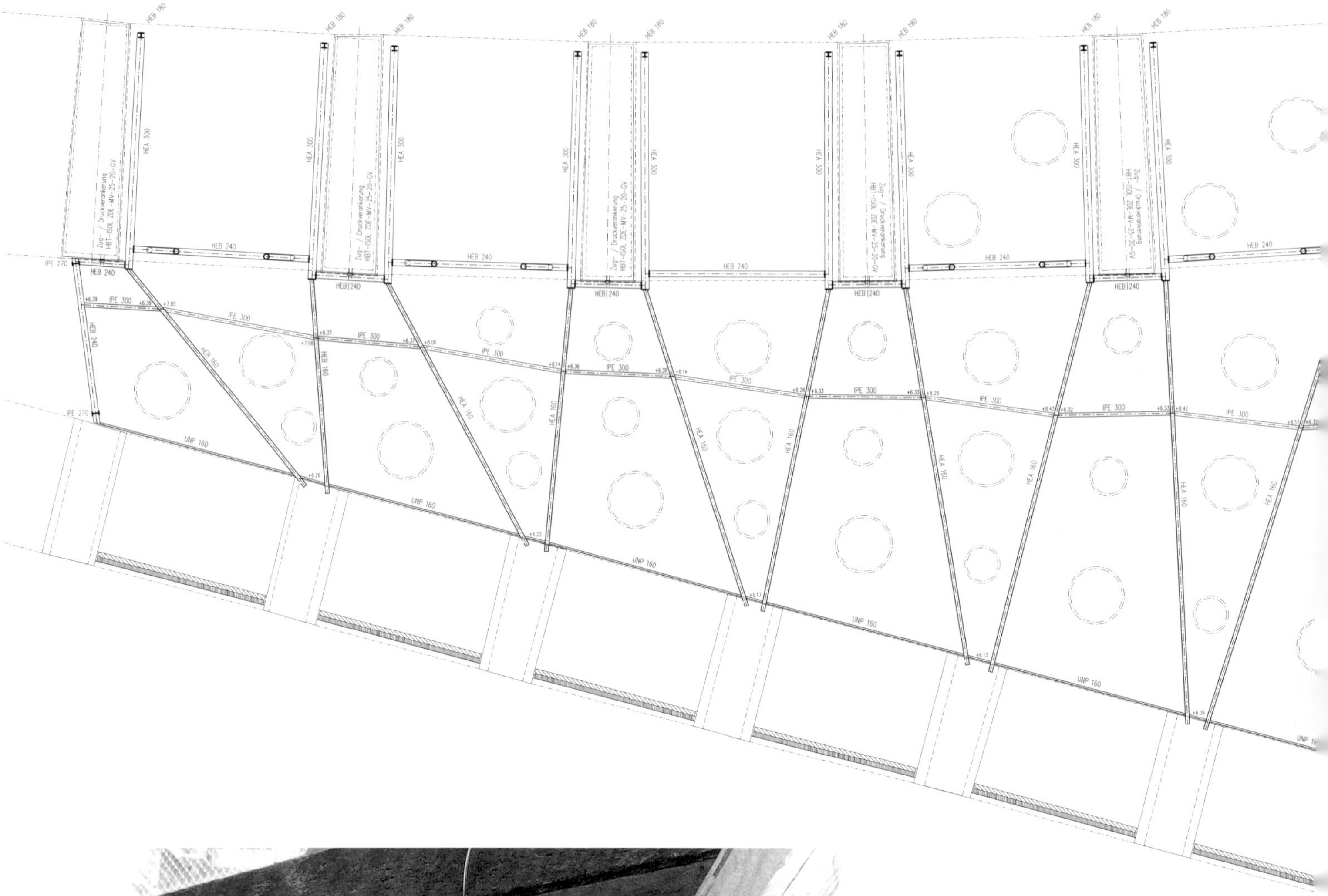

←← Insulated steel trusses prior to being clad with wood elements
↑ Plan indicating the individual steel members of the roof structure
← Bearing detail of Lettenviadukt prior to recess being filled with concrete
→ Cross-section with steel structure of the market hall

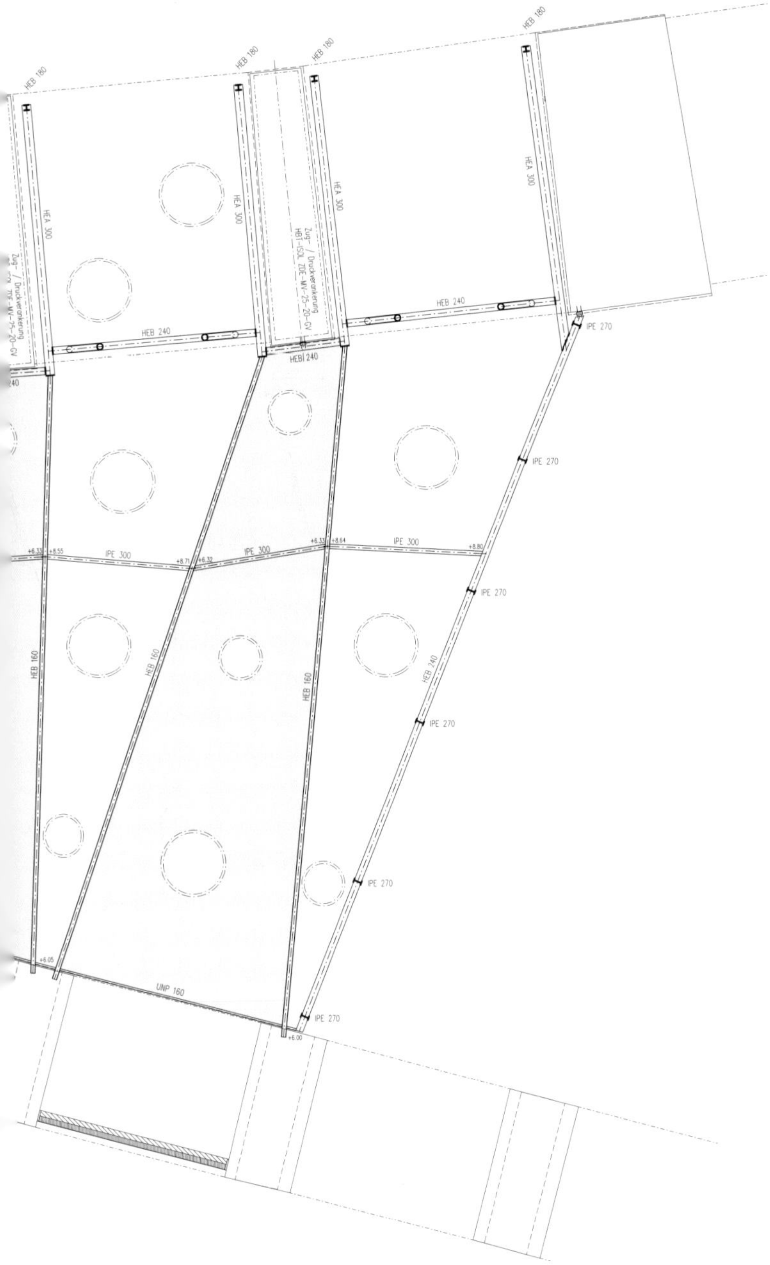

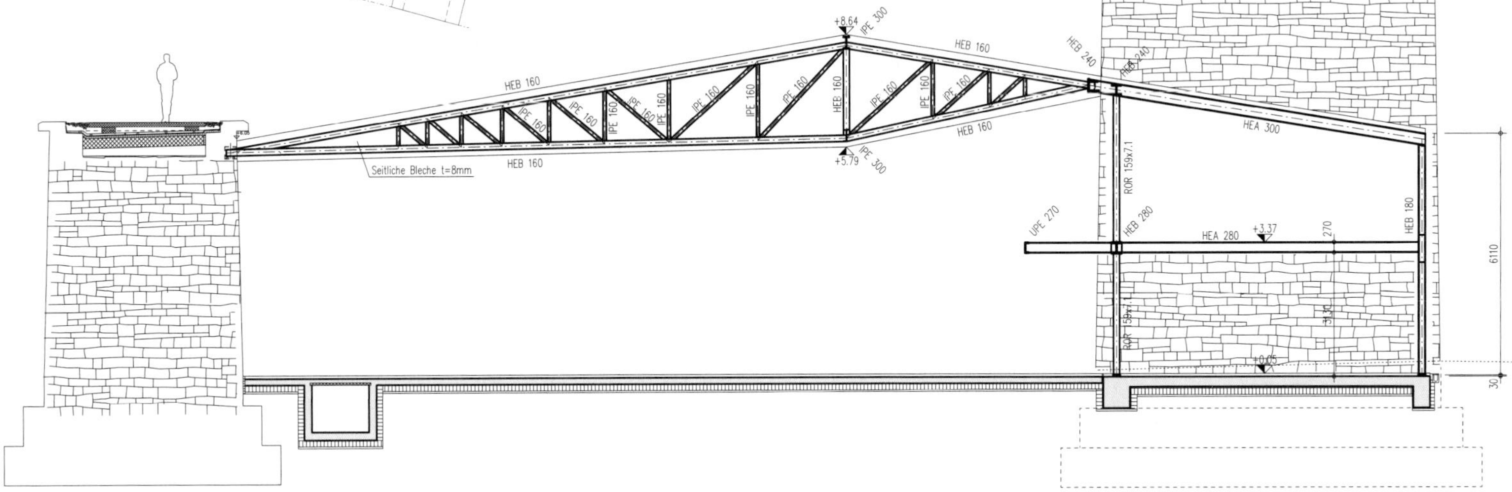
+8.64
IPE 300
HEB 160
HEB 160
HEB 240
HEB 240
IPE 160
IPE 160
HEB 160
HEB 160
Seitliche Bleche t=8mm
HEB 160
+5.79
IPE 300
HEA 300
ROR 159x7.1
UPE 270
HEB 280
HEA 280
+3.37
270
HEB 180
6110
ROR 159x7.1
±0.05
30

↑ Underside of the finished roof
→ Finished roof with EPDM membrane
→→ Evening impression of the illuminated Wipkinger Viadukt

amalgan
...30
...29
FASHIONSLAVE
...28

Competition for Hardturm Stadium
Zurich 2012

← Exterior view (visualization)
→ Cross-section of grandstand structure at VIP box area
↘ Prestressing concept for the grandstand structure
→→ Juncture between roof and grandstand girder, typical condition (left), VIP box area (right)
↓ Cross-section (visualization)

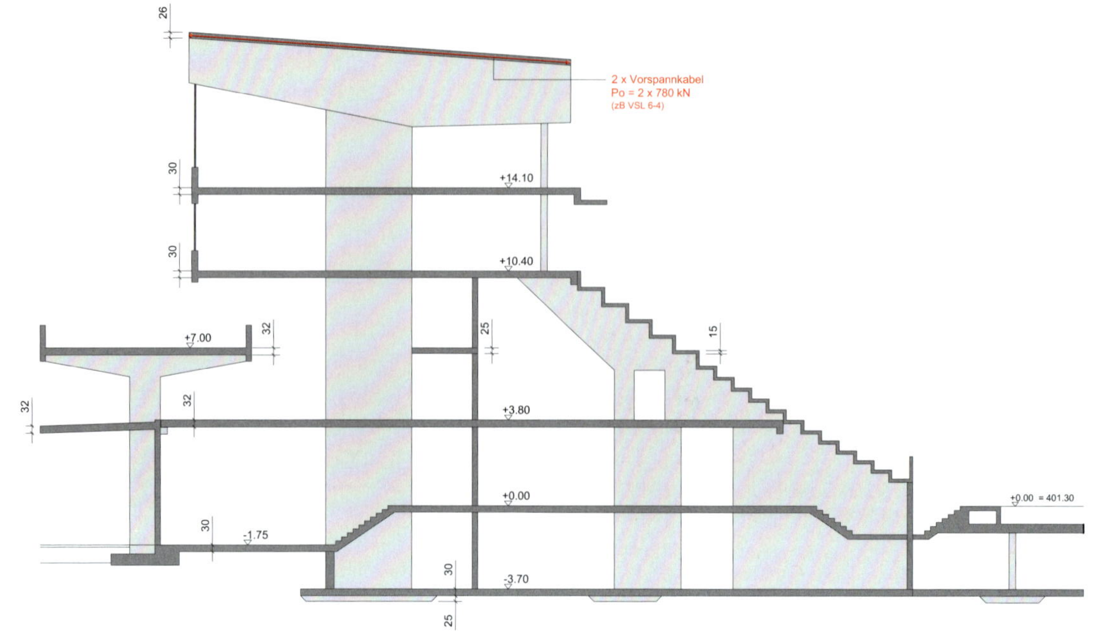

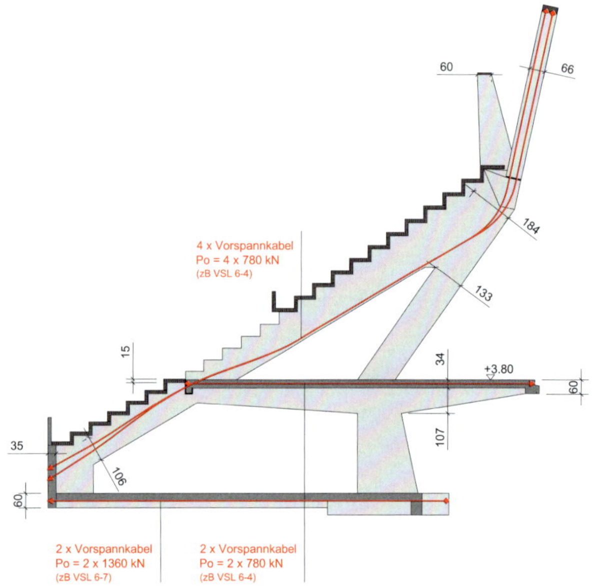

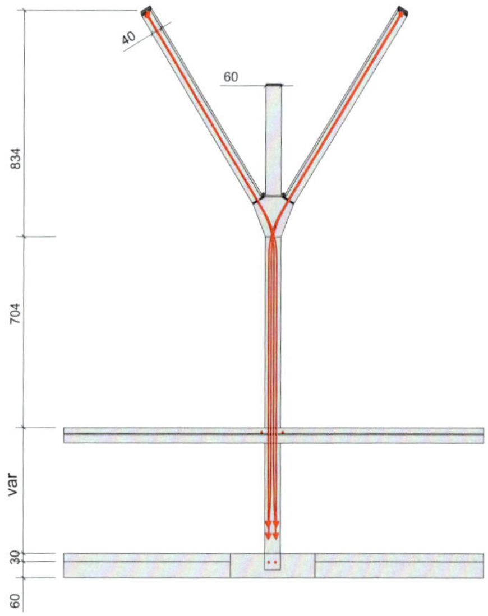

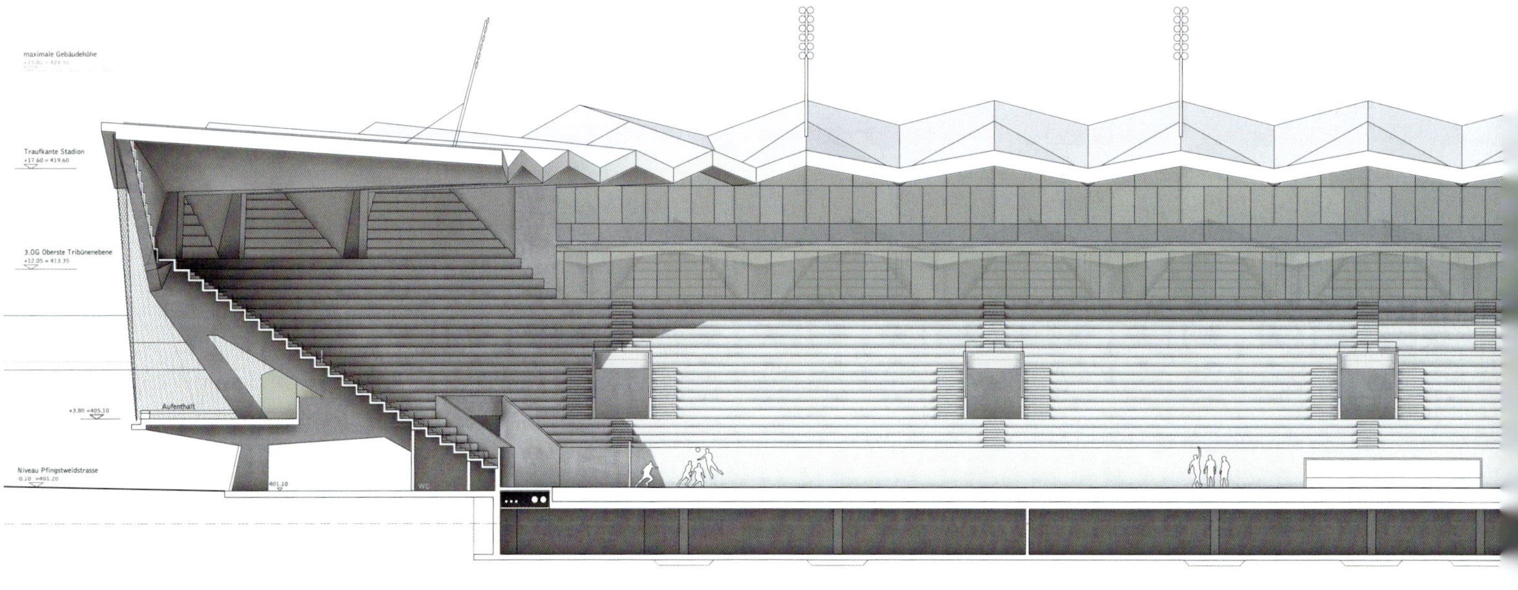

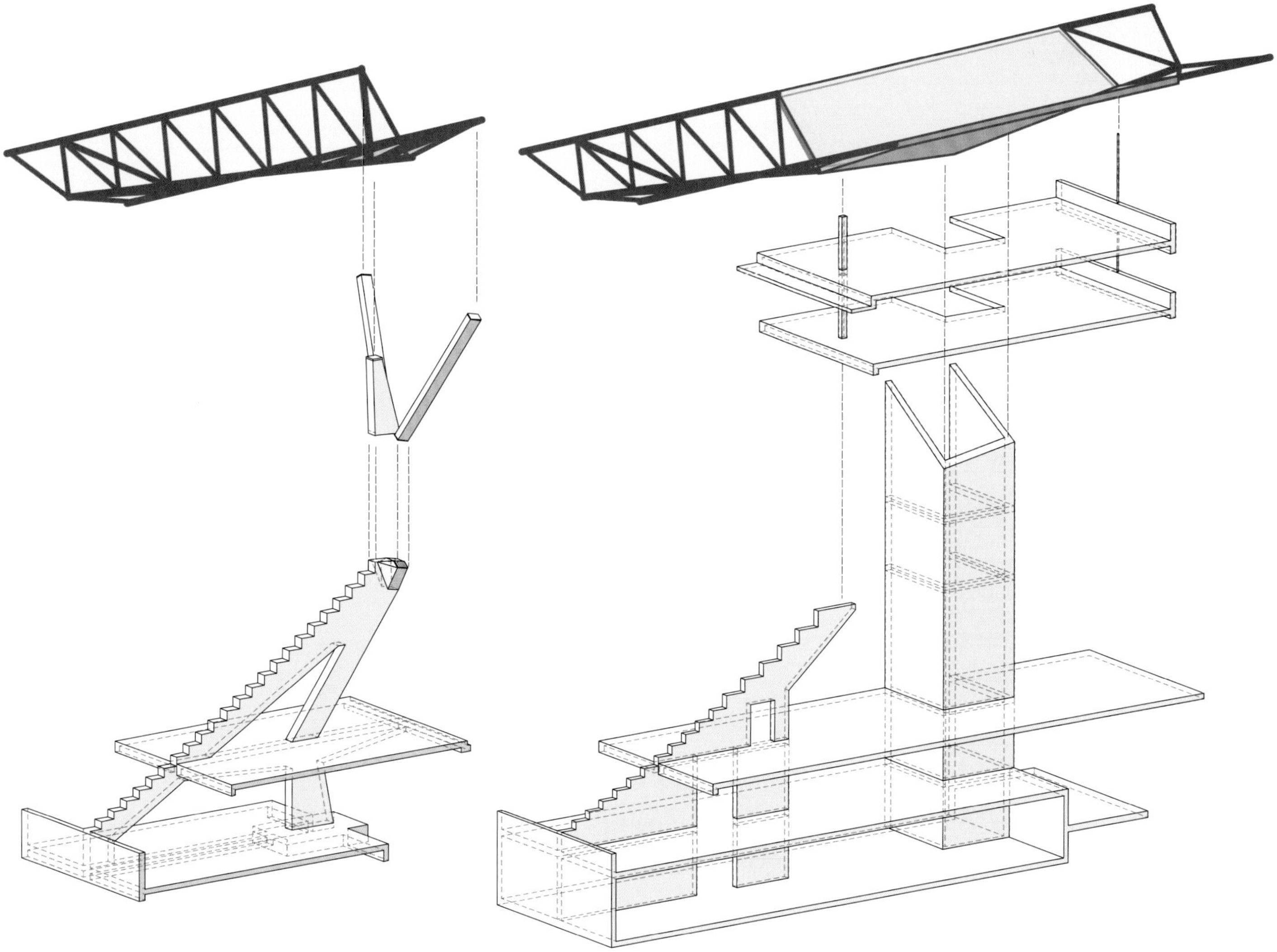

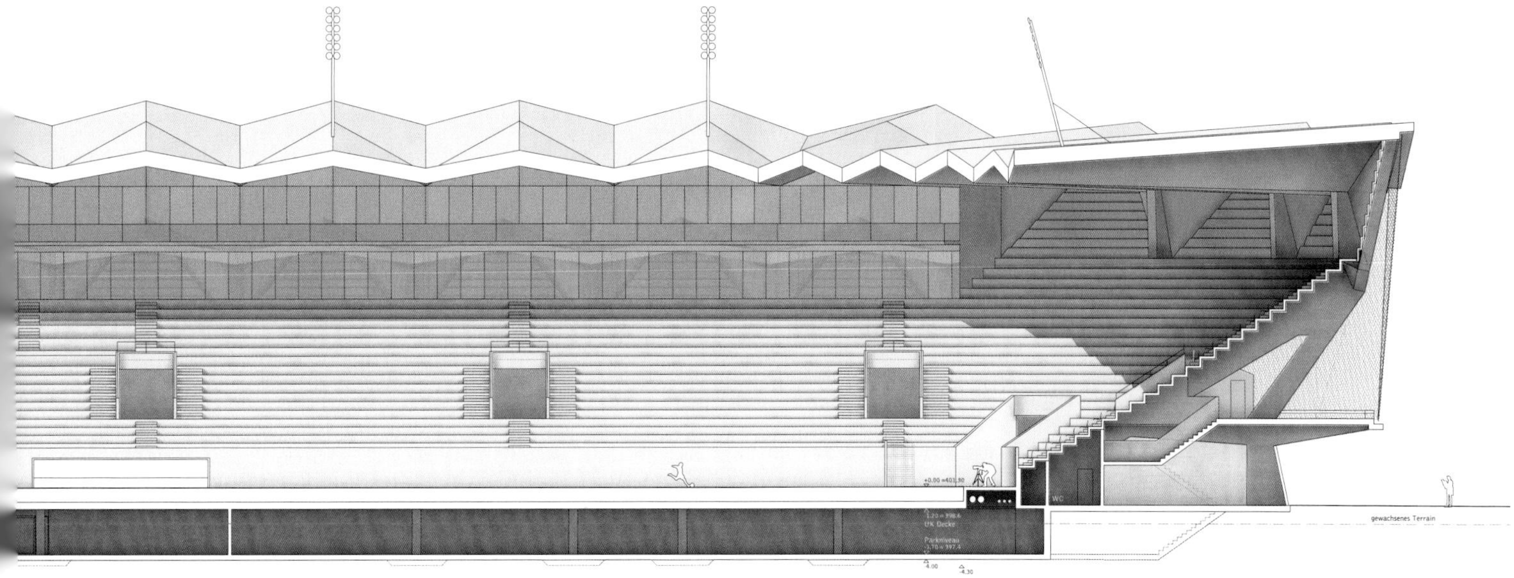
WC
UK Decke
Parkniveau
-4.30
gewachsenes Terrain

5 Structure and Facade

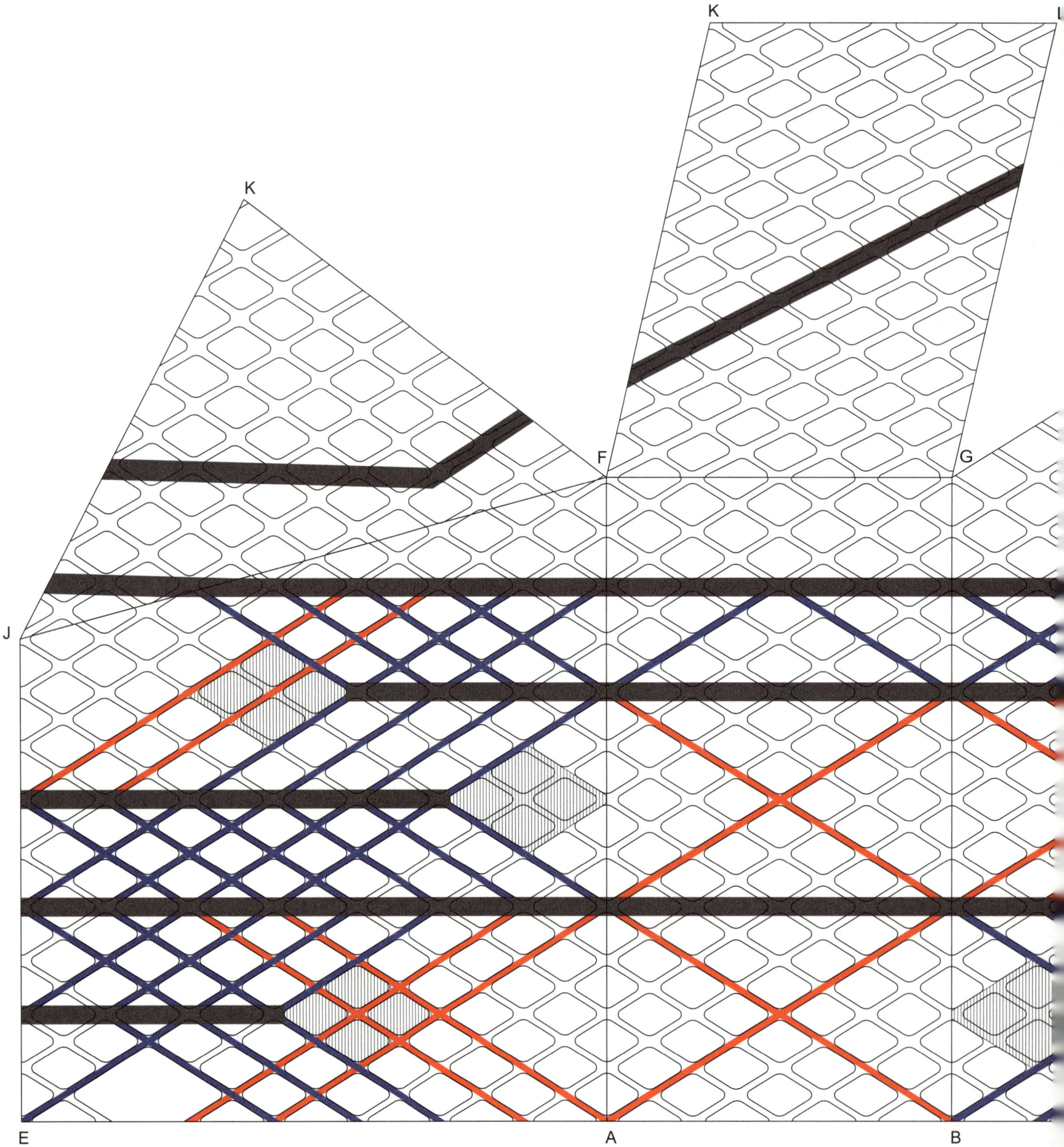

K
K
L
F
G
J
E
A
B

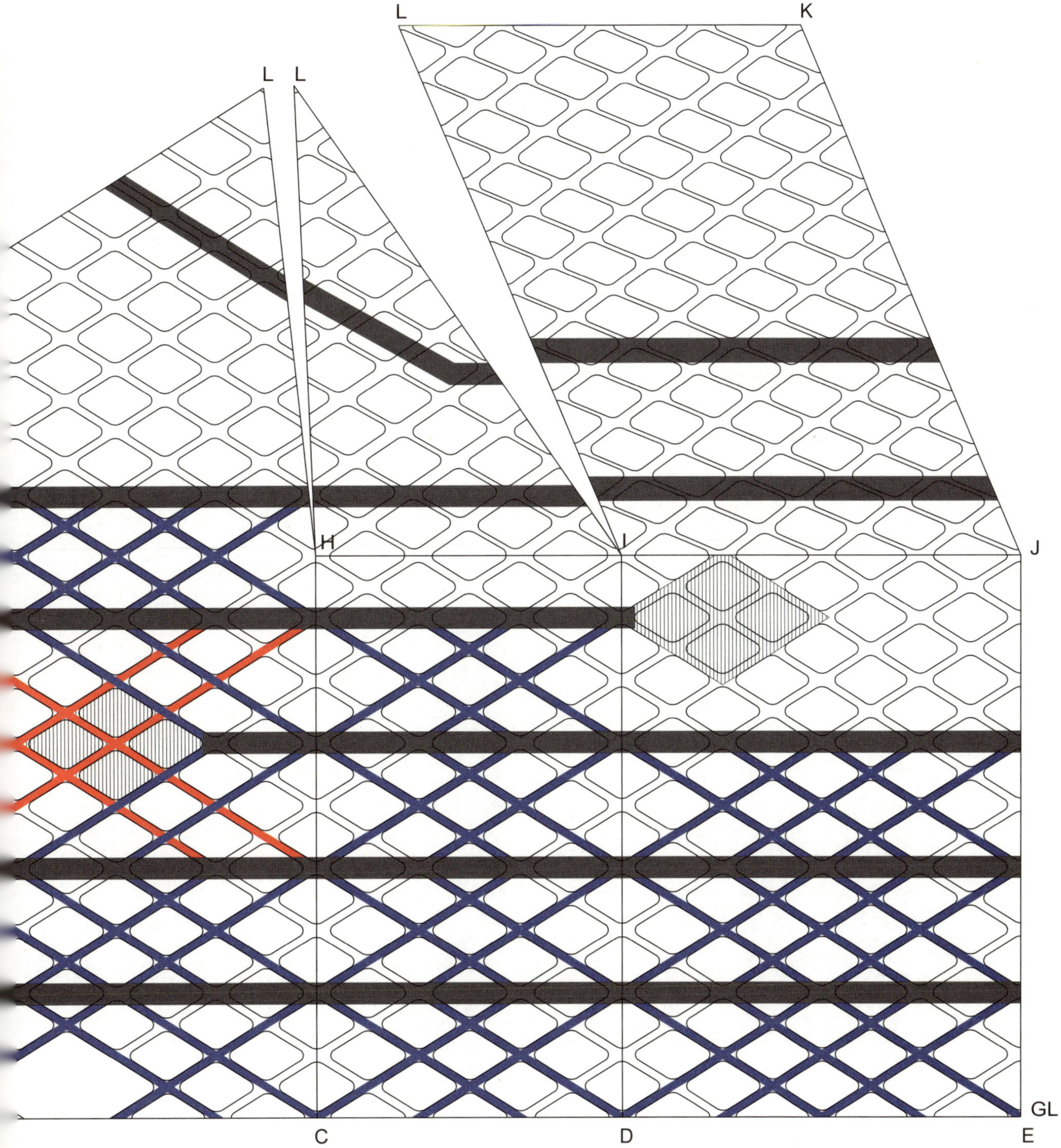
L
K
L
L
H
I
J
GL
C
D
E

← Developed facade,
Prada Aoyama, Tokyo

Conversations V

Learning from Nature?

AF Throughout architectural history, nature has repeatedly served as a model for construction and design. To what extent do structural concepts or construction in general have something to do with imitating nature – with learning from nature?

HS That depends on the scale you're looking at to learn from nature. A blade of grass that bends in the wind and whose deformability enables it to absorb smaller forces certainly tempts me to make a comparison with earthquakes and skyscrapers – since a principle that is valid in nature has been adopted in this field. Bone structures, like the familiar model of stress trajectories in hip bones, which, however, are much more complex than buildings, certainly serve as another example. If we examine the bone itself, we discover that, for the sake of lightness, it is hollow inside – these are in turn construction features or principles that engineers have adopted. A one-to-one translation of such principles, like what Santiago Calatrava does, for instance, seems to me to be more of a gimmick – we engineers are not nearly as good as nature itself! The way we use analogies from nature is perhaps well illustrated by the example of Actelion Business Center. The building consists of individual, layered beams stacked on top of each other. During the development process, we sought the optimal arrangement of the truss members. We considered how they deform and realized that the analogy of a tree seemed to fit the structure fairly well – but we had not sought a tree structure from the beginning. The process let us discover that the tree analogy would be suitable for the principle.

TP Certainly, engineers try to derive things from nature. Woven or tree structures come to mind, like those used by Professor Jörg Schlaich, for example, or the branchings of lightweight structures that disperse toward the top – principles that may have been used even more often previously – when materials were more expensive than labor. There are different aspects where we look to nature for guidance, be they vortex shedding, core analogies, or even the oscillation behavior of a building. All these things have to do with natural phenomena.

HS As already stated, it all depends on the scale. At the level of the load-bearing structure, for example, physics compels you to look for solutions that are similar to nature. At a small scale, however, we are confronted with materials and the production process – and that's where we very quickly reach certain limits, because we simply don't have the possibilities that nature does. Take cobwebs, for example – we would love to have that available as a building material!

AF What do you think of analogies that aim at achieving highly efficient structures, structural elements or node typologies with as little material as possible and reduced resources?

HS We are clearly rooted in our time. For the bridge construction project in Bhutan, we investigated the relationship between hourly wages and the costs of materials. It is roughly equivalent to the situation here [in Western Europe] in 1900, as engineers were still compelled to develop extremely material-efficient structures. We see that in industrial halls from this period and in Maillart's bridges. These days, efficient building here at home means low labor costs. The challenges of material economy will not again become relevant until the relationship between materials and labor costs again go out of kilter.

TP In designing ribbed slabs, slabs with integral beams, or decomposed structures in general, we also take a cue from nature in an abstracted way. But we always seek a balance between economic and material efficiency. Today, we can build an eight-centimeter-thick concrete wall that costs twice as much as a fifteen-centimeter-thick wall. That's a phenomenon of modern times – the resources are too cheap!

HS That's a fundamental dilemma in which engineers find themselves – there are no incentives for economy of material, which would actually be an important guideline for the design.

TP When we were young engineers, slab thicknesses were sixteen centimeters or less. Today, slabs are crammed with MEP systems; together with the sound insulation requirements, slab thicknesses are being driven up, so that twenty-eight centimeters are now common. It's a paradox: although we can build more delicately, we live in a massive world!

HS At the same time, sustainability is on everyone's lips.

Sustainability
High Tech/
Low Tech

AF What do you think about high and low tech – and which do you prefer?

SB I'm a low-tech guy. As an engineer, you can plan the most complex things on paper, but it must also be possible to build them – one needs to be conscious of that. Deliberations about a construction's susceptibility to flaws and how it can be tested are extremely important. High-tech constructions are, of course, much more susceptible. Personally, I get the most satisfaction when I can tackle a structurally complex problem with a solution that is as simple as possible to carry out.

TP Whether low tech or high tech is a contextual question – a question of how the problem is defined. We are not averse to low technologies, which are more likely to be used in places like Bhutan. A temporary exhibition pavilion, conversely, can be built perfectly well using high tech. Fundamentally, though, I believe that low tech is more sustainable.

HS Our construction sites are outside in the wind and rain. Then there's the training of construction workers – the trades are quite simply low tech. With high-tech designs, we can contribute ideas – but that's just the preparatory phase. Thinking about how the materials interact could be considered high tech, whereas the production process, and hence the final product itself, is low tech.

AF To classify things properly – Actelion was therefore high-tech planning, but a low-tech construction site?

HS Yes – that's exactly what was the big challenge for the contractor! We were able to cope with the geometry and the calculations for Actelion with the aid of the computer; the truss members were cut to length and drilled in the workshop by computer, too. But then these truss members had to be assembled in an ordinary, low-tech process. Setting the elements on top of each other and aligning them in the correct position on site were low-tech activities, and we quickly reached certain limits. In the end, the low tech on the construction site is the crucial factor: with steel construction, Actelion was at the limit of what is feasible!

TP We've also built some pavilions and additions using high-tech construction, such as this carbon-fiber box as an extension to a single-family house. High-tech connections, special adhesives, and glued elements were used – so in that respect, the actual construction work was also high technology, although here, too, the manual assembly is again low tech. These adhesive bonds are aligned in a protected space, making it possible to verify that they fit together properly. The materials, even the adhesive, is, in other words, very high tech, but as soon as a person comes into play to assemble the parts, we are again dealing with low tech.

SB We're not mechanical engineers – we always build prototypes. In other words, we have to get it right the first time!

AF Do the aesthetics of engineered structures come from the laws of nature and does efficiency lead to elegance?

TP Definitely. We are trying, for example, to create a roof structure by abstracting the elegance of a tree and its branches. Material efficiency is always linked with delicacy, which is, for me, associated with elegance.

HS Material efficiency can, but need not necessarily lead to elegant structures. The Leutschenbach school is certainly efficient in its load transfer and also elegant – yet presumably, nature would not set a large mass atop four small legs. So it's clearly an efficient support system that's not necessarily "natural."

Technology

AF How would you describe your basic understanding of technology and the application of technologies – can technology be appropriated limitlessly? Does it dominate or enable?

HS Those are loaded questions!

TP Technology is more of an enabling element that we take advantage of in our designs. It's just one factor; people and their perceptions also come into play. After all, we aren't building machines!

AF But as engineers, surely you are extremely interested in technology, aren't you?

HS We are interested in technology, but we're not devout believers in it. I must say that my view of technology has changed over the past ten years – I was much more of a technophile than I am today. Today, I see the limits of technology in a different light; they begin earlier than you think, and that's a good thing.

SB Through digitization, a major technological development has taken place in the planning process; we're pushing technology to the limit. But the work done on the construction site is still relatively archaic. As a result, the contrasts between designing, planning, and building have intensified.

HS Perhaps once again in fundamental terms: nature is more powerful than technology and, over time, it always manages to reconquer its territory. A powerful image for me is the street in front of my house: it was paved six years ago, but the first weeds have already begun to sprout through the asphalt. I also keep this image in mind with regard to our buildings.

AF What role does mathematics play for you? And abstract axiomatic thought? Are you mathematically minded engineers or do you work with graphical statics?

Mathematics
Mechanics

SB As an engineer, I've undergone development: During my training, mathematics was an important tool that aided my ability to visualize, and the longer I've been working and the more experienced I am, the more intuitive my approach becomes.

AF For helping me to visualize forces, however, the Cremona diagram was more useful than mathematics ...

HS You probably equate mathematics with analytical statics. Statics is similar to music: you can learn how to listen to music and develop an understanding of it. Similarly, you can learn to develop a sense of the forces, for the structure in general. Stefan undoubtedly meant approaching structural analysis intuitively as opposed to analytically.

AF But I suppose you were pretty good in mathematics ... what must you master mathematically?

TP We have to understand what an integral is and what a differential is, we have to be able to solve an equation with multiple unknowns, and we have to master today's tools, in other words, computers. But those are all merely facilitating elements: mathematics is an essential tool – but a good mathematician is not necessarily a good engineer!

SB The most important tool for us is mechanics. It is, of course, linked to mathematics: without mathematics, you cannot understand mechanics. But the mathematician does not necessarily understand mechanics.

TP Nor is the good mechanical engineer automatically a good structural engineer. Primarily, the latter must have a broad intuitive understanding of the forces involved so as to design an ingenious load-bearing structure.

AF If a structure is located on the exterior of a building these days, great efforts must be made in terms of building physics and construction – displaying the structure is no small act. In the Prada project in Tokyo, the facade and the load-bearing structure were combined into a single plane – was that solely to achieve a certain architectural expression?

HS Yes, definitely. The initial metaphor behind the concept was a Japanese bedside lamp – the paper and the sticks that give form to the paper are integral to each other. That led to the idea of combining the structure and the facade into a single plane, in order to create the image of such a luminous object.

AF Is the engineer's role in our decorative world simply to facilitate images?

TP Certainly that, too. If the ornamentation is load-bearing, the demands placed on us as engineers are great.

HS We engineers and the rest of the planning team are there to enable different systems. Your question is more about whether it makes sense today to design an integral, load-bearing facade. For me, it's still a valuable and sensible approach. It is, of course, associated with much greater complexity in the facade plane and, consequently, with a big effort from our side.

AF In a recent lecture at Architekturforum Zürich, Professor Antoine Picon of Harvard University identified digital technologies (after modern-era industrialization) as the locus of a renewed attempt to reconcile nature and technology: digital technologies now make it possible to model natural phenomena. As an aside and a transition to Forum 2004 in Barcelona: The ornamentation you developed there is no longer even structural. The engineering task was to develop a ceiling geometry that should suggest a randomly acting pattern. Does the engineer of today serve as an ornament provider?

HS The metaphor was that of a water surface, which inherently lacks repetition. The goal was a pattern that looks as if it were randomly created by the wind on a water surface. The ornamentation was developed with mathematical formulas that yield a different pattern with each touch of a button. The ceiling idea is a by-product of my dissertation. It was about random events when concrete fractures. We were able to use the models I used back then for predicting fracture processes in concrete to generate the desired, random-looking water surface. That's mathematics, of course, but the construction itself is also interesting: None of the 14,000 triangles is alike. Their positioning and production is elaborate, very high-tech CAD design.

AF It's an interesting development that engineers are becoming key figures in creating new forms of ornamentation.

TP When we think of Cordoba, with its mathematical, repetitive, and superimposed patterns, we are absolutely in the same tradition!

Remarks on the Structural Behavior of the Prada Aoyama Facade in Tokyo

Jürg Conzett

Standing in front of the Prada building in Tokyo, you might think you are looking at a beautiful, but purely decorative facade; after all, the corners of the building, where the latticework is bent in plan, seem quite unstable. Vertical corner posts would be needed to carry the loads, you think, because without them, a facade would bow outward. So how are the loads actually carried? The few small cores alone cannot do the job; the cantilevers to the corners of the building would be too large. Does the facade carry the loads after all?

The engineers' explanations provide insight: the facade has a force-locked connection to the floor slabs, and each facade has diagonal "struts" within the lattice that are connected to several slabs at once. These struts conduct normal (axial) forces; like inclined columns they bear the weight of the slabs in compression. Since adjacent struts generally take up some of the loads in the opposite direction, an equilibrium of the horizontal components results; the points of connection between struts and slab edges, however, are forced apart in the process. The slabs must hold these support points together, and this is quite possible with adequate reinforcing.

To maintain control of the flow of forces, it makes sense to allocate at least *one* pair of mutually supporting struts to each facade. Facade A–B, for example, has three such pairs (fig. 1): The two crosses in the two lower floors carry the slabs at the corners to the sides. The half cross above them supports the third slab in the middle of the facade as an A-shaped trestle; the corners are supported by the adjacent fields.

This simple principle yields a geometric order: the ratio of facade width to the height between floors corresponds to the slope of the struts. A narrower facade with the same floor-to-floor height would not permit supporting struts. And indeed: facade C–D, which is one rhombus field narrower, has half the floor-to-floor height of side A–B, which makes it possible to have four struts per floor.

So in this building, there is a constructive condition that governs the relationship between the facade width and the number of floor slabs. Even for an observer interested in such matters, this relationship is not readily apparent, but requires attentive thought. That makes it all the more interesting to discover that the putatively decorative facade follows the rules of equilibrium in a very fundamental way – it is a *constructed* building.

Can other structural properties be assigned to this facade? If we consider the facade lattice as a construction element, the column-free corners are reminiscent of the intersecting stress fields of a box girder subjected to torsion. If the tensile and compression forces in the struts that meet at a corner node of a facade are of equal intensity, the outcome at that corner is a vertically oriented force. Similarly, this vertical force can be split and taken up by the two coinciding struts of the adjacent facade (fig. 2 below). Imagining this effect distributed over many pairs of struts, the column-free building corners can transmit a so-called *shear flow*.

When a horizontal force acts parallel to the facade on every node of the facade lattice that intersects a horizontal plane (e.g. of fictive eaves), then a tensile force develops in all the diagonals running in one direction and a compressive force develops in all those running in the opposite direction (fig. 3). In this case, the external application of force corresponds to a continuous line load running around the entire building. In other words: a torsional moment about a vertical axis, which subjects the building to twisting, generates uniform tensile and compression forces in the facade lattice. In such a case, the facade becomes an extremely efficient structure.

During an earthquake, however, such torsional action is dwarfed by comparatively large horizontal forces acting strictly in parallel. Under such conditions, the behavior of the facade is more complex than under rotation, since not only

1

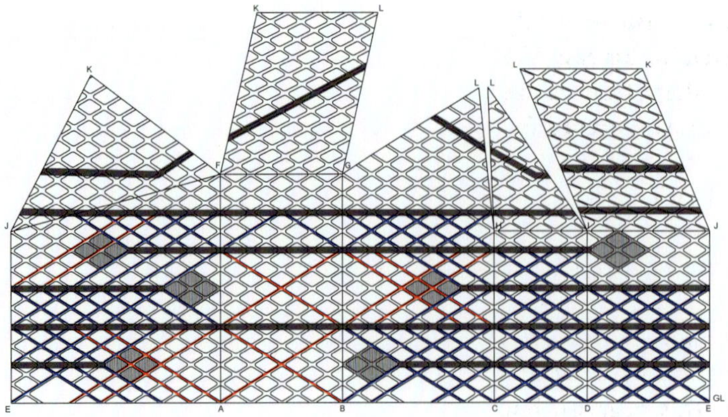

2

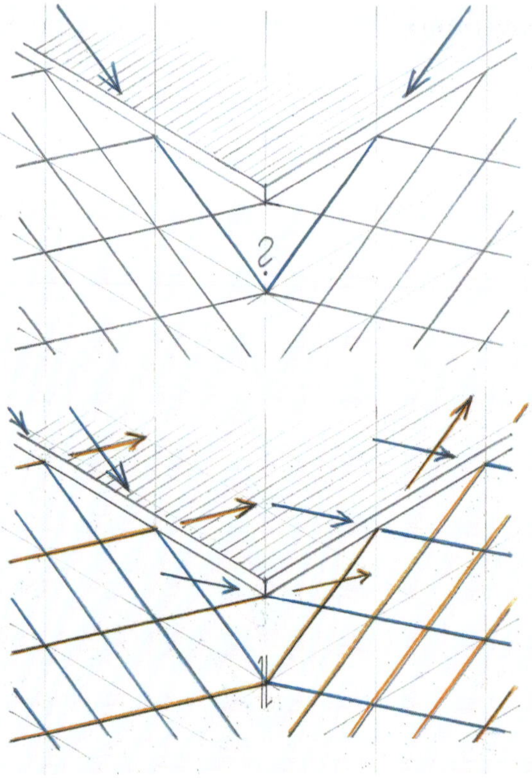

3

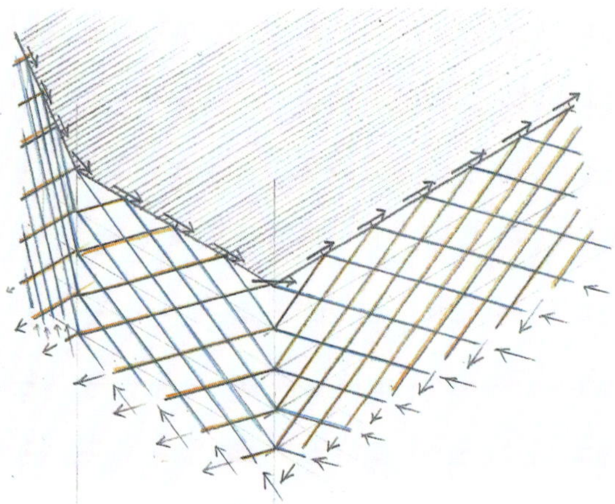

1 Developed facade. The members shown in color act in compression and take the vertical loads from the floor slabs. Blue: Compression struts over one story. Red: Compression struts extending over two stories.

2 Building corner: Equilibrium is impossible with two converging compressive forces (above); equilibrium is possible with alternating tensile and compressive forces (below).

3 Developed view of facade in response to horizontal forces, with compressive forces shown in blue and tensile forces in orange.
The facade would most effectively resist moments rotating about a vertical axis.

thrusting forces are generated, but also vertically acting overturning forces. But even in this case, the ability of the facade to transmit shear flow at the corners is helpful. It is the prerequisite needed for the individual parts of the facade to work effectively together in distributing and absorbing the horizontal seismic action.

Is the Prada Aoyama facade lattice a vivid constructive response to seismic activity in Japan? Coping with vertical force effects would not be the main focus (even though, as mentioned, the building structure is interesting in this regard), but rather the horizontally acting earthquake forces. Considering this aspect, the facade's architectural expression receives a profound importance that goes far beyond the decorative pattern of a fashion house.

Structure and Facade

Stefan Bänziger

Originally, solid walls meant the load-bearing structure and the facade were united. A steady increase in the specific requirements for thermal insulation, impermeability and transparency meanwhile led to a distinct separation of the load-bearing structure from the building envelope.

Prefabricated wood construction, thermally insulated single-wythe masonry and insulating concrete are indeed modern building systems and materials that now enable the construction of monolithic, single-wythe facades meeting current requirements. But they are limited in their structural performance and quickly reach their limits with respect to the dimensions, design, and complexity of a facade structure. Therefore today, larger buildings usually have load-bearing structures that are separate from the building envelope. From the viewpoint of building physics, the obvious approach is to design a building that has its load-bearing structure on the warm, inner side and an insulation layer on the outside. As soon as such an arrangement does not meet the design and use-specific requirements, because an additional layer or glazing is desired as the enclosure, the question of the appropriate and optimal sequence of layers arises. For selecting the system, moreover, minimal construction thickness is often crucial because more usable floor area can then be generated. For every building, the ideal sequence of layers must always be sought anew: whether the load-bearing structure should be located on the cold outer side, on the heated inner side or even within the facade construction must be decided anew each and every time.

Expansion Town Hall , Regensdorf, Switzerland, 2010

The town hall in Regensdorf has an exterior load-bearing structure. The building is a distinctive cubic monolith in the center of the town. The competition concept envisaged a three-story skeletal frame with central, stabilizing circulation cores and internal facade columns. The closed facade surfaces were to be clad with non-load-bearing concrete elements. Working together with the architects, this assembly was revised during the planning process toward a single-wythe load-bearing facade of varnished cast-in-place concrete, which markedly strengthens the intended architectural expression while minimizing the thickness of the overall construction.

There are no expansion joints in the entire cast-in-place concrete envelope, which is thermally separated from the inner structure near the slab edges with a continuous, eight-centimeter-thick layer of mineral wool insulation. The floor slabs, which span approximately seven meters, bear solely on the facade via lateral, sliding shear force dowels, and the facade is, in turn, secured horizontally to the slab edges with in-situ tie rods. The monolithic exposed concrete facade rests on the ground floor on a peripheral sliding bearing. This permits horizontal movements between the inner and outer construction, which arise due to differing thermal expansion and differential shrinkage contraction. This bearing concept enables the exposed concrete construction to remain virtually crack-free.

Double-sided cantilevers at the entrances to the building and the need to straddle the existing underground garage in some areas lead to a Vierendeel effect with considerable bending stresses on the facade lattice. Adequate additional reinforcement absorbs these forces. With an appropriate component thickness of twenty-eight centimeters, the builder achieved a perfect exposed concrete surface quality despite the existence of thick reinforcing that intersects vertically and horizontally in some areas.

Windows, interior insulation and all interior cladding of the facades are rigidly connected to the concrete facade. Flexible transitions to the floor coverings, interior walls, and ceiling linings absorb the differential movements, thereby simplifying the process of sealing the facade surfaces. Only the window sills and tops of the walls received a coating of synthetic liquid sealant.

ABR 5 Administration Building, Roche Diagnostics, Rotkreuz, Switzerland, 2011

The nearly seventy-meter-high administration building for Roche Diagnostik AG in Rotkreuz is built in a classic manner with an inner load-bearing structure and an outer facade. The supporting structure is isolated from the floor-to-floor glass curtain wall facade and set in front of it. The inclined, load-bearing columns of the facade are arranged in diamond shapes extending over four stories, dominating the building's architectural expression and providing most of the needed horizontal stabilization. In combination with the floor slabs, the columns form stable triangles that each extend over two stories and which function together as a rigid, tube-like mesh. This system is so efficient that it transmits about seventy percent of the horizontal loads from earthquakes and wind. Since the cores must only absorb a small portion of the horizontal loads, their wall thicknesses can be minimized and, thus, additional usable area can be generated. The load-bearing structure's functionality and the forces acting upon it are directly legible and tangible. Concrete as a high-performance building material is literally put on center stage, thereby conveying safety and visual stability to the building's users.

Due to the high aesthetic demands, the required yield strengths, and repetitive geometric forms, prefabrication of the columns was an obvious choice. The facade columns up to the thirteenth floor were prefabricated as V- and A-shaped column pairs. Only the two-story columns in the fourteenth and fifteenth floors were produced and installed separately for technical reasons. The design and execution of the details turned out to be tricky: To keep the reinforcement cross-section as small as possible at the complex areas where the reinforcing bars intersect, for instance, a higher strength reinforcing steel was used. In addition to compressive forces, the column couplings also had to absorb shear forces and, in some cases, even tensile forces, which is difficult to engineer and construct.

Prada Aoyama, Tokyo 2003

In the dense urban fabric of Tokyo, Prada Aoyama appears like a solitary crystal. Its sup-

1

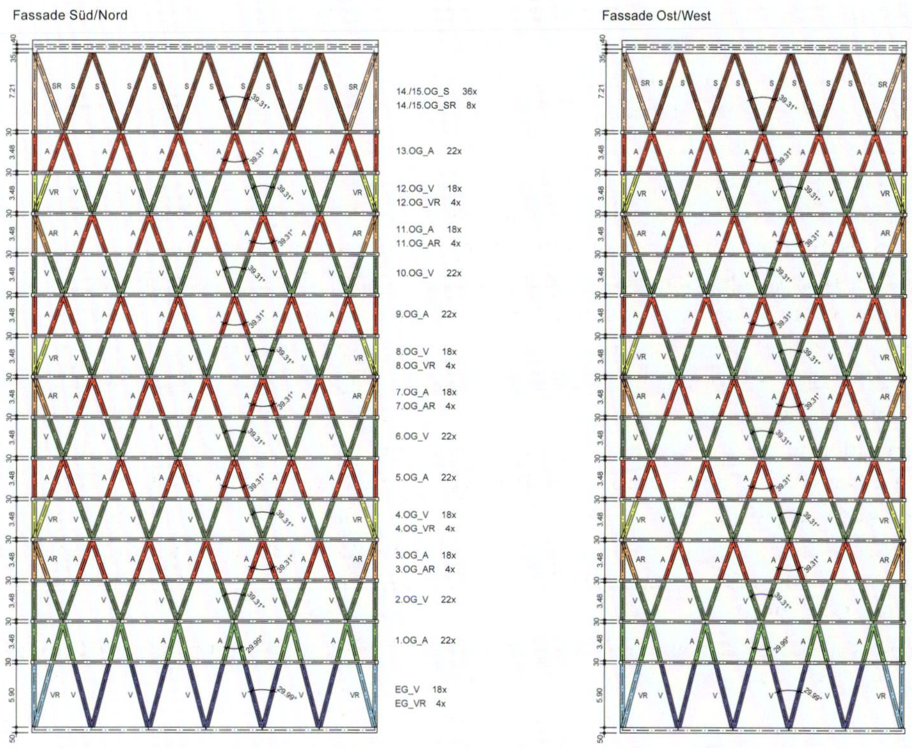

2

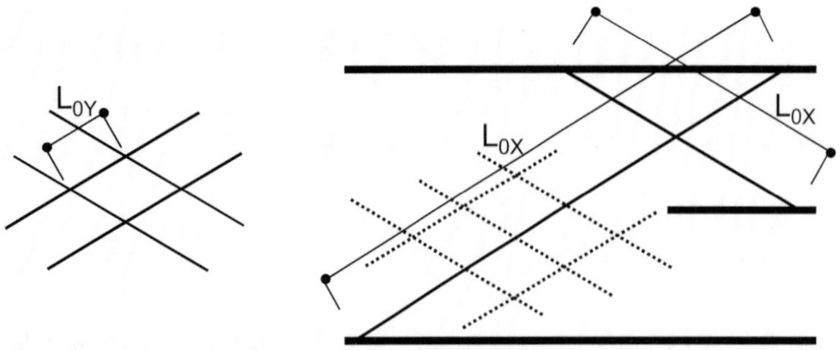

1 Classification of facade columns, ABR 5 Administration Building, Roche Diagnostics
2 Variations of the buckling lengths in- and out-of-plane, Prada Aoyama, Tokyo

porting structure that is united with the spatial structure is an integral part of the overall concept. Load-bearing steel sections are integrated into the delicate honeycomb pattern that uniformly surrounds the building. At first glance, the design suggests a classic, self-supporting curtain wall facade. This impression is reinforced by the horizontal orientation of the diamonds, which would have carried the gravity loads considerably more efficiently if oriented vertically. In addition, the floor slabs are recessed, so they end at the inner surface of the facade structure, even though they are needed as integral tension elements to complete the structural system's trussed framework.

The facade's load-bearing honeycomb structure consists of rolled sections with a double T cross-section, which are welded to one another at the nodes. The outer dimensions of the members are constant. The plate thicknesses vary with the different loading conditions. For individual nodes with very high loading, cast iron was actually used. Single honeycomb rows were prefabricated and fitted with temporary bolted connections to facilitate their installation. Two connecting plates at the sides of the sections enable the member to be bolted in place while leaving sufficient space for the welds. After welding, the bolted plates were removed. The honeycomb structure is connected to the slabs – and hence, to the tension elements of the closed triangles – with a hollow section that is welded to the node. The actual tension element is a steel plate that runs along the slab edges.

Due to the basic geometry and the way the load-bearing structure wraps around the corners, the facade's stability is ensured at the corners of the building. In the straight sections, however, the facade's stability problems are much greater perpendicular to its plane. Here, too, the floor slabs provide help by reducing the buckling length of the columns to the distance between the floors. In addition, three tubes that contain changing rooms penetrate the internal space of the building. Their cross-sections each encompass four honeycomb sections of the facade structure and also serve as bracing.

Prada Aoyama
Tokyo 2003

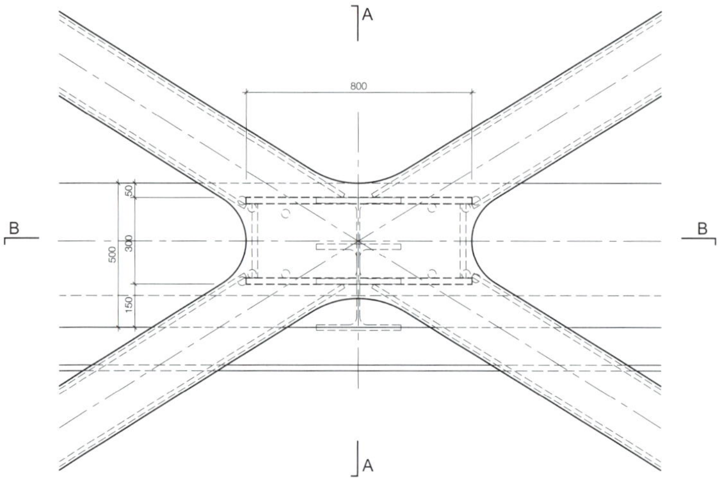

← Assembly of the steel structure at the roof
↑ Construction of the cast-in-place concrete slab in concert with the steel structure of the facade
↗ Depiction of deformation for part of the facade
→ Details of slab juncture at the facade nodes

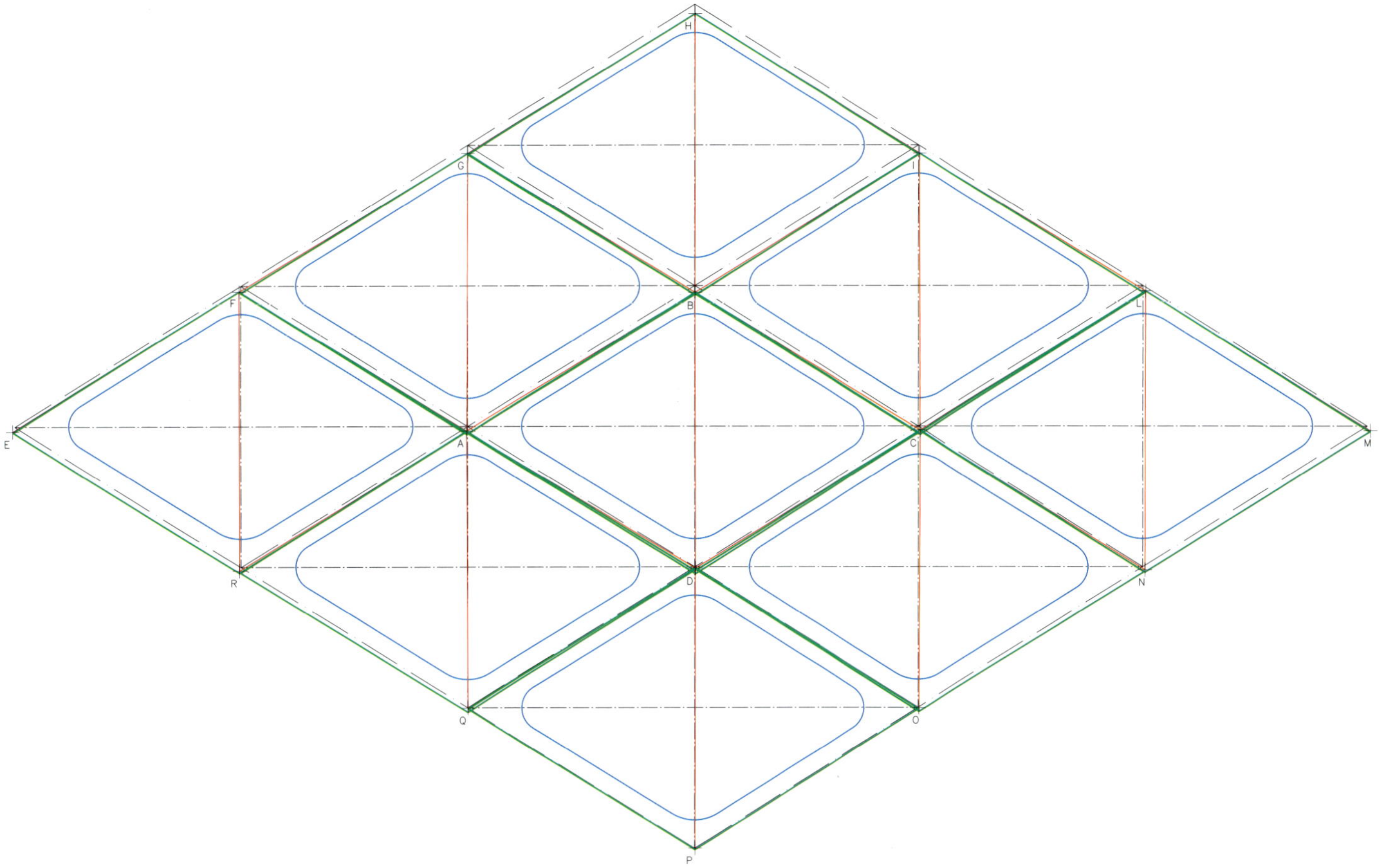
H
G
I
F
B
L
E
A
C
M
R
D
N
Q
O
P

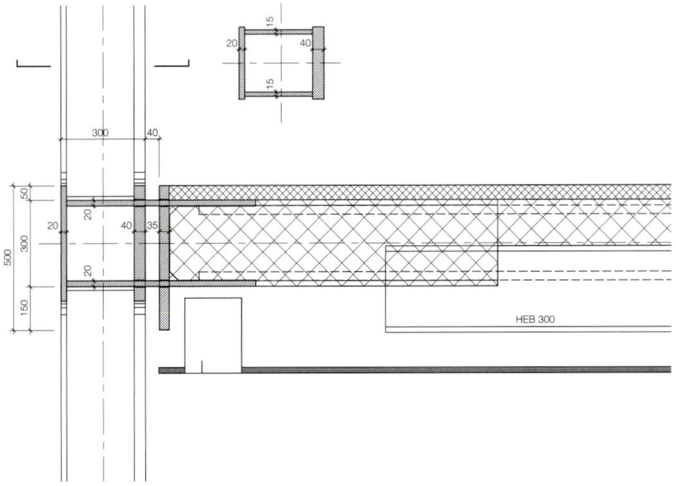
15
20
40
15
300
40
50
20
20
40
35
300
500
20
150
HEB 300

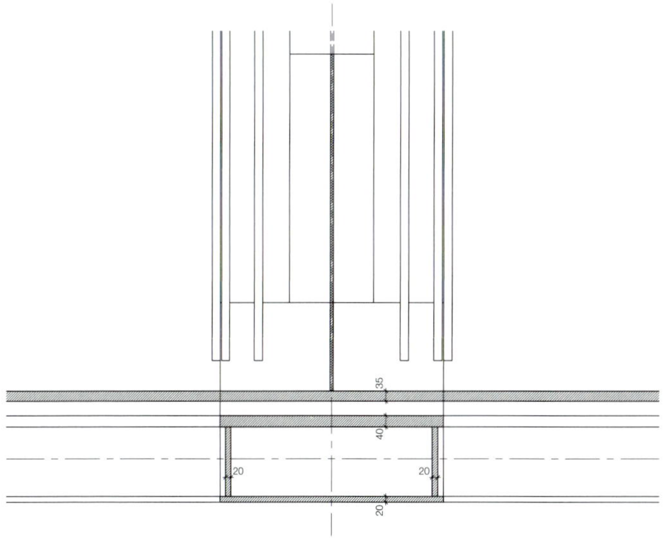
35
40
20
20
20

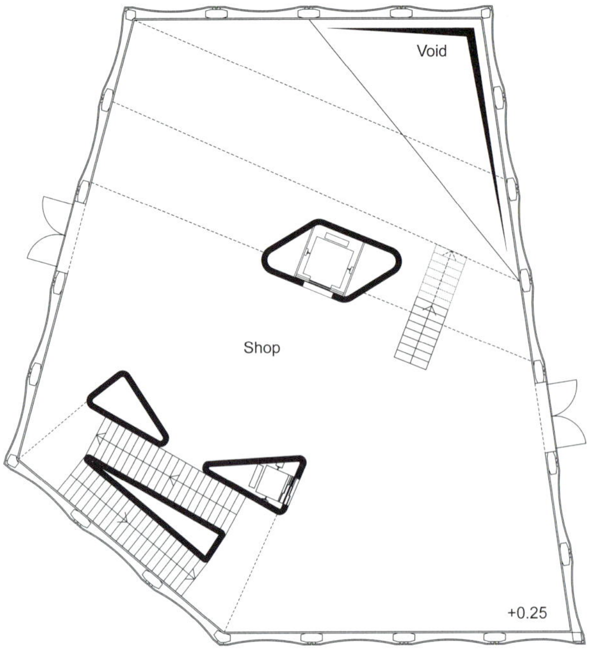

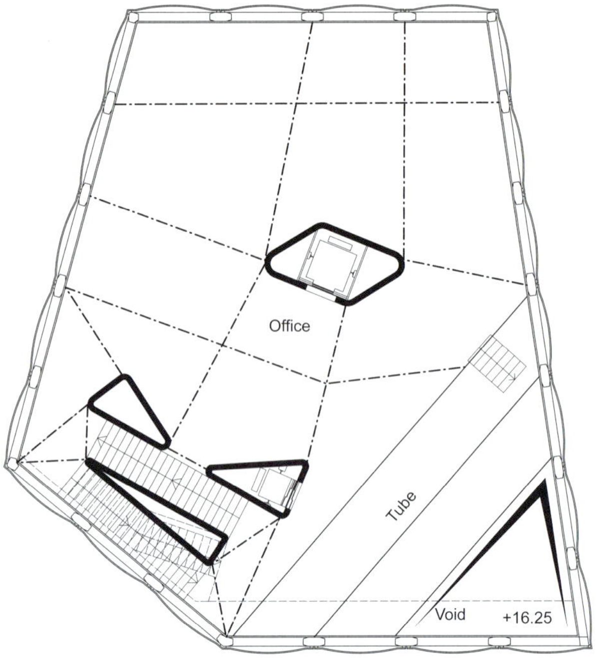

↑ Entrance level plan (left), upper floor (right)
→ View through the finished facade
→→ Exterior view

PRADA

Expansion Town Hall
Regensdorf, Switzerland, 2010

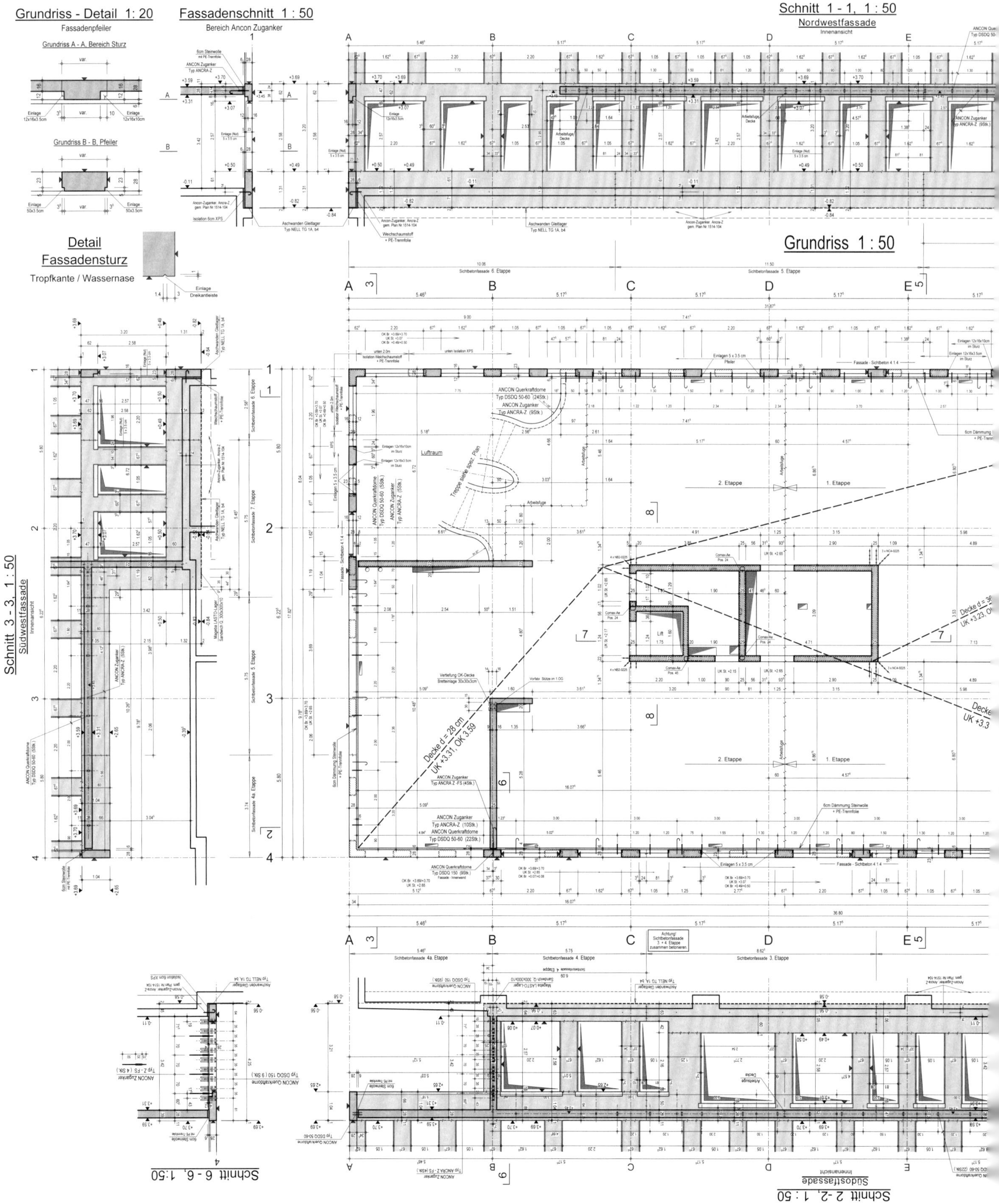

Grundriss - Detail 1: 20
Fassadenpfeiler
Grundriss A - A, Bereich Sturz
Grundriss B - B, Pfeiler
Detail
Fassadensturz
Tropfkante / Wassernase
Einlage
Dreikantleiste
Fassadenschnitt 1 : 50
Bereich Ancon Zuganker
ANCON Zuganker
Typ ANCRA-Z
Aschwanden Gleitlager
Typ NELL TG 1A, 64
Isolation 6cm XPS
Schnitt 1 - 1, 1 : 50
Nordwestfassade
Innenansicht
Grundriss 1 : 50
Sichtbetonfassade 6. Etappe
Sichtbetonfassade 5. Etappe
ANCON Querkraftdorne
Typ DSDQ 50-60 (24Stk.)
ANCON Zuganker
Typ ANCRA-Z (9Stk.)
Luftraum
Treppe siehe spez. Plan
2. Etappe
1. Etappe
Lift
Decke d = 28 cm
UK +3.31, OK 3.59
ANCON Zuganker
Typ ANCRA Z -FS (4Stk.)
ANCON Zuganker
Typ ANCRA-Z (10Stk.)
ANCON Querkraftdorne
Typ DSDQ 50-60 (22Stk.)
6cm Dämmung Steinwolle
+ PE-Trennfolie
Schnitt 3 - 3, 1 : 50
Südwestfassade
Innenansicht
Sichtbetonfassade 4a. Etappe
Sichtbetonfassade 4. Etappe
Sichtbetonfassade 3. Etappe
Achtung!
Sichtbetonfassade
3 + 4. Etappe
zusammen betonieren
Schnitt 6 - 6, 1:50
Schnitt 2 -2, 1 : 50
Südostfassade
Innenansicht

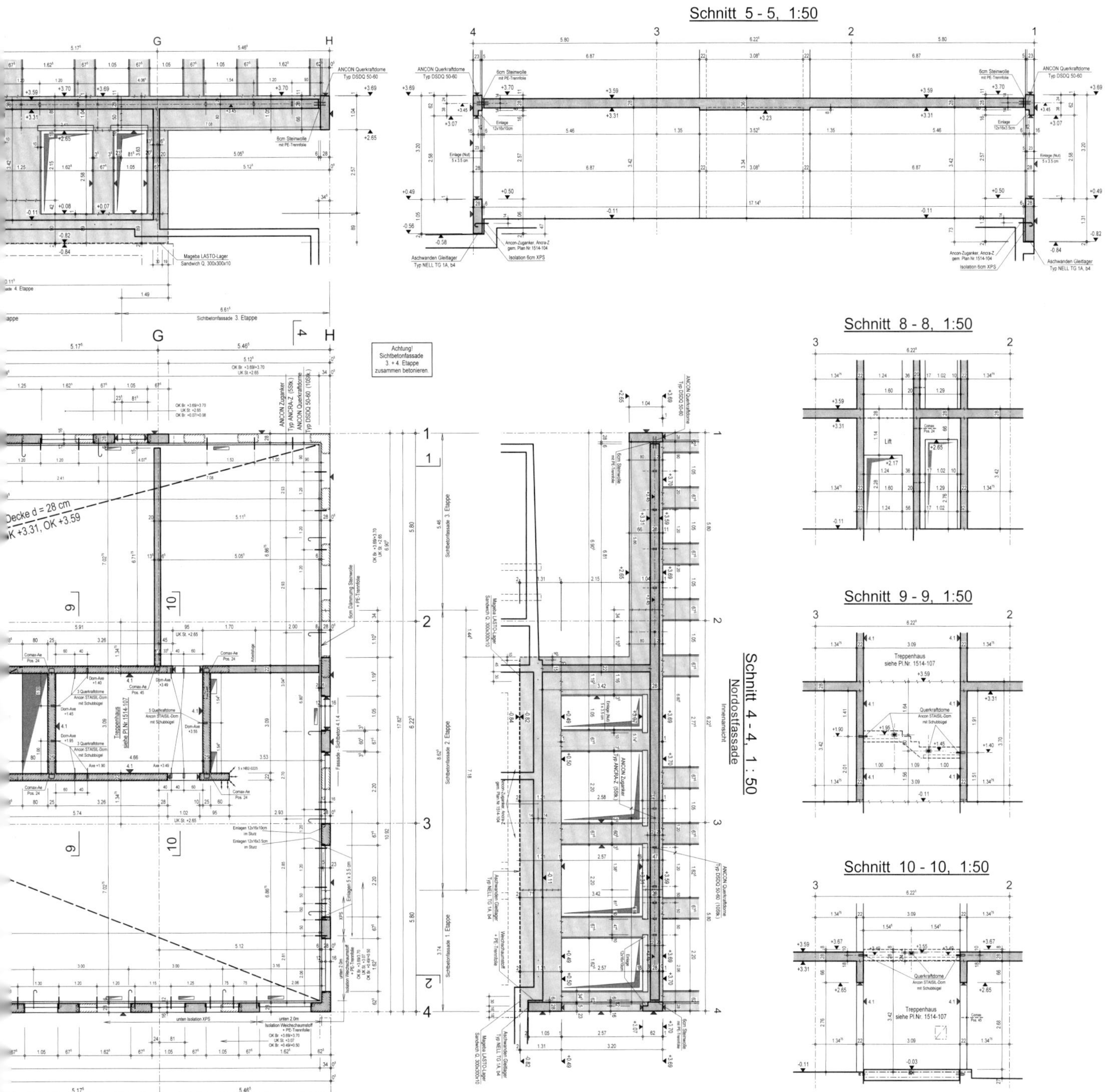

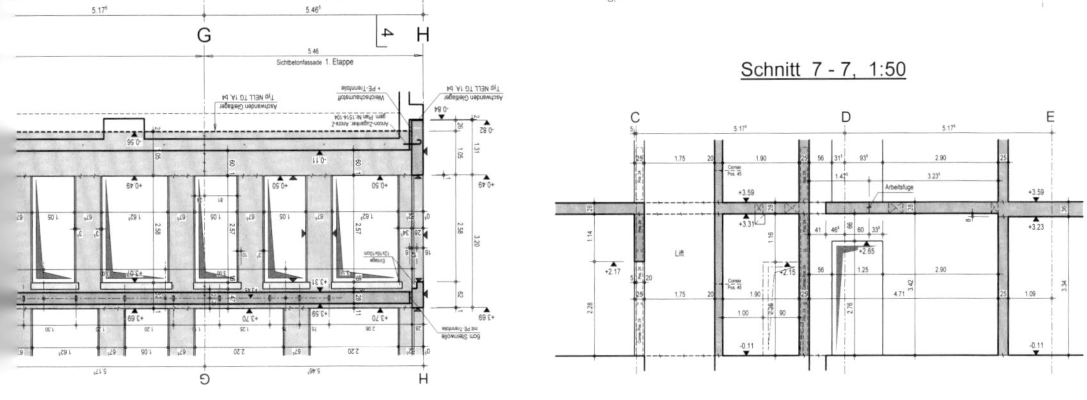

←← Shear force dowels in the facade prior to construction of the floor slabs
↑ Formwork drawing – ground level

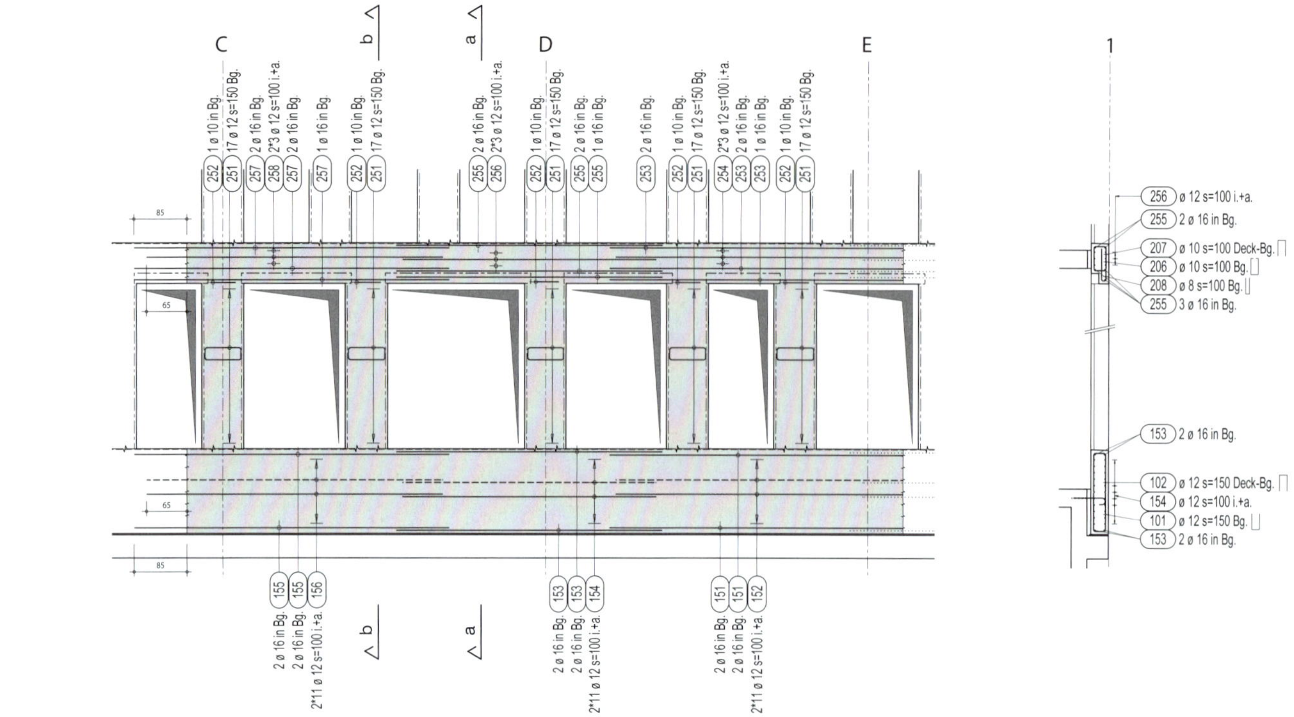

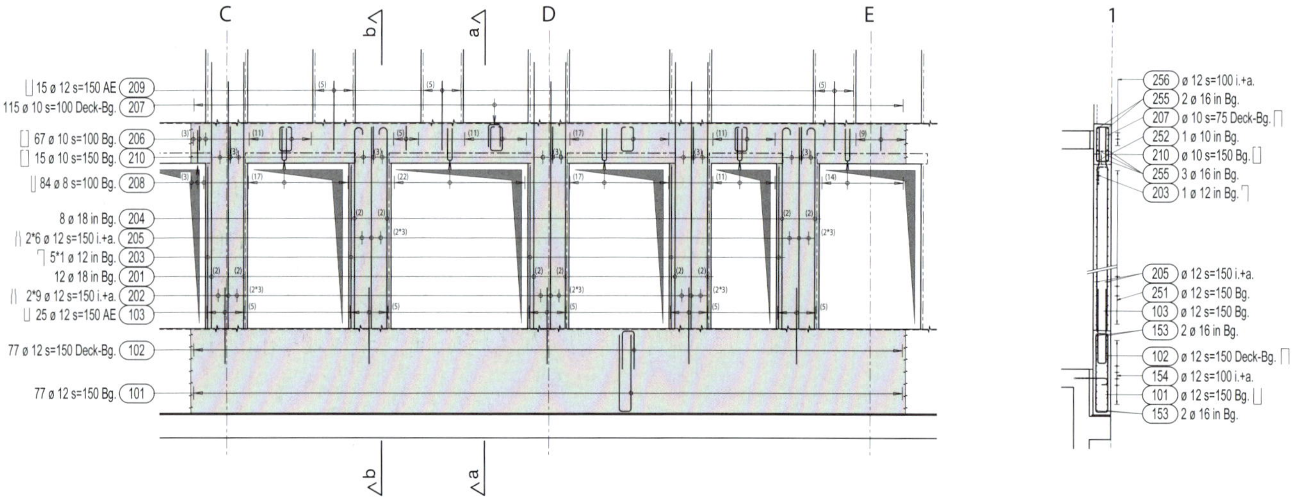

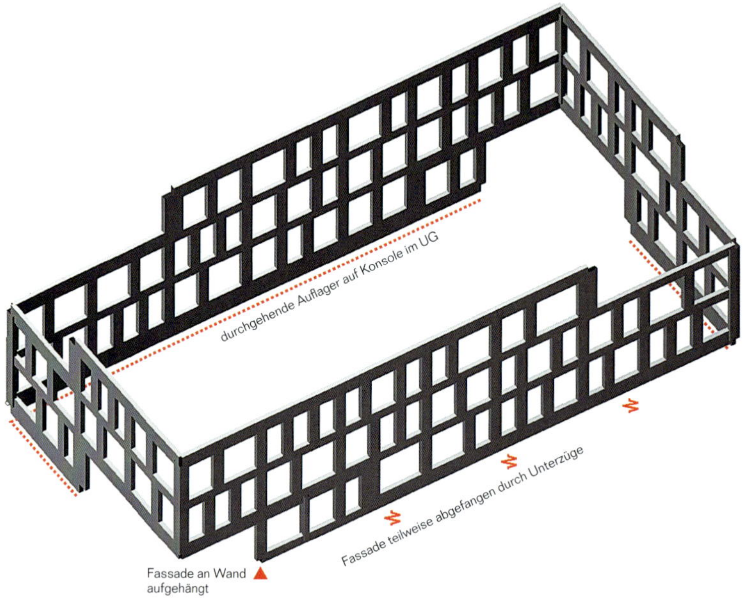

↑↑ Horizontal reinforcing of the facade at ground level
↑ Vertical reinforcing of the facade at ground level
→ Structure and bearing concept for the exposed concrete facade
→→ Exterior view

ABR 5 Administration Building, Roche Diagnostics Rotkreuz, Switzerland, 2011

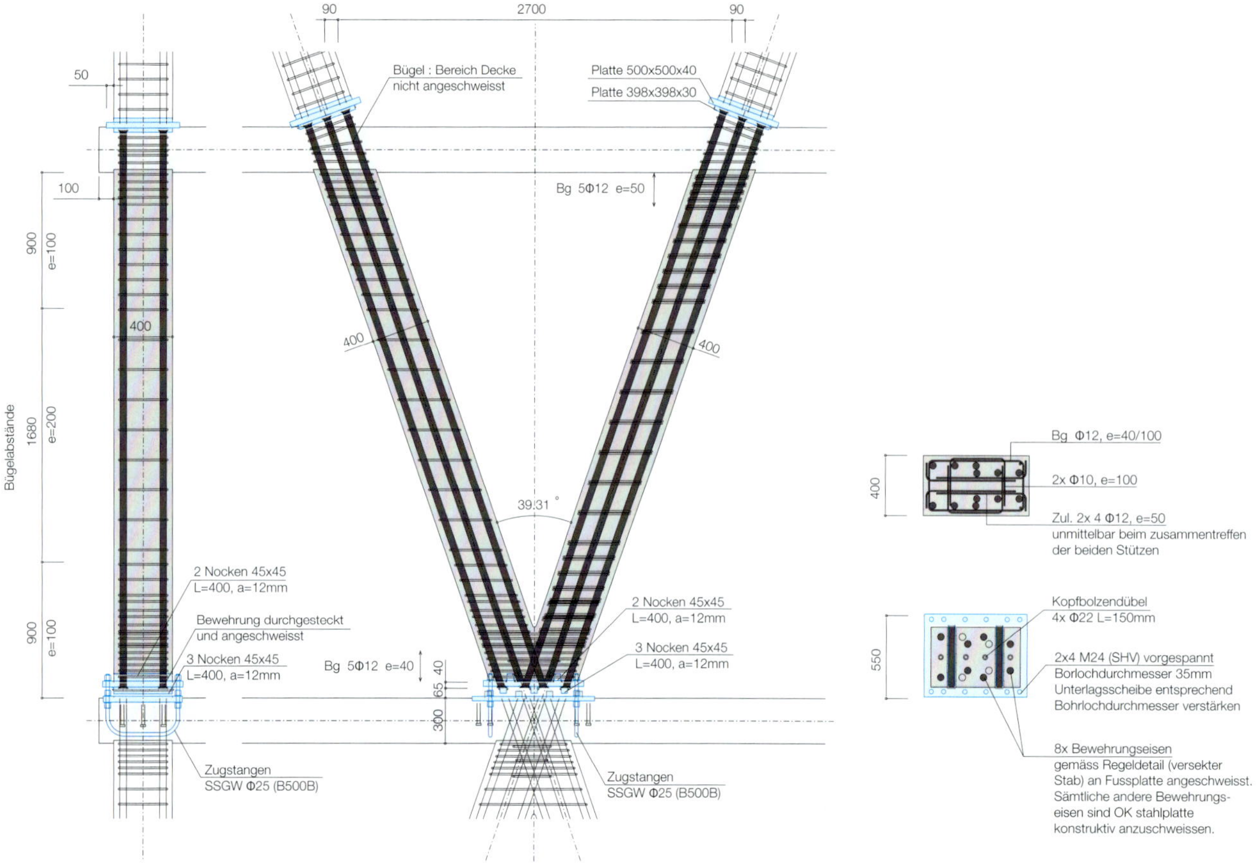

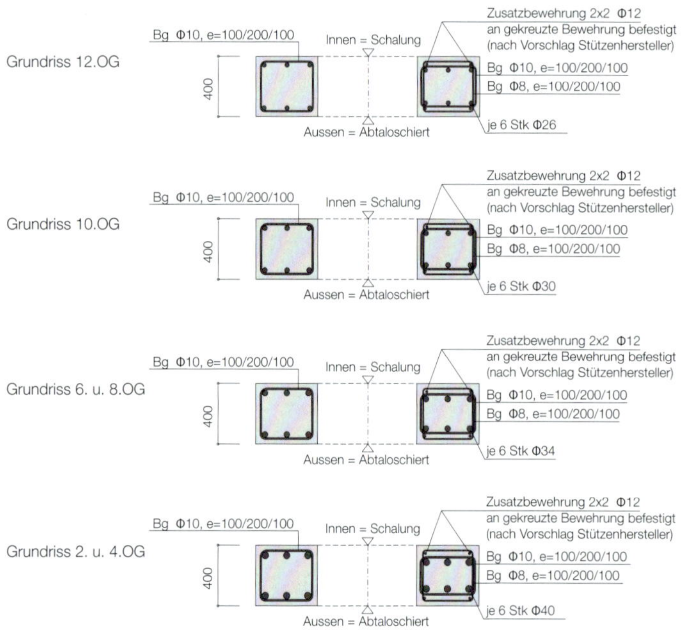

←← Construction in progress, typical floors
↑ Reinforcement of a typical V-column, with column base details
← Reinforcement of the V-columns varies as a function of the story
↗ Soffit formwork drawing, eighth floor
→ Horizontal deformations with only core walls (left), with only facade columns (center), and with facade columns together with core walls (right)

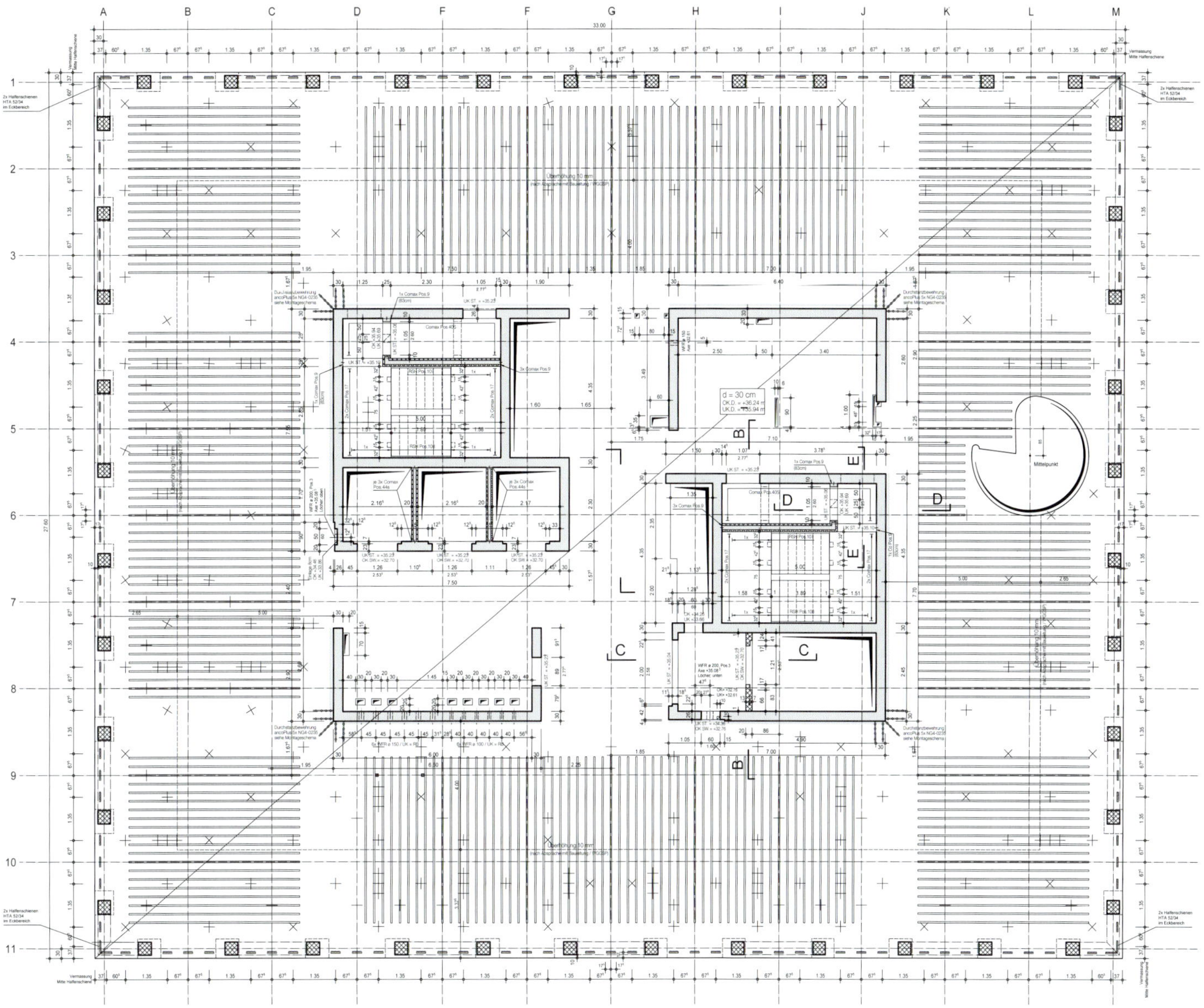

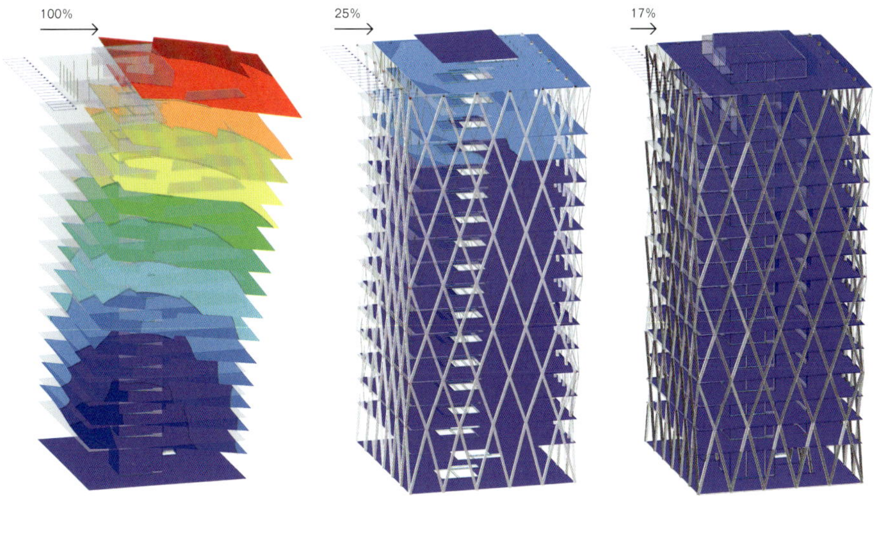
100%
25%
17%

↑ Corner situation on the top floor
→ Ground level during construction
→→ Exterior view of finished building

6 Tall Buildings

← Model of Claraturm in the wind tunnel at Wacker Ingenieure in Birkenfeld, Germany

Conversations VI

AF What do you find interesting about high-rise projects? High-Rise

TP As the name says, for one thing it's the height of the buildings that makes high-rise projects special. For another, the repetition of building elements that you get is appealing, for example with the slabs: when a floor slab is repeated thirty times, the economic and structural efficiency of the system plays a significant role. Other interesting aspects include the vertical circulation, meaning the arrangement of the cores, and the stabilization, that is, the question of how the vertical cantilever is stabilized.

AF Structures have become lighter in recent years. Consequently, structural damping is an issue: buildings have become more susceptible to vibrations – and this applies exponentially for tall buildings. The importance of dynamic responses to wind forces and earthquake safety has increased. How much influence does vibration damping have today as a design parameter?

HS The higher a building becomes, the more the technical aspects move to the forefront at the cost of design concerns. With tall buildings, there's increasingly little design latitude for the architecture.

TP That applies to all aspects, such as fire protection, etc., but also to the vertical circulation. Recalling the genesis of the skyscraper in Chicago, which was made possible by the invention of the elevator, it's interesting that even today, the maximum travel height – currently 557 meters – still defines the upward limit for now. The vertical circulation is key, and Frank Lloyd Wright's dream of the mile-high skyscraper remains, for the time being, a vision. Vibration damping mainly has to do with comfort requirements, so that a building's occupants don't become seasick. For all intents and purposes, however, that's not really an issue until you reach a height of one hundred fifty or two hundred meters. We are not working at such heights – at least not for now ...

HS When we speak of high-rise buildings in Switzerland, then we are talking about "tall buildings." Here, we are dealing with heights of up to one hundred twenty meters, and thus we can still build them conventionally. Beginning at a height of two hundred meters, earthquakes and wind demand other responses.

AF So building a tall building does not seem to be a formidable challenge?

TP I would not want to generalize that much. It is, of course, always a new stimulus. We are currently working on Claraturm, which is only one hundred meters high, but stands crosswise to the wind and has very slender cores. So there, we must indeed be aware of the susceptibility to vibration. An agreement needs to be reached with the users, acknowledging that the tenants might feel the effects of a one-in-ten-year wind. But I want to emphasize: it's not the size or height of a building that represents the complexity of its structure.

Computers

AF Thanks to the computer, it is possible for engineers today to dominate the complexity of the systems designed. To do so, the building's behavior is simulated with modal analyses. These require the creation of a model and its interpretation by a programmed set of instructions. Can you briefly describe how you approach the structural behavior and what enables you to correctly evaluate the calculated results?

SB To start with, we always make a preliminary analysis by hand and have done rough dimensioning. Computer simulations are, for instance, important tools for evaluating vibration behavior, earthquakes, or wind.

HS Before I bring in the computer, I usually have an idea of the dimensions. It's a matter of estimating things from experience.

AF And how does the young, inexperienced engineer go about it?

HS That is, of course, a process. Today's young engineers can build up their experience by using the computer to check the assumptions of seasoned engineers.

TP In the past, we sometimes also had to repeat calculations three or four times to reach the goal. Maybe today we are guided by something like artistic intuition, which is, of course, based on experience. The assumptions that have been made are verified by modeling the structural system on the computer. Either the results correspond with the expectations, or we need to check the assumptions used for the modeling. Then we can change parameters, such as the material properties, the load assumptions, the influences, etc., and observe how the system reacts to that. Then we need to narrow things down – at this point we find ourselves in an area of our discipline where things are not black and white, but consist of many gray zones.

HS Your question focuses, for one thing, on the use of computers in the planning and, for another, you want to know specifically how we model a problem on the computer in order to obtain the correct result. We've already discussed the first aspect. In the initial dimensioning, experience makes it possible to recognize the critical points that need to be checked by computer. That leads to a correction of the design if need be. And when the construction documents are needed, all the detailed dimensions and junctures are calculated with the computer. The broader question is, in which phase of the planning process is a problem modulated? In this case, experience plays a crucial role, especially in deciding what simplifications should be considered. After all, we have to abstract; we cannot accurately model reality. Ultimately, the hardest part is interpreting the results of the modeling.

AF If the result does not meet your expectations, you simplify different aspects, or adjust the parameters?

HS Of course, you have to question why the result doesn't match what was expected, and whether there was a fundamental error in reasoning. That's where a simple assessment by hand helps again. Or you recognize in this way that the error was actually made in the modeling.

AF If I understand you correctly, the engineer's intuition thus decides when enough calculations have been made.

HS Calculations can be made to an arbitrary level of detail and carried out for an arbitrary amount of time. At some point you have to reach a conclusion, even knowing that you don't know everything. You must, as it were, estimate the truth.

TP We must be able to live with something that's ninety to ninety-five percent true – the absolute truth doesn't exist.

AF So your discipline is not an exact science.

SB Engineers' digital structural analysis programs can be compared with architects' physical models. Using our digital models, we can envision a structure very precisely – similar to how an architect might accurately visualize a space by using a 1:20 scale model. Most conclusions about the correctness of a structure come by studying the deformations. These can also usually be easily recognized in physical architectural models – when, for instance, a cardboard corner begins to sag.

Standards

AF Your opinion of standards interests me: Are these ultimately nothing other than a convention that is based on intuitive experience? Shouldn't the terms intuition and calculation be seen less as a pair of contradictory elements and more as ones that interact?

TP Standards are not to be understood as an interaction of intuition, experience, and calculation. The standard is a fixed method of calculation.

AF That was not developed intuitively?

HS Interesting question. In a recent publication, Professor Menn revealed that in the various concrete standards that have been in effect in Switzerland over the past thirty years, the steel and concrete properties were always determined arbitrarily. An important constant, thereby, has been the safety factor of 1.8 between the rupture of a building element and its complete failure. In other words: even with a load 1.8 times greater, the concrete must still hold.

AF So are the standards precise or not?

HS No. Sometimes they are actually arbitrary.

TP Buckling formulas are fairly accurate, but material properties are less so. The lobbyists for the wood or steel industries also help determine what factors shall be assumed.

HS Regarding "buckling," an interesting example just came to mind: Sometime around 1800, Euler developed the formula for buckling on a strictly mathematical basis. The supports of a bridge near Münchenstein were calculated according to this formula, but the bridge collapsed. It was then realized that the residual stress within the steel affects buckling. So the Euler formula was adjusted accordingly. But adjustments of this kind are arbitrary. Therefore, the standards must be interpreted.

TP Standards are something like a book of standard recipes. Whoever wants to test the limits must seek tests that go a little further.

HS Whoever keeps to standards remains safe. If you ignore the standard, you must know exactly what you're doing.

AF Are you permitted to ignore the standards?

HS Standards are not prescribed.

AF You are not obliged to adhere to the SIA standards?

HS Yes, of course. But somewhere at the beginning of the compilation of standards, there is a statement that deviations from the standard are possible if they are justifiable with theories.

TP ... or with trials. Just recently, a bridge made with a composite sandwich construction of synthetic fibers and balsa wood was inaugurated. There is no standard for that yet, but the construction was naturally tested in advance through trials at the EPFL.

AF So that means engineers can defy the standard by supplying their own proof.

HS Yes, through trials or analogies – proof at different levels.

AF Let's go back to the subject of models: Heinz Hossdorf's development path moved away from the physical model and headed toward a pure computer model that he valued "as a unique tool for people's cognitive dealings with their real, three-dimensional world." Is the significance you find in the virtual model that reflects spatial-material reality related to this office tradition? Models

HS It's probably more a sign of our times. Architects have physical models to evaluate spaces. But with paper or cardboard, engineers can only evaluate the surface, not the forces. In our field, digital models grant us a much deeper insight than physical models. Early on, Heinz Hossdorf took self-built Plexiglas models and loaded them to measure the deformations. From these, he drew conclusions about the flow of forces and, in turn, extrapolated them into reality in terms of scale. Today's computer models enable this to be done directly.

AF So physical models are used only for evaluating the design?

TP For prototypes – for the development of a new building material, for example – physical models are still needed in order to balance out the material properties.

HS For engineering purposes, physical models must be to scale not only geometrically, but also in terms of their materials.

Risk

AF Let's talk about risk. Structural engineers work "on the edge," as Jack Zunz formulated it, and that's their dilemma: The designing architect wants to stand out and thus moves very close to the edge. The limits of the feasible, however, are ultimately determined by you engineers, who calculate the risk. How do you deal with this balancing act?

HS Personally, I repeatedly go through different phases. Early in my career, I was restrained and careful. Then, with increasing experience, I wanted to explore the limits myself – until I had some experiences on the construction site that vividly showed me the limitations. That caused me to again become more careful. I've already gone through three or four such episodes. But as you've rightly noticed, it's indeed so that, often enough, those on the team who have certain desires are not necessarily the ones who are themselves liable for ensuring everything is implemented correctly. The responsibility we bear is not appreciated enough by society.

AF What about you, Tivadar, have you had sleepless nights for that reason?

TP Yes, I've naturally gone through that, too – for instance, when we built our first high-rise building, the Messeturm, with its enormous cantilever. Together with Stefan, we must have run the calculations at least four times. During construction, it was raised with jacks and then lowered slowly – such actions take a toll on your nerves. At the first strong wind, I immediately went to the construction site and climbed to the top to see how the building behaves in the wind.

HS Apropos risk and responsibility: At the beginning of the Actelion project, the architects and we presented the plans to the board of directors. The president was very interested in the details, and then finally asked: Mr. Schnetzer, will it hold up? I replied with a yes. In such situations, one must be aware of the weight of such a statement. For him, that meant the matter was settled.

AF You're simply trustworthy!

HS It's interesting, too, to observe young engineers sending their first plan to the construction site. Their uncertainty is practically tangible and you can sense that they also probably had a sleepless night.

TP Any engineer who takes the job seriously and also challenges the limits has sleepless nights.

AF In other words, you need strong nerves.

SB Sometimes you go to the construction site unsuspectingly and then, once there, you determine that something has been done incorrectly or the contractors for building equipment have added unplanned inserts. And the ready-mix truck is already standing by with a fresh batch of concrete. Such situations really require a clear head, because suspending the work or delaying it means immediate consequences for scheduling and costs. At the same time, as the engineer I must assume the responsibility. Such decisions are not easy. I have even given the O.K. to place the concrete, but then immediately went back to the office to spend hours doing calculations to verify that it works that way. We have the greatest responsibility if we also have to share the risk of a pre-existing construction. Planning documents for the existing building are often scarce and if they exist, it's usually unclear whether the building was also actually built that way.

HS In Bhutan, our local project partners informed me during a visit that knowledge of the old wooden bridges was unfortunately lost. I had already come up with some thoughts about the project and promised to send my ideas to them. On the return flight, I jotted down a few rough calculations; one of our draftsmen at the firm made a plan from that, which we sent to Bhutan. When I was there again one year later, they happily informed me that one week before, a bridge built according to our plan had just been inaugurated. I was quite shocked – I immediately sat down and carefully checked the calculations for the construction again!

Tall Buildings

Tivadar Puskas

"I hope you will understand that architecture has nothing to do with the inventions of forms. It is not a playground for children, young or old. Architecture is the real battleground of the spirit."[1]
Mies van der Rohe

It has been the engineers who, accompanied by developments in steel-framed construction and the potential of the elevator, have made pioneering achievements in high-rise construction since the end of nineteenth century. The Reliance Building in Chicago and the Flatiron Building in New York City are contemporary witnesses to the epoch of the first Chicago School. The architectural firm Skidmore Owings Merrill (SOM) was co-founder of the second Chicago School, and Fazlur Khan, partner of the firm since 1966, was the most important engineer of his time with regard to high-rise construction. With his structural typologies, such as "outrigger systems," "framed tubes," "tube in tube," and "bundled tubes," he set milestones in high-rise construction. The dream of building a mile into the sky inspired Frank Lloyd Wright's 1956 "Mile-High Skyscraper" project in the United States. Today, the technical capabilities that would allow such a tall building are still being discussed.

In connection with the height limit, questions about the rationality of skyscrapers and their relation to human scale play an important role. Technically, the limiting height for skyscrapers is determined by the space requirements of their vertical circulation and by the technical capabilities of elevators. By the same token, achieving sufficient comfort criteria pertaining to wind-induced vibrations and controlling the seismic behavior of a skyscraper are also important criteria. Information about the behavior of the structure is provided by wind tunnel tests, which are performed under statistically extrapolated wind speeds. At the same time, the structure should be formed in such a way that the wind does not even induce any disturbing vibrations. This is achieved with a rigid stabilization system, or with dampers and pendulums on the upper floors to limit the acceleration of the vibrations within a tolerable limit.

1

Densification, and thus also the high-rise building, will always have to represent an added value for it to be promoted at all. Thus in Switzerland, too, the high-rise building will make a significant contribution to the physical densification of the urban realm and constitute an important

2

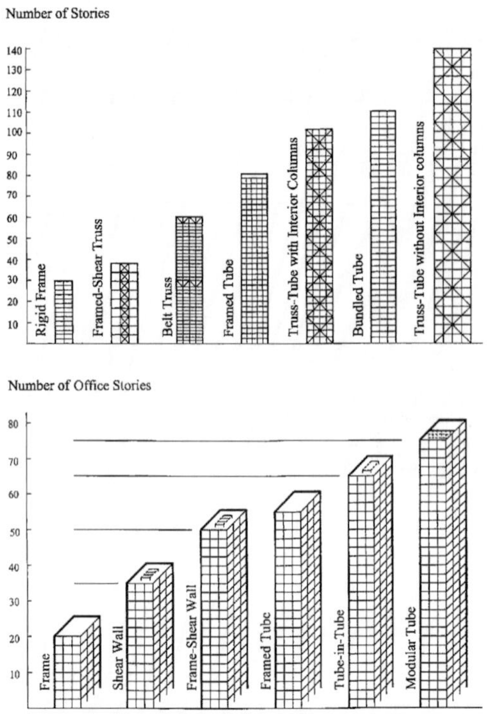

building contribution for sustainable urban design. Tall buildings of the kind we are planning and building also feature densification in the vertical dimension inasmuch as the floor-to-floor heights and the slab thicknesses are reduced to a minimum. Simultaneously, the flexibility of use is also maximized while the circulation spaces are minimized. These structural steps are necessary to ensure that high-rise buildings are cost-competitive with low-rise buildings. With tall buildings, the structural engineering is an important and integral part of the architectural statement. Thus, the design of high-rise buildings is becoming an ever more lengthy and ever more highly interdisciplinary and iterative planning process.

Messeturm [Trade Fair Tower], Basel 2003

Modeled on Lever House in New York City, the 105-meter-high glass shaft of the Messeturm dominates Basel's cityscape.

The starting point for the tower was the strategy set forth in the "Messe Basel Plus" project that was adopted in 1994, which foresaw keeping the trade fair in the city and expanding its traditional location. The city of Basel, together with the Swiss trade fair organizer MCH Group, held an international, two-stage architectural competition, which resulted in the selection of the design by the planning team Morger & Degelo / Marques Architects, in cooperation with our firm.

A formally restrained built form with a clear, cubic geometry characterizes the project. The implemented structural concept addresses this aesthetic premise inasmuch as the load-bearing structure is an integral part of the architecture.

The Messeturm consists of two basic parts: the actual tower itself and a flat, rectangular section that protrudes at its base. Both have transparent facades, floor space that can be used flexibly, and compact, central circulation areas.

The tower's structural system consists of a concrete core in the middle, delicate steel columns directly behind the facade plane, and flat concrete slabs with ten-meter spans. The core incorporates stairs, elevators, and utility lines and it carries sixty-five percent of the vertical loads. In addition, along with the outrigger system on the top floor, it stiffens the tower against wind and earthquake loads. The columns are made of solid round steel bars that are sheathed with a concrete shell to protect them in the case of a fire. Commensurate with the varied loading, the cross-sections of the column are biggest in the basement and decrease up to the twenty-ninth floor. The concrete slabs, which are simultaneously designed as chilled ceilings, distribute the loads to the core and the columns. The slab's edges are stiffened with integrally cast steel sections. These continuous, composite upstand beams reduce the slab deformations and enable the structure to have a delicate appearance on the facade.

The time needed to construct a typical story (casting the core and slabs and erecting the steel columns and beams) was one week. For the first time in Switzerland, work on the building's slabs and walls progressed simultaneously as the formwork was prepared, the MEP systems were integrated, reinforcement was placed, and the concreting work was carried out.

The building volume appended to the lower floors is a ten-meter-high space frame. It protrudes eccentrically from the tower's ground plan: twenty meters on the short side facing the square and ten meters on the long side. The column grid is the same as in the tower, thus it appears uniform over the facade of the entire building. The two volumetric forms interlock elegantly and efficiently. The loads from the tower provide the needed re-

3

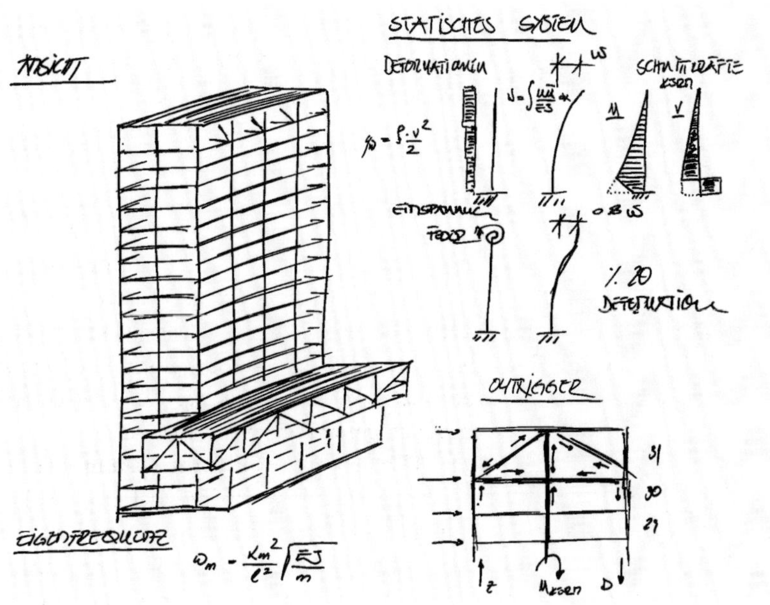

1 Skidmore, Owings & Merrill Architects (Fazlur Khan, structural engineer), John Hancock Center, North Michigan Ave., Chicago 1969 (height = 344/457 meters)
2 Development of load-bearing structures for high-rise towers as a function of the number of stories
3 Structural system with estimate of the influence of the outrigger on the natural frequencies of the building

straint – after the building carcass had reached the twenty-first floor, the auxiliary shoring beneath the trussed framework was removed using jacks.

Wind forces and seismic actions are the decisive, albeit difficult to assess horizontal effects acting on the building. At the beginning of the planning, wind tunnel tests on a 1:400 scale model helped to determine the complex behavior under the influence of wind. By connecting the core with the facade columns on the top floor (outrigger system), the deformations and accelerations were able to be reduced by about twenty percent, which meant a significant increase in comfort for the building.

All the loads from the tower's load-bearing structure flow into the subterranean box girder forming the building's basement. It is ten meters high and has a two-meter-thick bottom plate that rests on 102 bored piles. The bottom plate and the deep foundation jointly restrain the tower, since, in structural terms, a high-rise building is a more or less rigidly fixed vertical cantilever.

Aquila Commercial and Residential Building, Pratteln, Switzerland, 2013-2015

The sixty-six-meter-high building is the result of a competition that was won by the architects Christ & Gantenbein. The building, which is directly adjacent to Pratteln's newly designed urban square at the train station, responds to its context with a closed facade to the noisy north side and facades to the west and the south that open out to the town center and to the foothills of the Jura landscape. These local conditions result in a unique, rhombus-shaped floor plan. The structure supports this geometry in the truest sense of the word.

The pentagonal concrete core is shifted directly along the closed north facade. It forms the backbone of the tower and is flanked by two wing walls that can absorb the torsional loads from wind and seismic actions. The flat slabs span from the core and the wing walls to the prefabricated facade columns.

From the fourth to the nineteenth floor, apartments are oriented to the south, east and west. Shops and offices are in the four-story, wing-like base, and storage rooms and parking spaces for cars and bicycles are located in the basement levels. Large panoramic windows yield apartments and workplaces awash with light. The winged annex has its own entrance, separated from the residential tract, and two elevators of its own. These elevator cores are in turn coupled to the tower's core by the concrete slabs. The architectural and structural concepts combine to form a single entity.

Claraturm, Basel (in planning)

The area around Messeplatz has been undergoing a dynamic process of development and densification since the 1920s. The primary catalyst for this is the growth of Basel's popularity for conventions and exhibitions. With the Messeturm and the new exhibition center, a new scale has been estab-

4

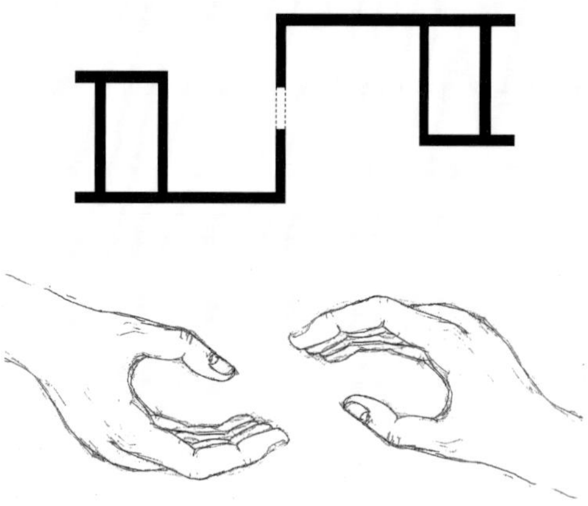

4 Visualization of interlocking hands (freehand drawing by Giotto Messi, project manager for Claraturm) as approach to the structural model for coupling the two cores

lished in the vertical as well as the horizontal. This steady development and the aspect of densification formed the basis for the urban spatial concept developed by Morger + Dettli architects. One hundred eighty apartments close to the old town, as well as offices, two restaurants, and two stores on the ground floor, will enhance the district for residents in the neighborhood.

The elegant, twenty-nine-story building with its distinctive horizontal stripes will have a total height of approximately ninety-six meters and will taper towards the top in order to reduce the shadows cast onto neighboring buildings as much as possible: Starting from a footprint of about forty-eight by twenty meters, the tower's floor area continuously decreases from the fifth floor upward on its one narrow side, and from the twenty-second floor also on one long side, so that the floor area on the top floor only measures about forty by sixteen meters. The wraparound bands on each floor level express the stacking of the twenty-ninth upper floors. Through the omission of spandrels, the residents have a direct "high-rise feeling" with long-distance views as well as a view downward.

The load-bearing structure is designed as a skeletal frame and consists of three different structural elements that transmit the vertical loads into the ground via a deep foundation: point-supported flat slabs, prefabricated facade columns, and two interlocking cores.

The non-structural soffit bands prevent the spread of a fire from one floor to another. The facade columns are aligned above each other over the entire height of the building. Where the floor plan diminishes in size, the columns are slanted. The column grid is a result of the interplay of slab specifications, structurally required member thicknesses, punching shear forces, deformations, and column diameters. The two cores, which are coupled to the slabs and optimally integrated with the utility shafts, not only transfer the vertical loads, but also ensure the building's horizontal stability against wind and seismic actions. Located in the middle of the building, their shape is reminiscent of two interlocking hands, symbolizing the architectural and structural requirements placed on them: horizontal bracing, vertical load transfer, stiffening against oscillation, integration of MEP installations, and functional access. Entirely in the spirit of Aristotle, this complementary core arrangement is a whole that is more than its parts.

1 Ludwig Mies van der Rohe, "Technology and Architecture," in: *Programs and Manifestoes on 20th-century Architecture,* Ulrich Conrads (ed.), Cambridge, MA 1971, p. 154.

Messeturm [Trade Fair Tower]
Basel 2003

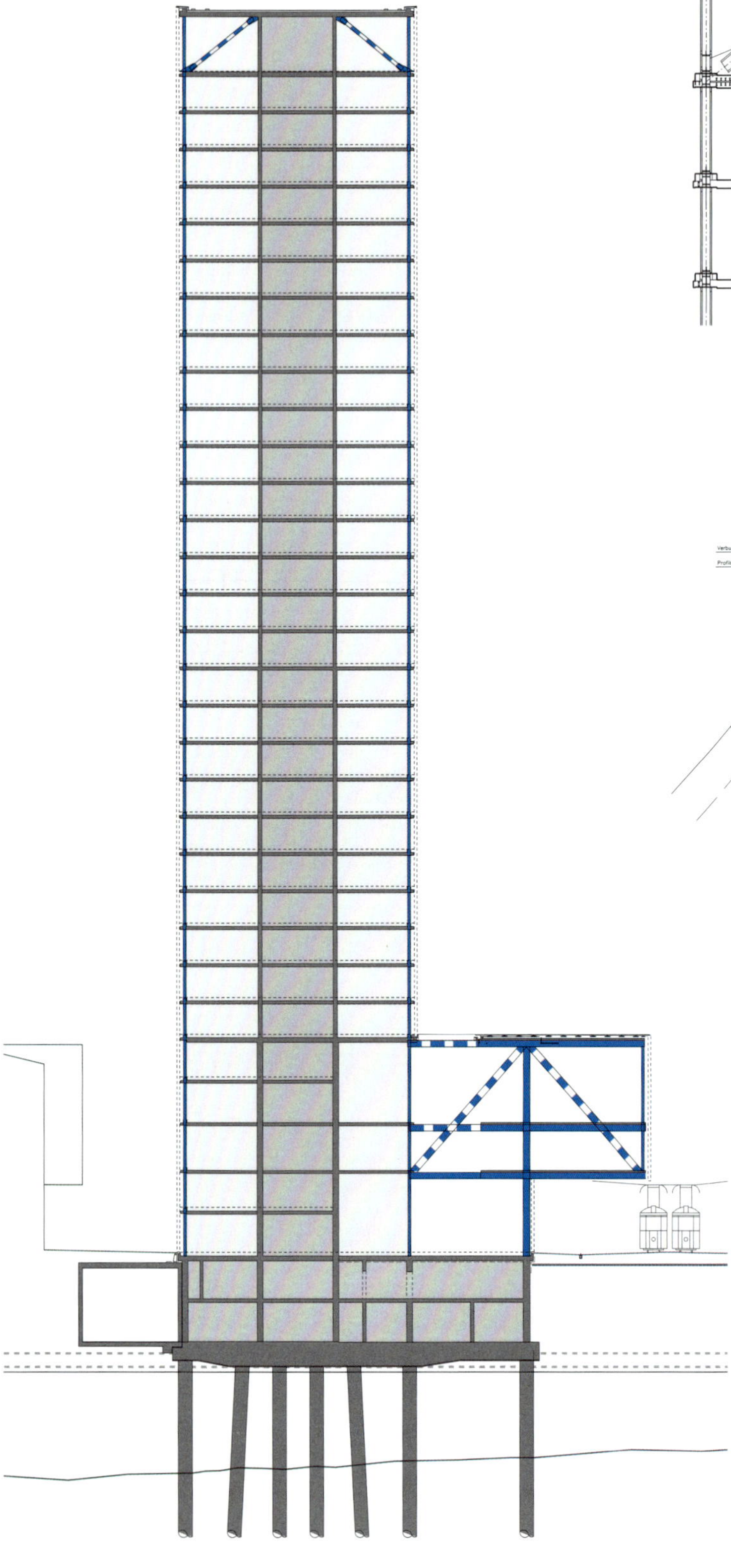

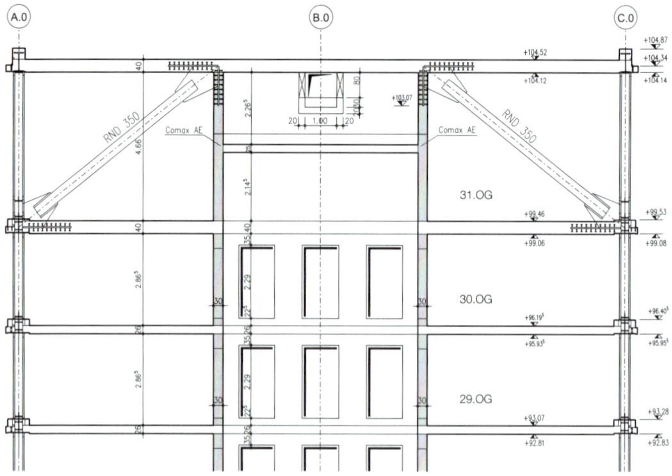

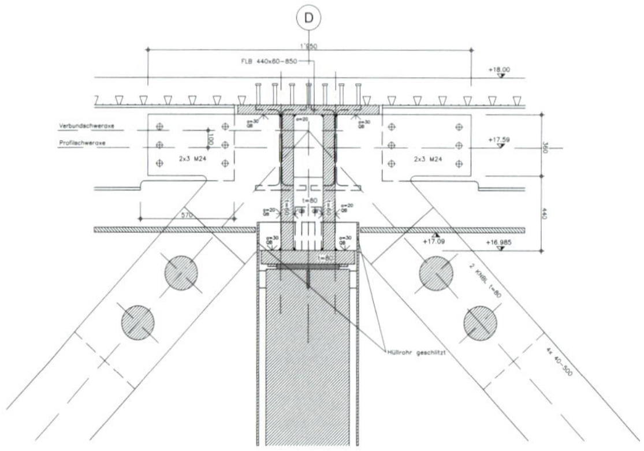

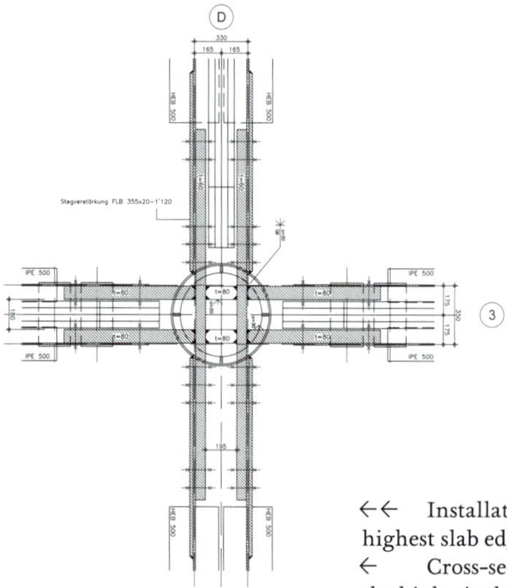

←← Installation of the highest slab edge beam
← Cross-section through the high-rise building with core, outrigger, and cantilever
↑↑ Outrigger system in the top floor
↑ Spatial node element in the cantilevered section: elevation (above), plan (below)

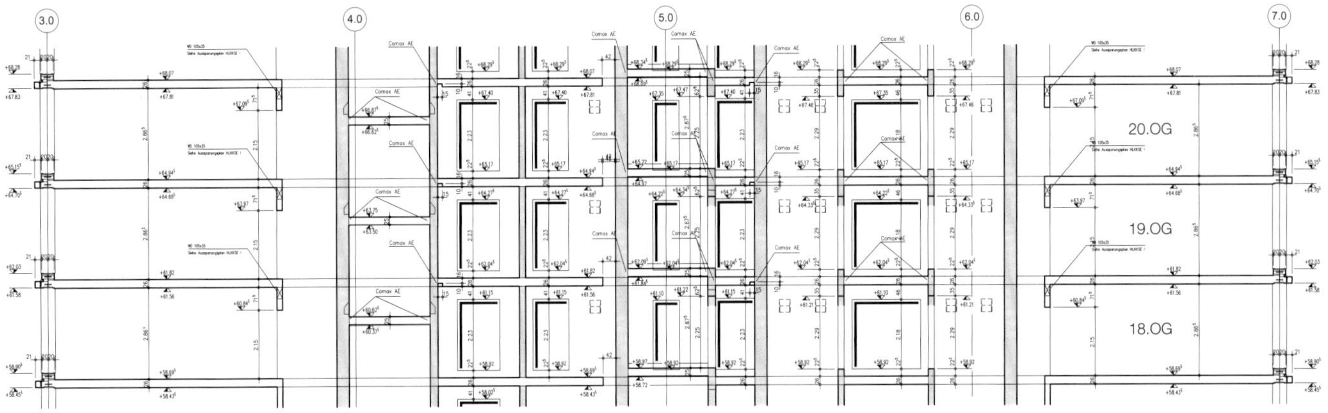

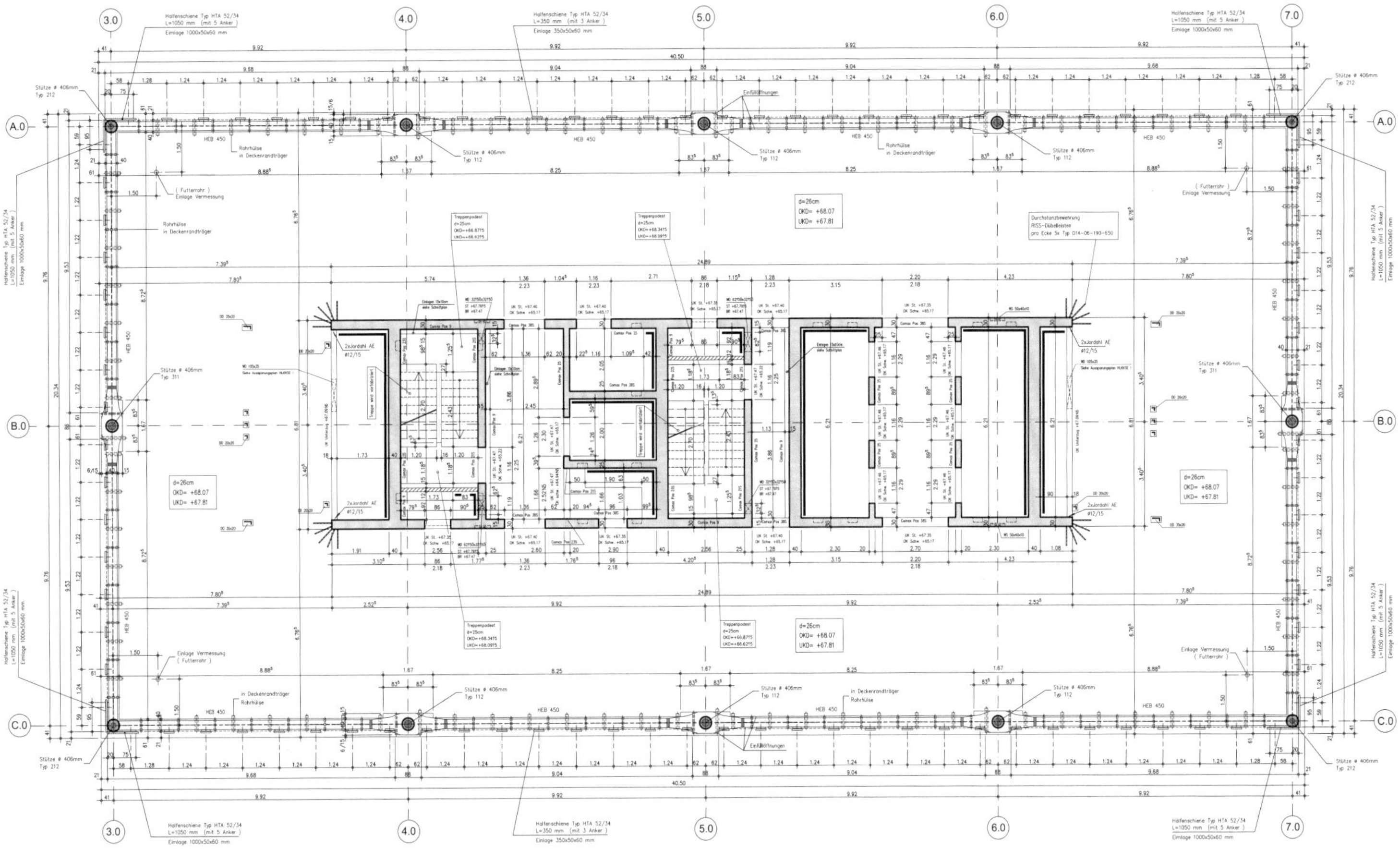

↑↑ Longitudinal section with a height between floors of 2.86 meters
↑ Plan with span lengths of 10 meters at the slab edges

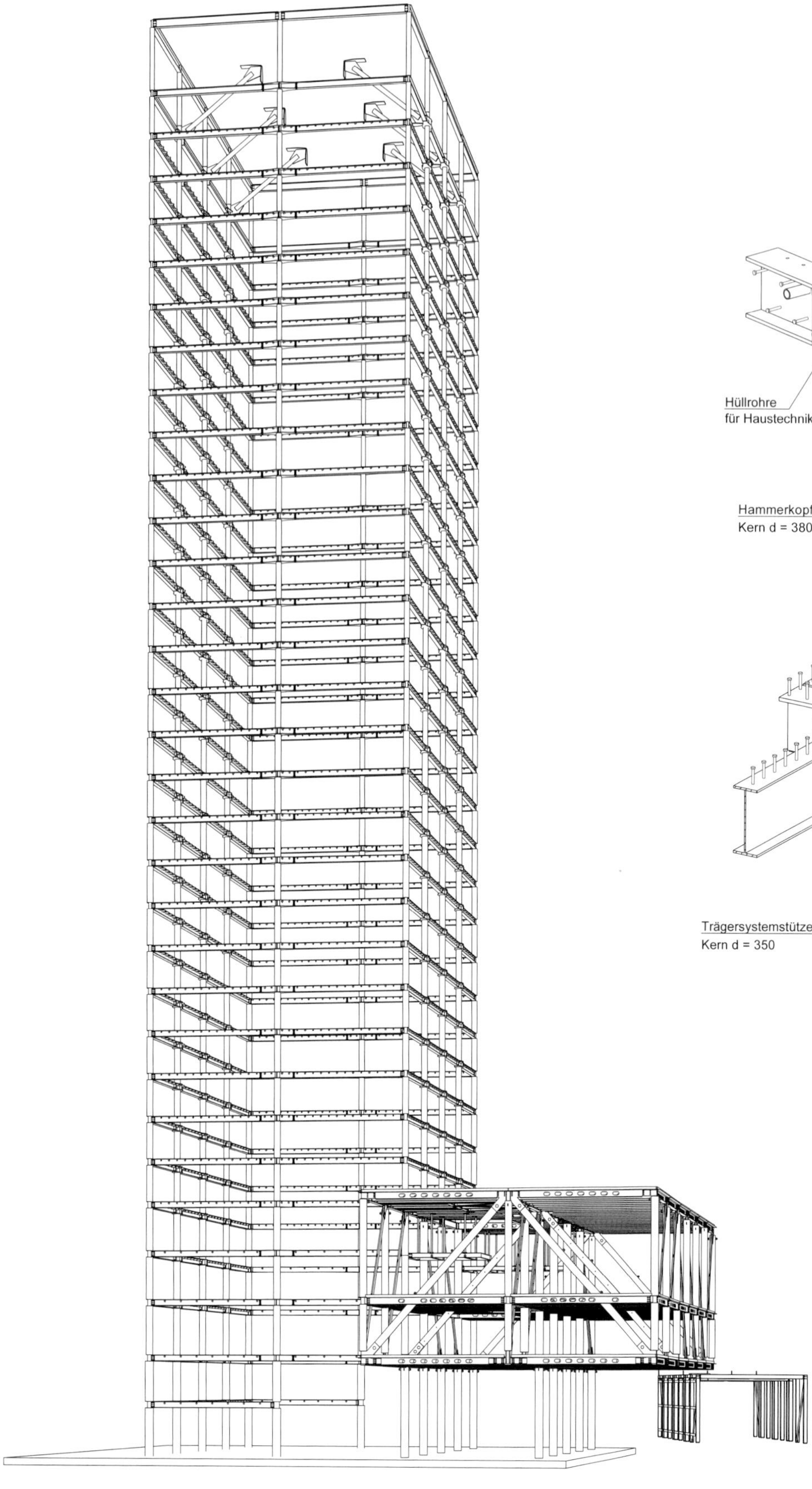

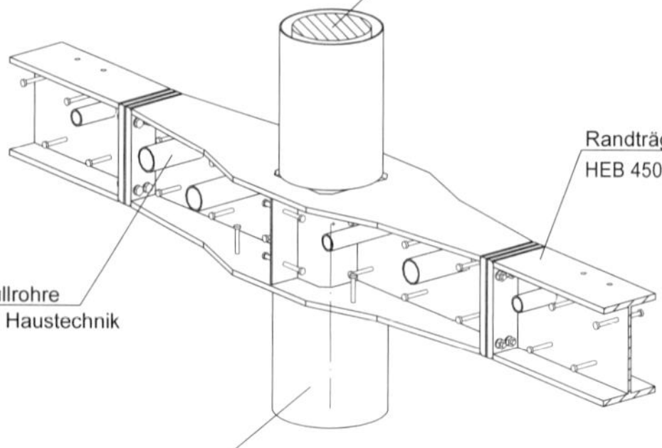

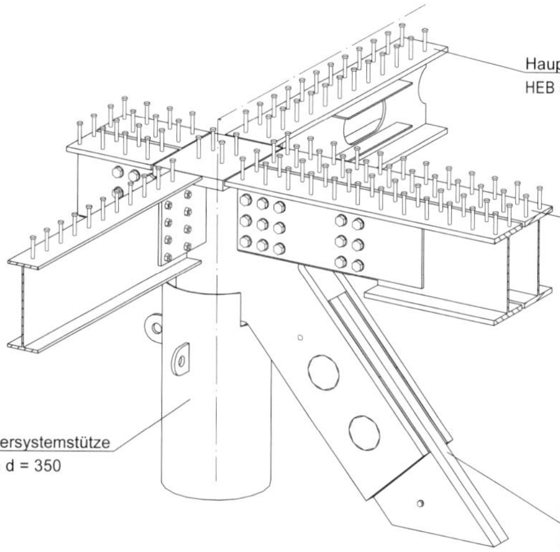

← Steel frame of the Messeturm (Josef Meyer Stahl und Metall AG, Emmen, Switzerland)
↑↑ Structural design of the hammerhead column with spliced girder connection for continuous slab edge
↑ Detail drawing of spatial truss node in the cantilevered section

↑↑ Welding together the hammerhead columns at Josef Meyer Stahl und Metall AG in Emmen, Switzerland
↑ Installation of the node element, with corrosion protection
→ Under construction, with climbing formwork for the cores and slabs at the twenty-second floor
→→ View of the building's cantilevered base, with hotel driveway
→→→ View as seen from Messeplatz

Aquila Commercial and Residential Building
Pratteln, Switzerland, 2013–2015

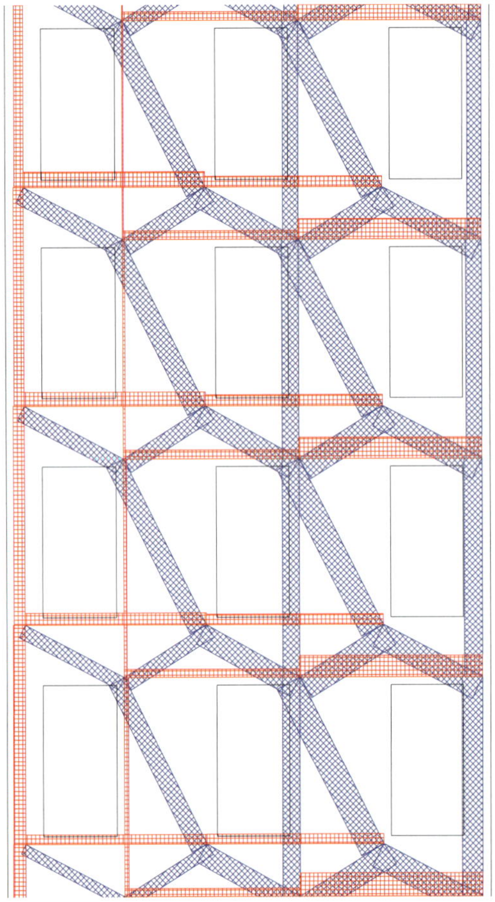

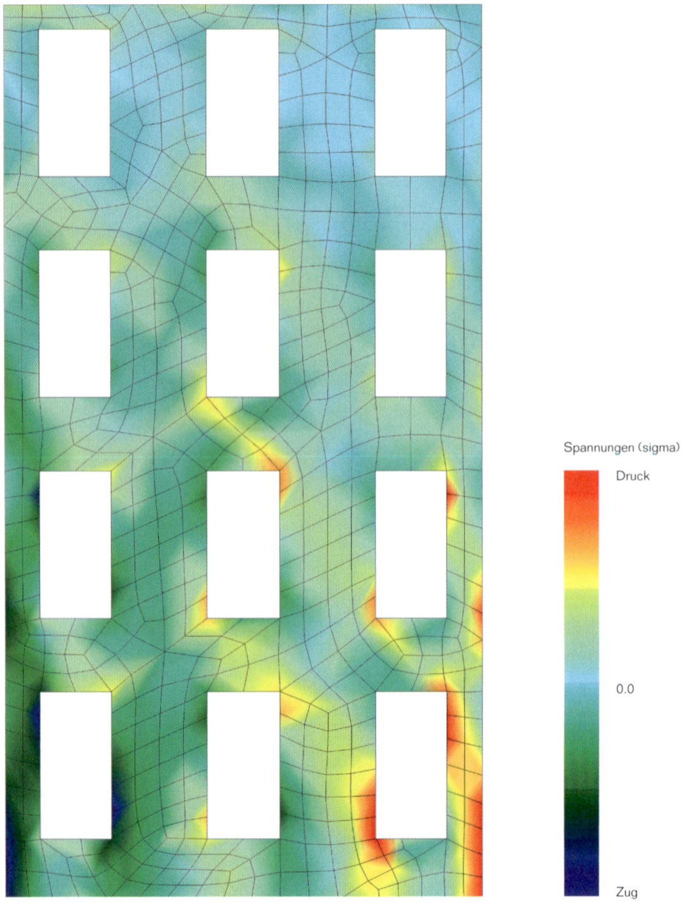

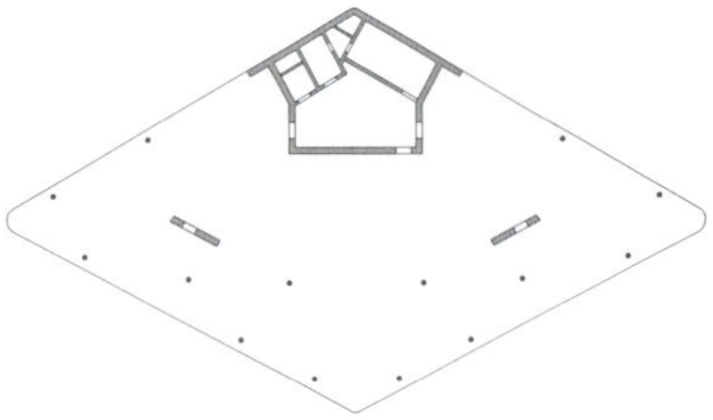

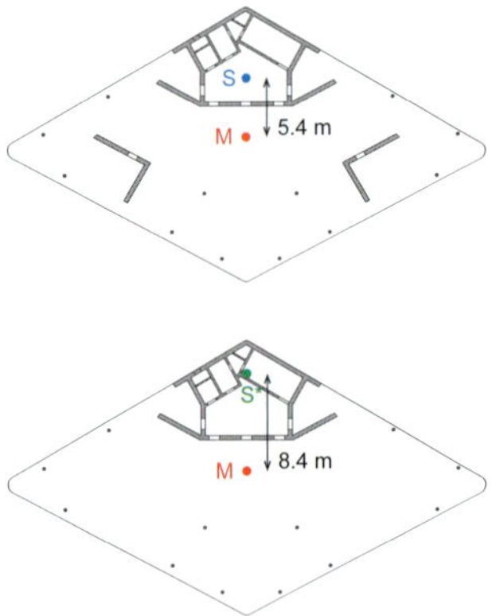

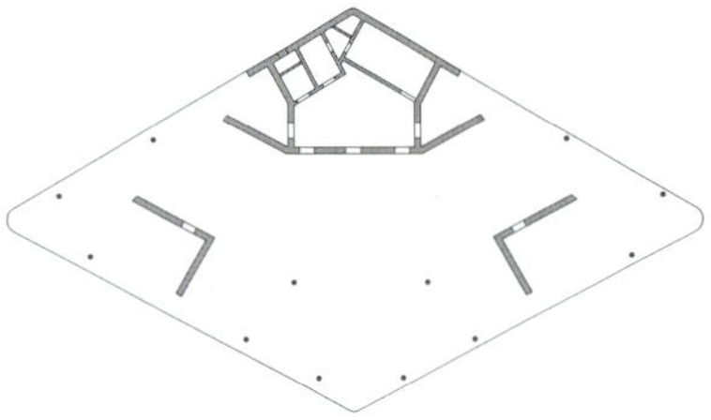

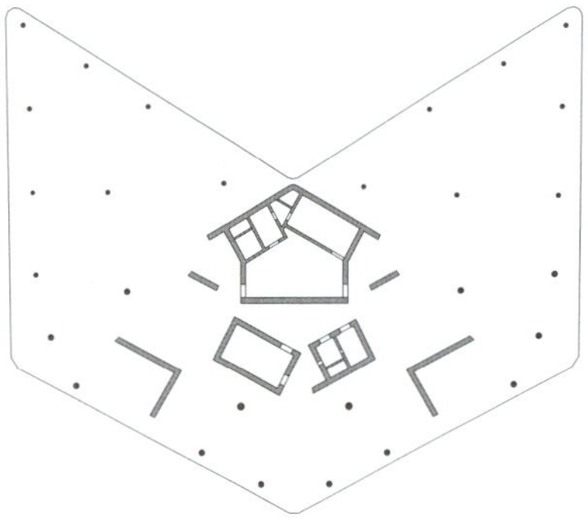

← Architectural model of the high-rise
↑ Horizontal flow of forces (left) and principal stresses (right) through perforated core wall
↗↗ Plan of the penthouse levels (eighteenth and nineteenth floors)
↗ Typical floor plan (fourth to seventeenth floors)
→ Plan at the base (ground level to third floor)

↑ Torsional stiffness increased by adding two angled walls (shear center shifts three meters toward the center of mass)
→ Spatial structural system with slabs, columns, and stabilizing walls

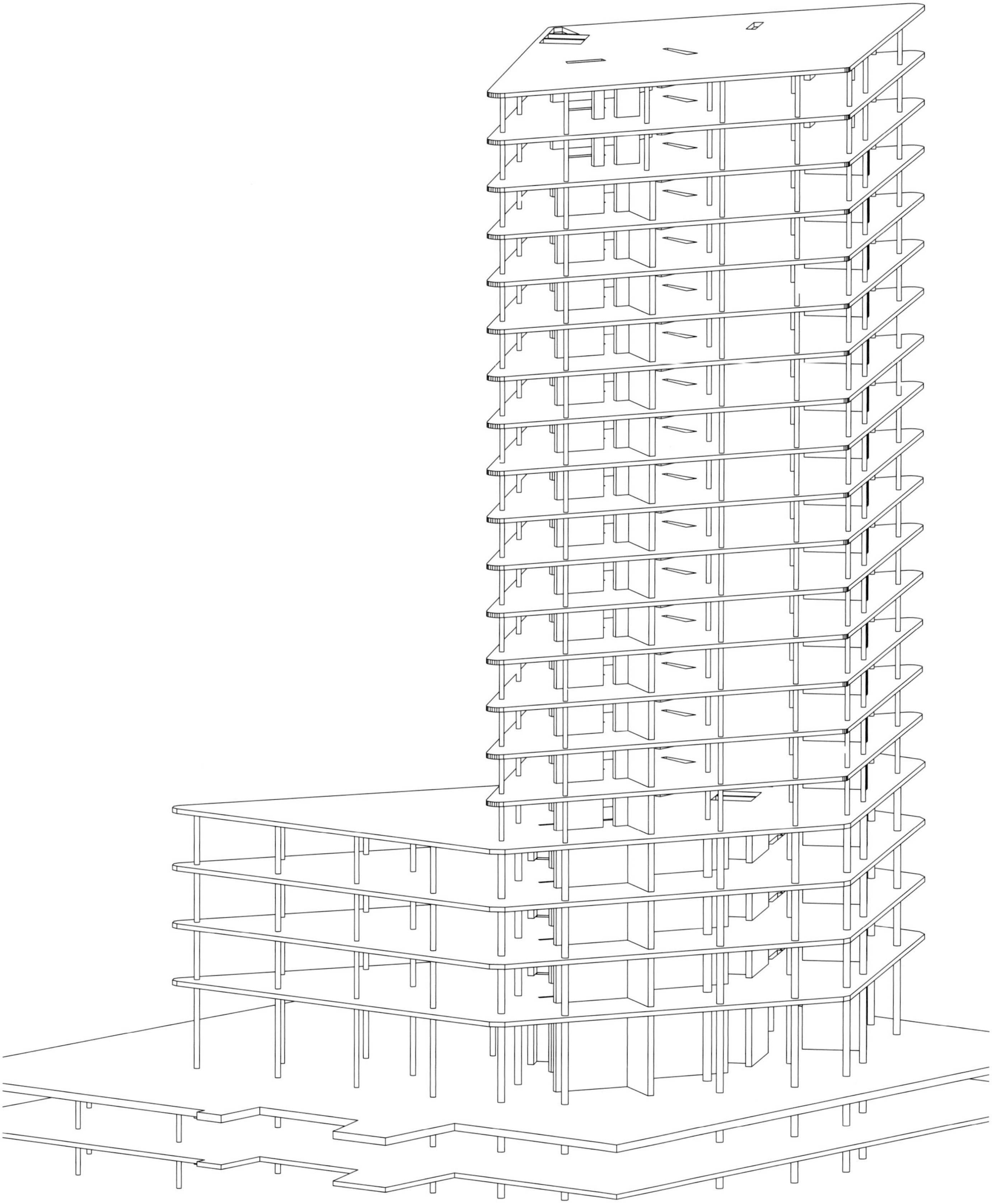

→ Core tube with angled wall
→→ Moment diagram of the decisive wall in the core

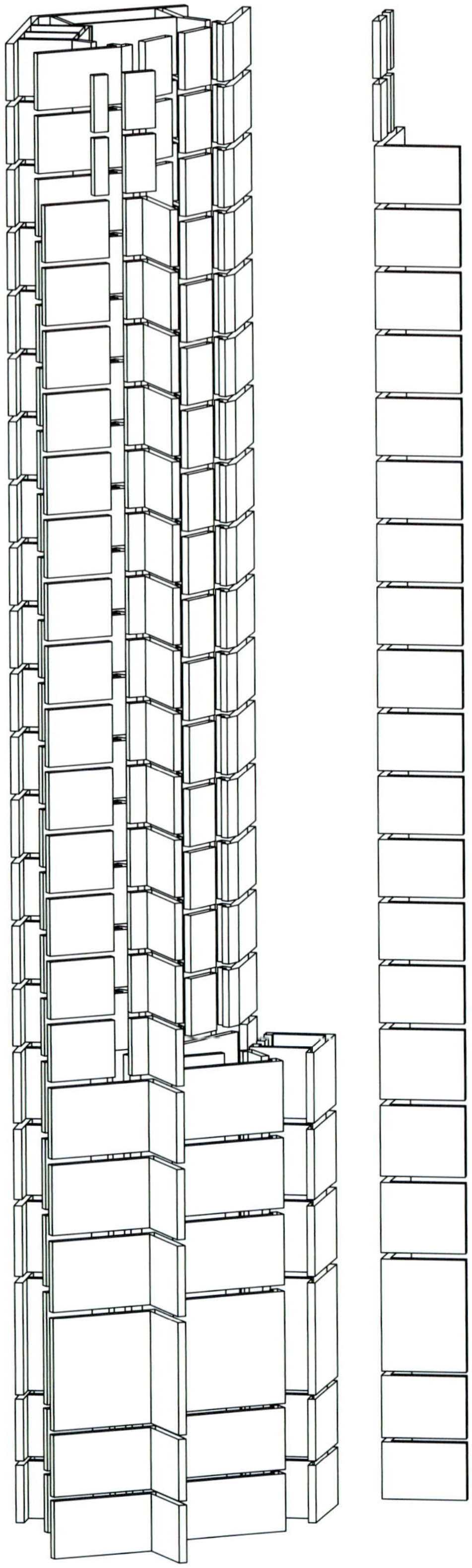

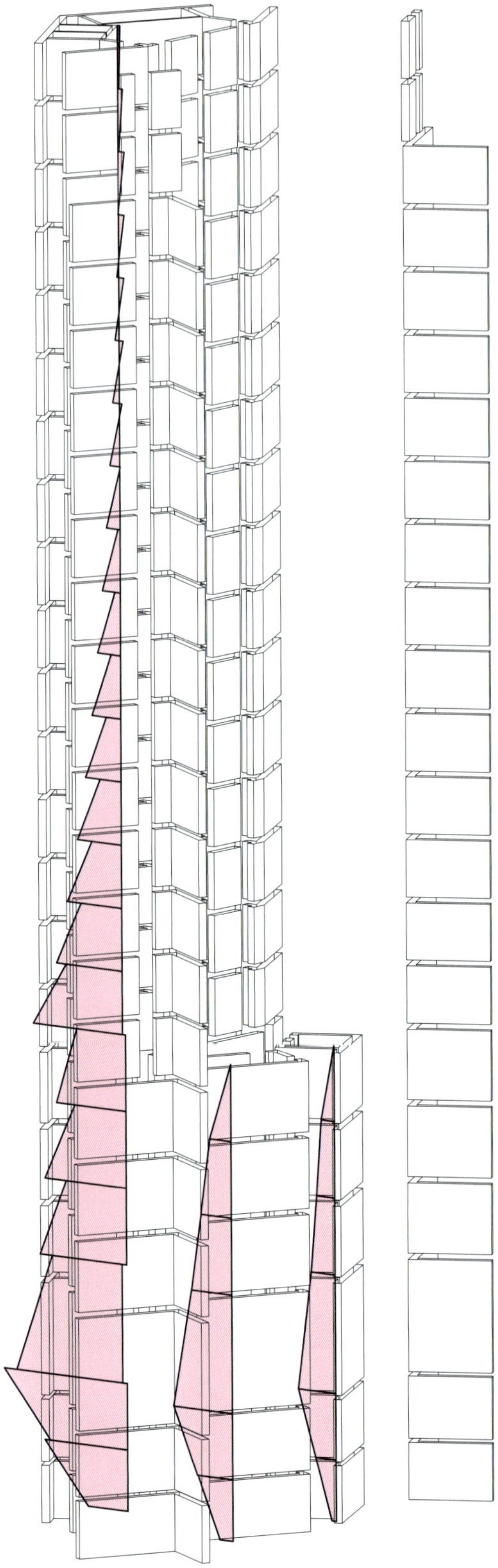

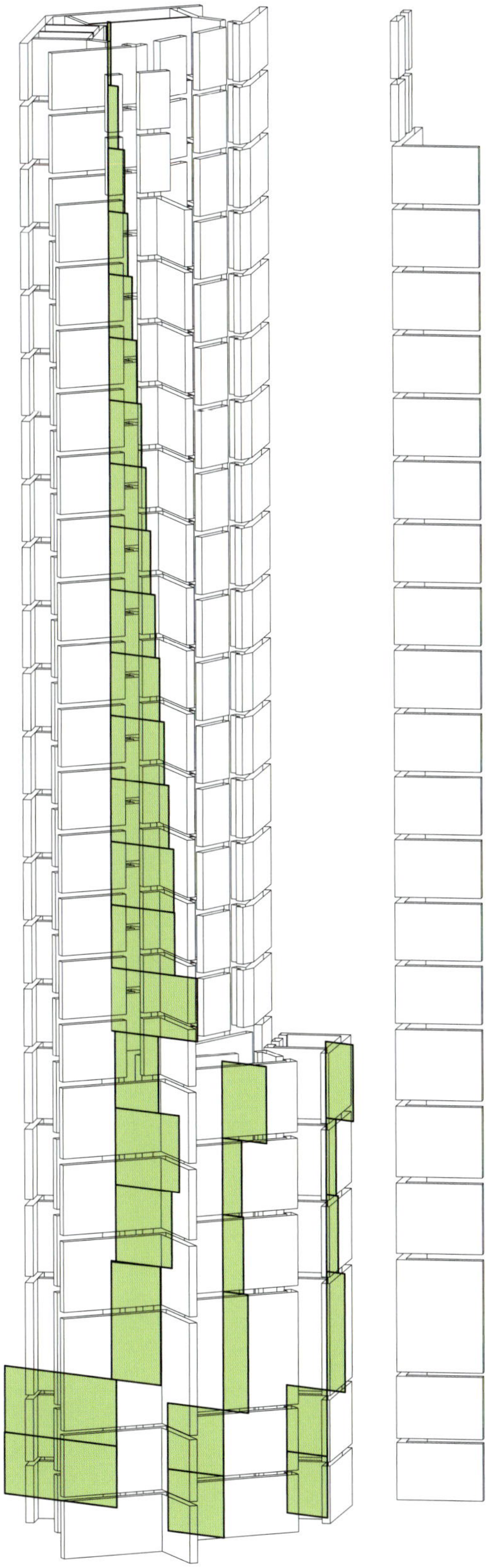

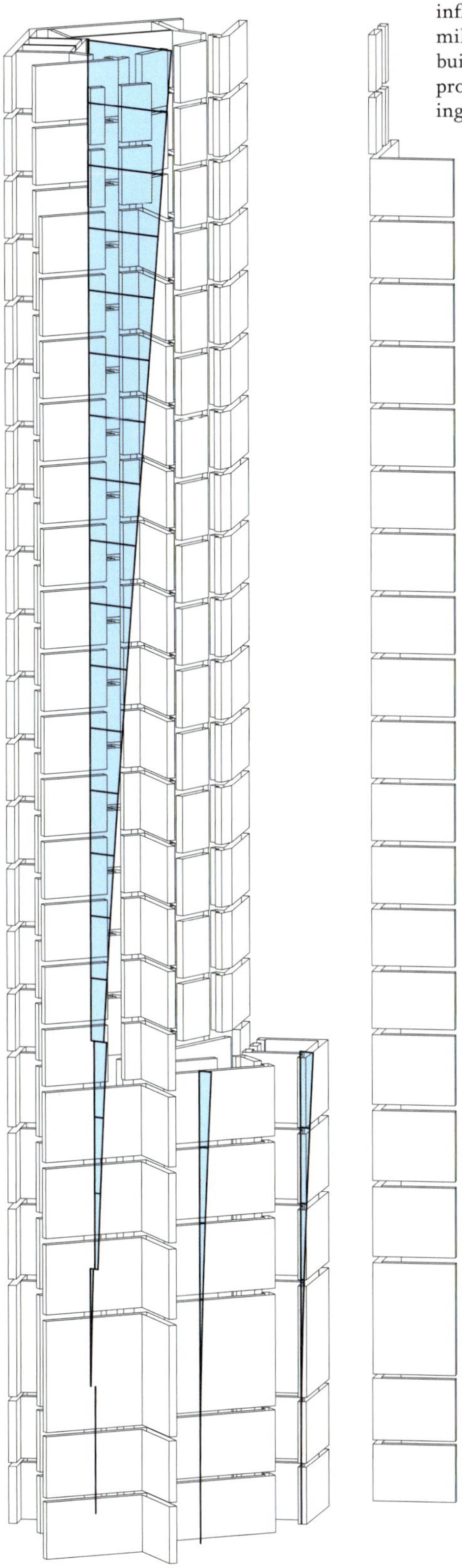

← Shear force diagram of the decisive wall in the core
←← Horizontal deflection of the entire system due to the influence of wind. Eighty-five millimeters at the tip of the building correspond to approximately $^{1}/_{750}$ of the building height

Claraturm
Basel (in planning)

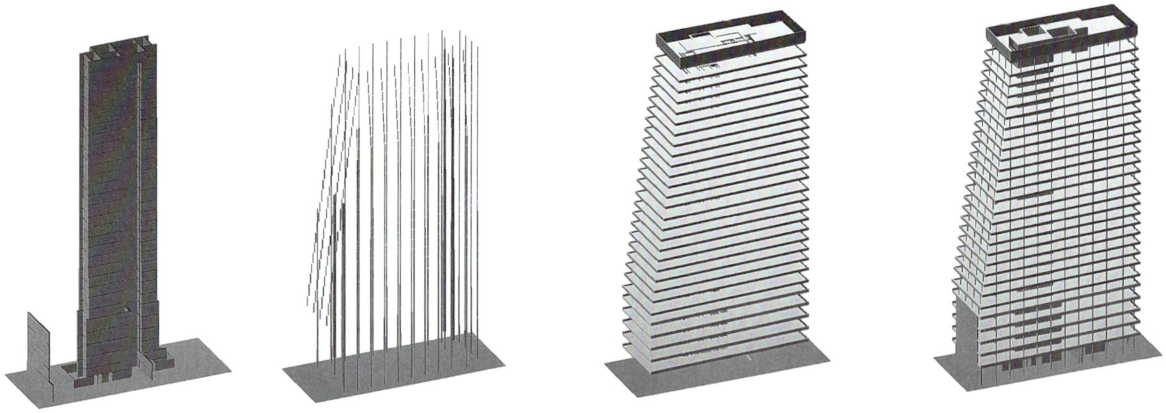

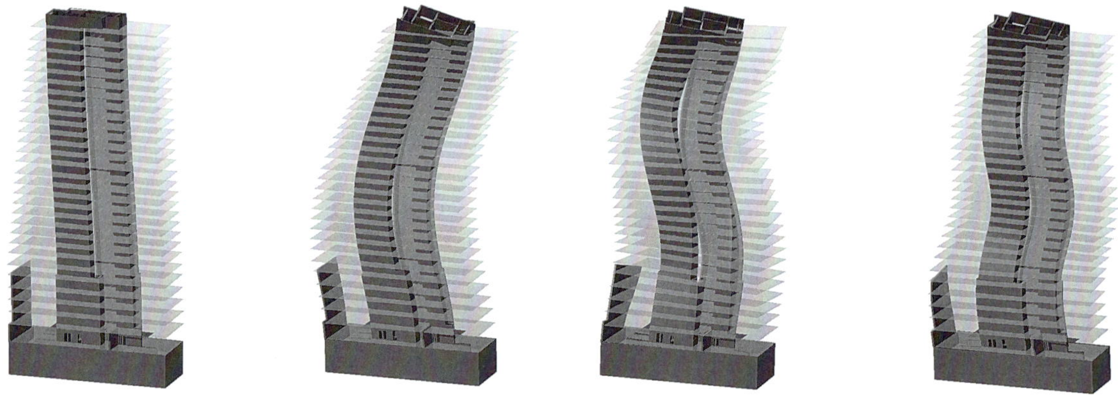

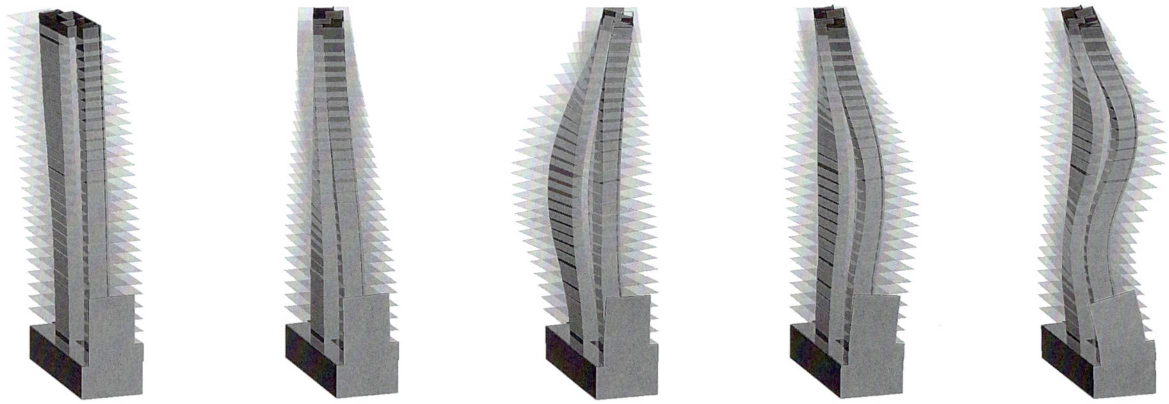

← View as seen from Riehenring, (visualization)
↑↑↑ Overview of the structural elements of the high-rise
↑↑ Initial natural frequencies in longitudinal direction
↑ Initial natural frequencies in transverse direction
→ Horizontal deflections from 50-year wind

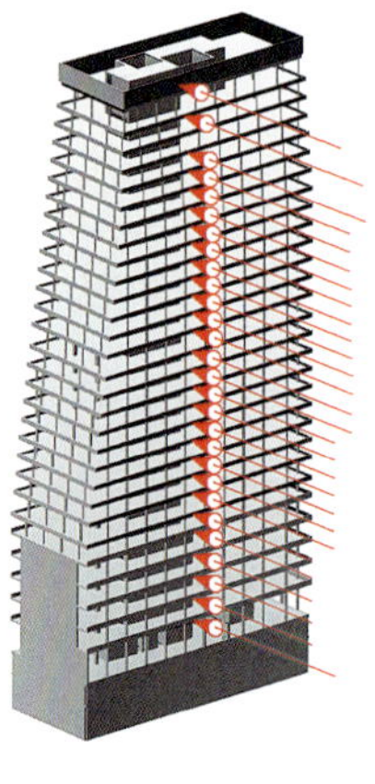

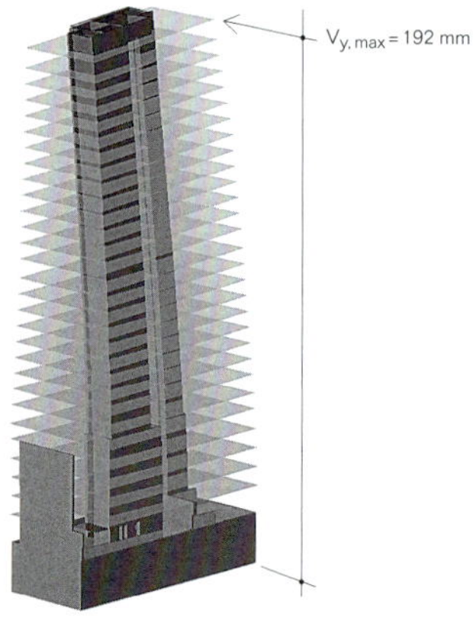

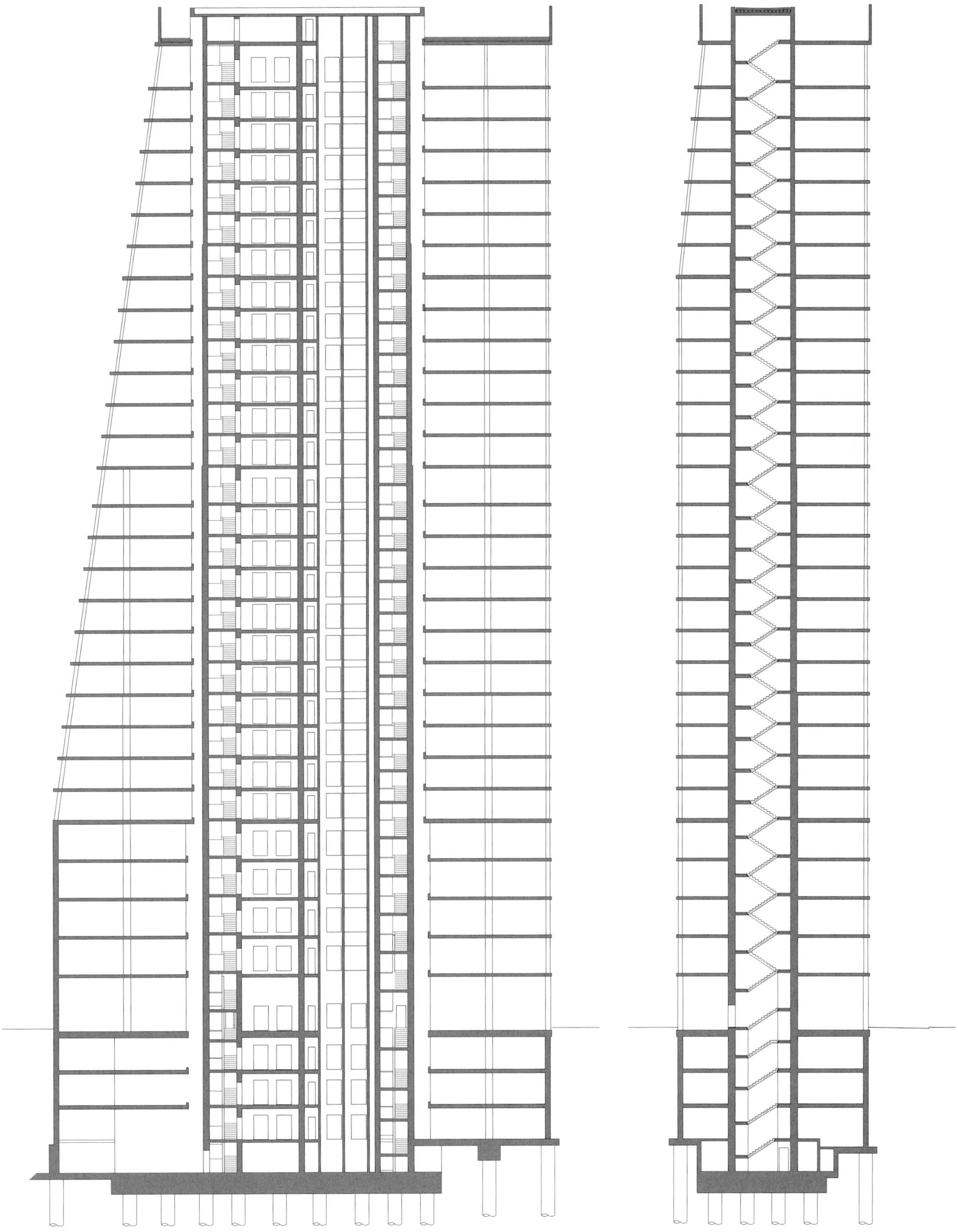

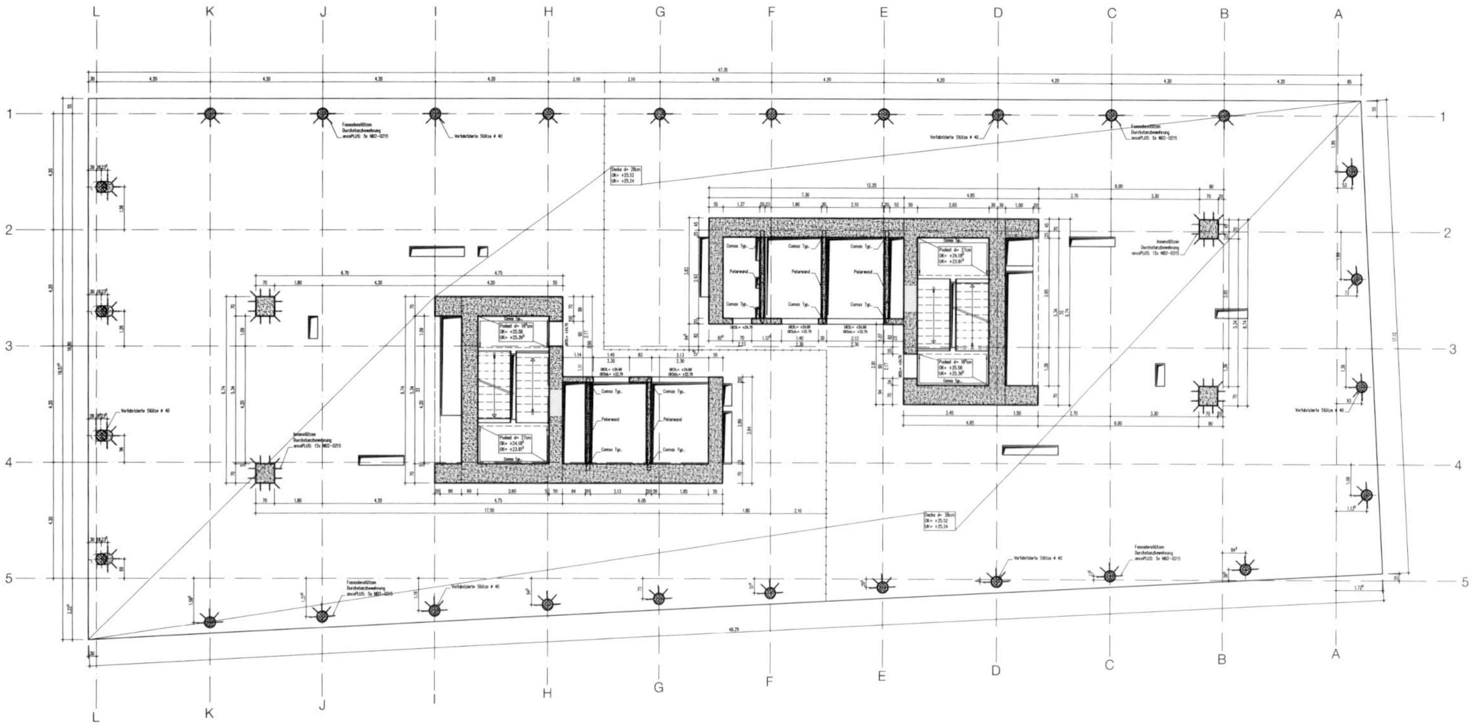

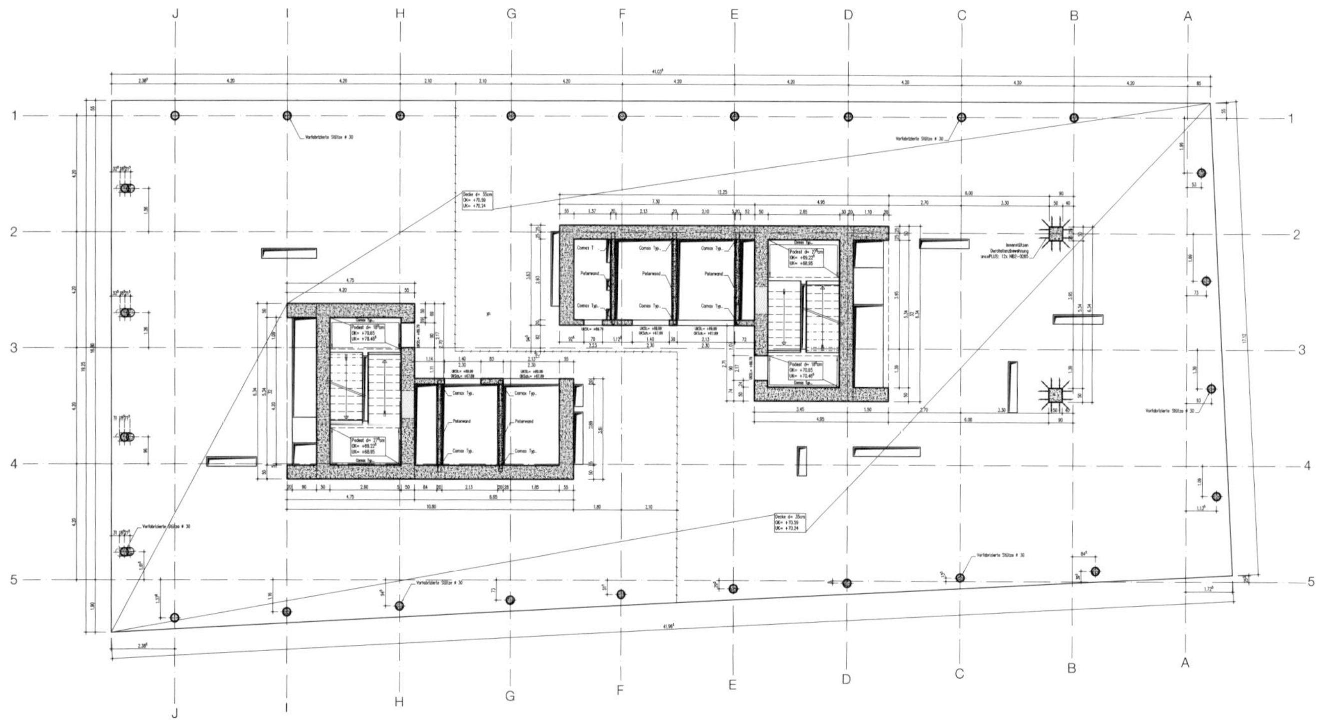

← Longitudinal section with slabs, core, columns, and foundation (left); cross-section with slabs, core, columns, and foundation (right)
↑ Fourth floor plan (above); twenty-second floor plan (below)

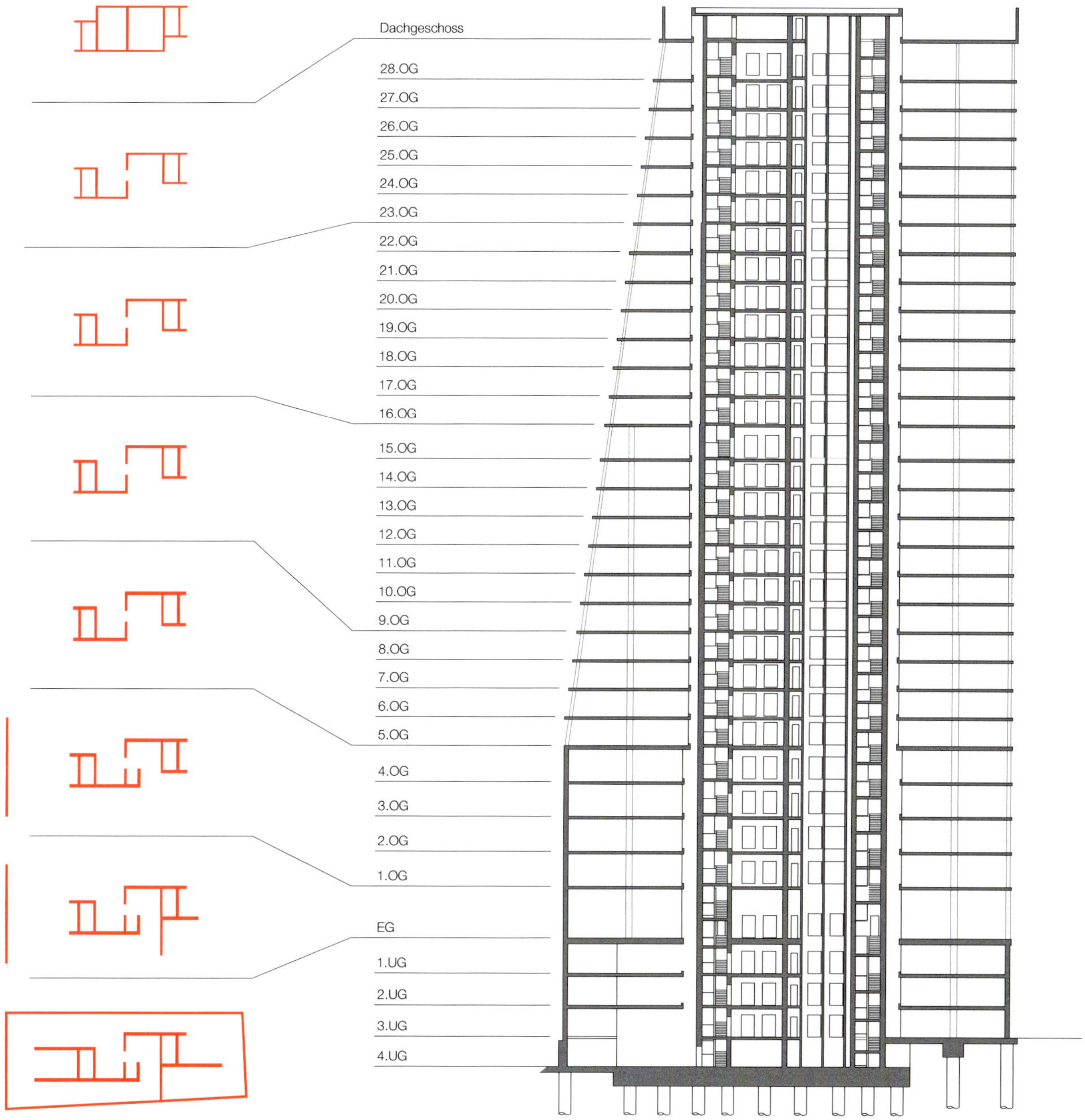

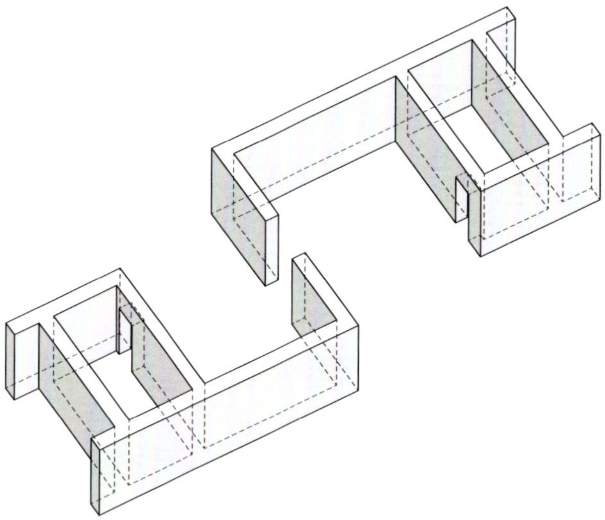

↑ Development of the building core at different floor levels
→ Two complementary cores that are coupled in the top story to resist shear. “The whole is greater than the sum of its parts.” (Aristotle)

7 Transformation

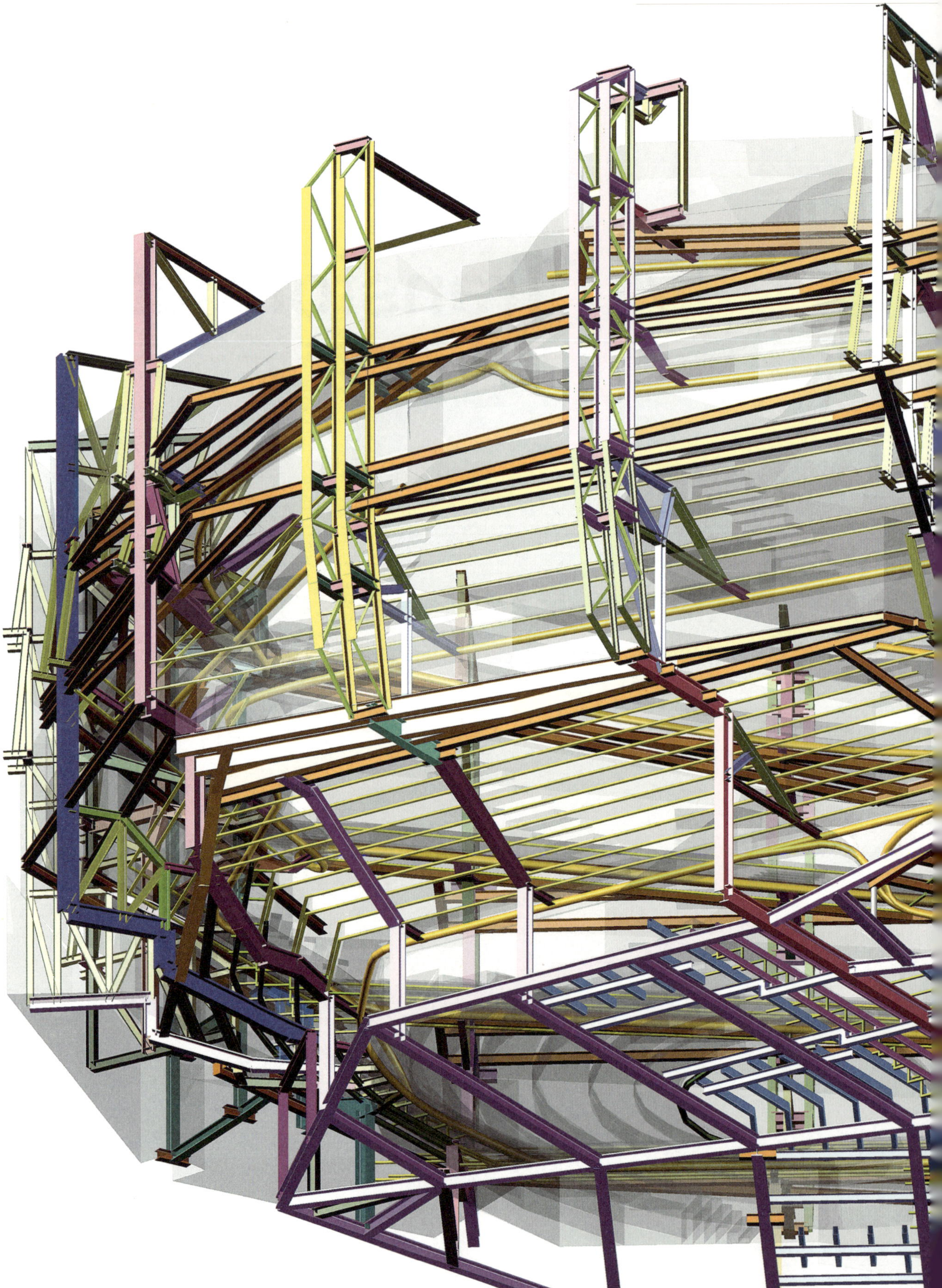

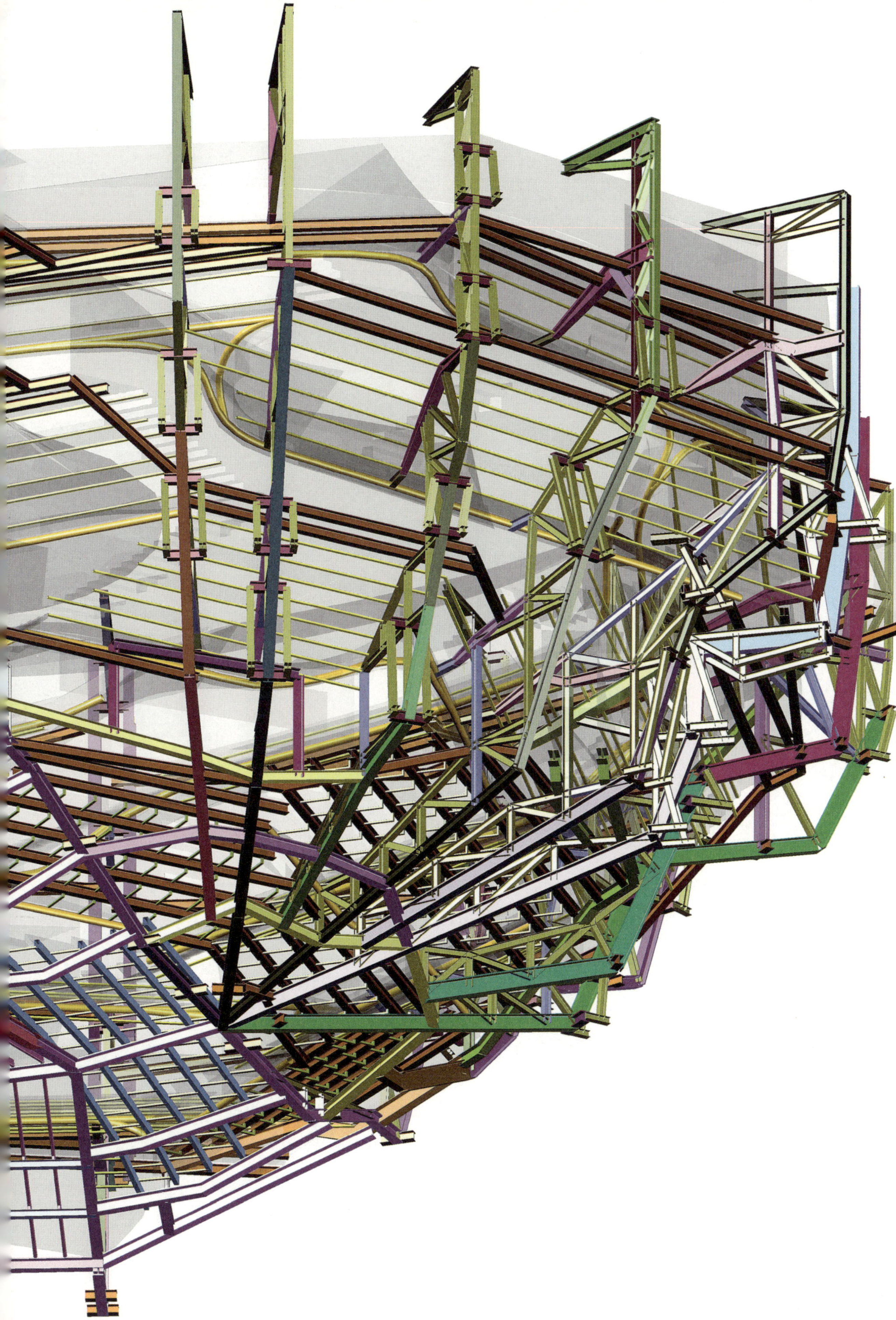

← Inner shell of the Grand Hall of the Elbphilharmonie Hamburg (computer model)

Conversations VII

AF For CaixaForum Madrid, you took an old power plant, rigorously gutted it, underpinned it, thrust it up in the air, and added new floors – probably one of the most radical conversions of a historical building ever made. Likewise with the conversion of the warehouse for the Elbphilharmonie in Hamburg, the original building was completely gutted and new levels were added on top. To what extent were you involved in preservation issues on these projects and how would you describe your fundamental position in dealing with existing buildings? Are there moral, structural, or transformational limits?

Radical Conversions

HS Neither of those buildings are really conversions. At Caixa, the historical facade has simply been re-used: otherwise, it is a new building. So I would not talk at all about transformation and conversion in that case.

AF Are you advocating a redefinition? Is CaixaForum Madrid a new building?

HS It's definitely not an example of a so-called conversion. The former building no longer exists and it could equally well have been replaced by a new one. From a landmarks preservation perspective, it is a new building.

AF A new building with a historical facade?

HS The building makes no claim of representing a way of dealing with existing built fabric.

AF This does seem somewhat contradictory to me – the facade is indeed still there and it is probably very important for the identity of that place. A new building would also presumably have been less expensive.

HS It would not have been less expensive.

AF All right then, let's agree on a new building with an old facade. Does that also apply to the Elbphilharmonie?

HS There's an interesting story behind that project. Originally, a competition for two high-rise buildings had been held there. Owing to a lack of investors, however, the high-rises could not be built. So a "temple to culture" was envisioned for where offices had initially been planned. The design draws its energy from the interaction between old and new. For economic reasons, the concept was initially based on using the existing quayside warehouse and its foundations, consisting of 1111 piles with ample load-bearing capacity. The warehouse was supposed to mutate into a parking garage. But the client believed that parking inside it would not be convenient enough, so a decision was made to gut the warehouse. I wouldn't cite this project as a shining example of conversion, either.

AF You cannot simply alter the definition of transformation: these are absolutely radical interventions!

TP Yes, I see it that way, too.

AF Apparently, you often work with architects who – at least on the well-known projects – take radical approaches to stripping away and gutting.

TP We are engaged just as much with other, less extroverted projects, like our current work on a bonded warehouse in Münchenstein. The old warehouse is being converted into a school, and it involves minor interventions that have relevance to landmarks preservation, where we remove columns or ceilings, for example, in order to create higher spaces. On another current project, a warehouse is being converted into an office building. Since goods are no longer stored there, there is less load, which makes it possible to add three additional residential floors to the building. So here, the existing internal structure is precisely what is used, perhaps in exactly the opposite way to what was done at CaixaForum Madrid. Inside, the built substance is retained, while outside a new facade is added. In our work, it is extremely important to be able to read and understand how existing structures work, be they bridges or buildings, in order to make continued use of them.

HS The question was actually about how we deal with buildings that are protected as historic landmarks. The two projects that we discussed at the beginning had no preservation-related requirements.

TP For the jazz school in Basel, we had to retain an entire building and underpin it with two new levels below. We supported the building with micropiles and built beneath it. This was done for reasons of landmark preservation, because the building contains important wood panel paintings dating from the fourteenth to the sixteenth century.

AF As mentioned earlier, I am interested in whether you yourselves get involved in the preservation debate and take a stance on it.

TP Preservation officers assume a hegemonic position on projects with historical built substance. We take their demands seriously and try our best to meet them.

HS An architect who is involved in the design of guardrails or noise barriers is not also involved in deciding the course of the railway line.

AF That's exactly what I see as a huge problem!

HS In transformations, legislation and the preservation aspects are the hard criteria.

SB But we can facilitate a lot.

HS The question is, though, whether that's all it should be about.

SB At the Stadttheater Solothurn (City Theater), soundings revealed the presence of old murals on the theater's balconies. My confirmation that they would hold up and could thus be left in place was crucial to the design.

AF By "facilitating," do you mean facilitating conservation measures?

SB Yes, that too.

HS It's no different for us than for the architects, to stay with my example of the noise-barrier...

AF Let's stay with transformations. I am firmly convinced that architects ought to develop firm positions and not allow the agenda to be set only by the landmarks preservation officers. I can think about whether I want to spatially extend something or simplify it, and whether I should didactically separate old from new, blur their boundaries, or clear away as much of the old as possible. These considerations can and must but be applied equally to the structure and the construction itself – indeed, they are integral to any reflection about it! But where should the engineer gain knowledge of the historical constructions?

TP We don't get this knowledge from our training, but we have books.

HS One thing is learning how something is built. That's more easily accessible for us. The other would be an education in the history of art and culture, in the history of structural engineering and construction: our training shows undeniable deficits in that regard.

TP A fatal flaw – we don't learn the roots of our own history! You can certainly hold that against our universities.

AF I remember a debate on engineering and preservation at the Swiss Society for the Art of Civil Engineering, during which tempers rose about whether to remove areas of brick infill that had been subsequently inserted into Maillart's Schrähbach Bridge in Innerthal, given that they are structurally irrelevant and were not part of Maillart's design. The engineers pleaded for the construction to be returned to its original state, while the landmarks preservation officer argued that the infilled parts were worth keeping because they chronicle a moment in time.

TP I agree with the engineers. The bridge should be stripped down to Maillart's original construction – away with the infill!

AF I'm interested in the general question of who should actually discuss the authenticity of a construction.

TP Surely the landmark preservationists and the engineers, together.

HS Of course, those bits of infill were not envisaged by Maillart, but now they are there – why should they be taken away? Why make the effort to remove them? I'm taking a purely pragmatic point of view.

SB When refurbishment is pending anyway, it can surely be considered; I wouldn't set up a construction site for it otherwise, though.

TP That bridge is a landmark monument! The monument has been modified, decorated – I can't judge the quality of the modification. But we would want to remove an added suspended ceiling if it were hiding beautiful timber beams from view!

HS Medieval houses were altered many times, too. How do you decide what to remove and retain in that case?

TP I differentiate between a Maillart bridge and a dwelling that has already been altered twenty times and has no special historical value in terms of design or culture. The Maillart bridge has that, though.

AF I wonder broadly whether engineers themselves shouldn't get more actively involved in debates on the history of construction.

TP Sure, we have a duty to do that.

HS But the competence and the willingness to contribute at all is something the engineer has to first develop. In Switzerland, things are certainly happening along this line; the Society for the Art of Civil Engineering also plays its part.

Civil Engineering Debate

AF Is there, in your opinion, a debate on the art of engineering? And if there is, are you involved in it?

TP The debate over civil engineering art does exist. Let's consider the current example of Isler's shell – the autobahn service station in Deitingen, which was targeted for removal. The engineers and architects fought that decision together.

HS We will increasingly be confronted by such questions. Many so-called "pure" engineering structures, such as roads or tunnels, are approaching the point when they need renewal or repair.

SB I don't see any examination led by engineers at the moment. The engineering journals usually restrict themselves to documenting a structure, which amounts to a one-sided report. There's no critical debate taking place!

AF What trade literature are you referring to?

SB *Der Bauingenieur* [The Civil Engineer] or *Tec21*, which used to focus more on engineering, but is now almost a purely architectural magazine…

AF So this kind of debate is not held in your trade journals. Is writing and talking about your projects relevant for you?

TP Yes, we've also published articles about our projects. They never get the same attention as architectural publications, though. But we're also addressing a much smaller audience.

HS Engineers are not accustomed to contemplating their own structures, so there is rarely any critical self-reflection taking place. Nor are we accustomed to talking about negative experiences, even though everybody has them.

Education
Training

I would like to add a general observation on the training that engineers get at ETH Zurich: The lack of instruction in the history of construction is only one aspect. The other point is that intrinsic engineering thinking, meaning the approach to a construction problem, is not given enough attention today. The professors and the chairs have a slant that tends to be too theoretical, which has disadvantages in the long term.

AF That's related to the requirement that the chairs in the Department of Civil, Environmental and Geomatic Engineering (D-BAUG), where the civil and structural engineers are trained, must all be appointed as full-time positions, which makes it impossible to run a practice on the side. As a result, engineers who actually build are excluded from teaching at ETH Zurich.

HS That is certainly one reason. But the real problem, in my opinion, is that because the candidates for the chairs are selected on the basis of doctoral theses and the number of published papers, only a certain type of engineer is considered anyway.

AF It's an absolute disregard for the relevance of practice.

HS It's headed the same way as in the United States, where the training is very theoretical. The strength of the courses at the ETH, or anywhere in Europe, used to be their close relationship to practice.

TP In our time, Professor Walther and Professor Menn were practicing bridge-builders who were simply among the greatest engineers of their day. They didn't just sit in the laboratory, conducting material research...

HS In the past, there was a balance between theoretically oriented professors and practically oriented ones. Today the theoretical side predominates.

AF How did this shift come about?

HS The current university administration only pays attention to the scientific side and that's not very expedient.

AF A sporadic dialogue between the D-BAUG and the Department of Architecture clearly seems to be called for...

HS Yes. Or a talk with the dean of the ETH, the President himself – he ultimately designates the professors!

On Dealing with "Wicked Problems"

Harry Gugger

Hasn't everything already been said about the collaboration between engineers and architects? Or do these domains change and develop so dynamically that there is always something new to report? The two disciplines have been separated now for a few centuries, so no one would still want to seriously bemoan or call into question their partition. And the planning and construction industry is not exactly known to attract attention for revolutionary innovations.

But nevertheless, we should stop for a moment and ask ourselves whether the collaboration between architect and engineer has changed under the influence of information technology (IT). In the building industry, as everywhere else, IT has been the most important driver of innovation in recent decades. Its use has led to massive changes in the design as well as the construction of buildings. Consequently, IT must be identified as a disruptive technology that, while not creating a paradigm shift, has provoked a clear bifurcation – a drifting apart of conventional building production, on the one hand, and building production that is dictated by information technology on the other.

Building production, after all, is at best managed purely rationally. In this field, IT has gained widespread general acceptance. The design process, by contrast, represents a *wicked problem.* It is, in other words, a question that is not definitively defined – one that has no unequivocally right or wrong solution, but only has solutions that are better or worse. A problem that, when working collectively – and the act of building is, of course, a collective good – cannot be resolved with purely rational means, but is instead dependent of rhetoric.

Here, I understand rhetoric as a method of finding truth. In debate, consensus is sought and truth is not necessarily found rationally. Now, it can be observed that architects and engineers react very differently to this situation. Generalizing, it can be said that engineers have difficulties with "wicked problems" and, for the purposes of their work, try to reduce them to rational relationships. In his essay "Infrastructural Ecologies," Pierre Bélanger describes engineers as follows:

"Their work, their language, their education, are determined by standards, specifications and systems. They relish in the isolation of variables and exclusion of social dynamics to reduce complexities down to verifiable quantities, certainties and solutions. Their work sits on a razor's edge of precision and a thin hairline of probability in order to produce exactitudes from indeterminacies."[1]

For the work discussed here, information technology is of great importance. Thanks to IT, ever more complex problems can be determined. At first blush, there is nothing objectionable about that. After all, we are all grateful when potential risks in the building process can be precluded with absolute certainty.

In the design process, however, relying too soon upon rational correlations can lead to dismal results and undercut motivation. Admittedly, the fundamental rejection of rhetorical debate, as practiced by some engineers, is the even greater evil, because then the architect, left on his own, engages in an unproductive mock debate that inevitably yields purely formal solutions.

With the help of information technology, though, more and more problems and also problems of a more complex nature can be calculated conclusively. It seems clear that in a world of ever greater certainty, the unknown is increasingly perceived as a threat. This suggests that the fundamental problem in collaborations between architects and engineers, namely dealing with "wicked problems," will become more acute in the future. But this evolution is not going to change the nature of this collaboration as such. From this I conclude that the collaboration between engineers and architects can only be assessed in specific individual cases.

This is best accomplished by examining deviations from the regularity briefly described above.

As we know, every rule has its exceptions – and I have had the good fortune to have experienced more than a few exceptions in my work. These include Heinrich Schnetzer and Tivadar Puskas. You can sit down together with them at the very beginning of the design process and be assured that an open and constructive debate will take its course. You feel you are among comrades. Not because they are longtime acquaintances, but because they are willing to engage in discussions when occasionally confronted with unknown territory. Working together this way, daring hypotheses can be formulated, which are then checked from different viewpoints as the work continues. You could say that the two engineers seem to work more like architects, except that what they bear in mind is not the architectural, but the structural potential of an idea.

This process can best be illustrated by looking at the CaixaForum Madrid project. An old, derelict steam-driven electric power plant on the Paseo del Prado was to be converted into an exhibition center for the Obra Social "laCaixa." This landmarked building was wedged into a small-scale residential neighborhood with narrow streets. The required spatial program was almost three times larger than the existing building's area potential. So creating public space was absolutely inconceivable at first. Since, for us, a purely subterranean expansion was out of the question, we had to challenge the rigid stance toward landmark preservation in any case.

1

2

1 The entrance stair of the CaixaForum Madrid, which leads from the plaza beneath the building to the first upper floor.
2 Schematic diagram of the construction measures at the CaixaForum Madrid

The audacious idea of removing the building's granite base and allowing the existing building to float in order to create a covered public space underneath came naturally.

In nine out of ten cases, an architect would reject such an idea immediately. Fortunately, we had consulted Heinrich Schnetzer, who happened to think the idea was not quite so reckless, and who was actually attracted to the structural challenge it presented. Consequently, he immediately adopted the project to also make it his own. The architectural, spatial idea became one with the structural idea. What followed was an exploration of this basic concept.

Then, we sought to develop the concept with extreme radicalness, both architecturally and structurally. For the public space, this required a column-free space on the ground floor, meaning that the building would only be supported by the three cores that were needed anyway (public stair with the public elevators, the freight elevator for transporting artworks, and the car elevator). A concrete ring girder is suspended from the three cores and interlocks with the existing brick masonry. Two additional concrete cross beams brace the structure and, as the primary girders, define the span length of the large exhibition halls. The rest of the structure is all built of steel and either erected on top of or suspended from this basic framework. The folded slab above the plaza is thus not a formal gesture, but a reflection of flow of forces in the suspended structure.

The architectural and structural concepts were so closely woven together on this project that it was impossible to work independently from one another. That could be perceived as a constraint. But Heinrich Schnetzer's and Tivadar Puskas's personalities do not even allow this perception to begin with. I don't know if there is such a thing as humorous sobriety. But that's how I would best describe their personalities. In a collaboration, this trait permits the "nature of the project" to come to the fore and causes personal convictions to recede. And that is, after all, the goal when designing a project.

1 Pierre Belanger, Infrastructural Ecologies, unpublished paper for the 4th Holcim Forum for Sustainable Construction, Mumbai 2013, p. 2.

Transformation

Heinrich Schnetzer

Cities develop not only toward the outside, but increasingly toward the inside. This is accompanied by densification of the core zones – meaning the treatment of existing building structures, which are often protected historic monuments, will inevitably be an ever more frequent topic of discussion. Including the existing building fabric in buildings that are being newly developed, often with completely different uses, requires intensive examination of the initial situation on both the urban and the structural level. Only an analysis of all the fundamentals can be the basis for conceptual deliberations. The aspects of construction and structure must receive the same attention given to the aspects of architecture and urban design. That makes it imperative for the close collaboration between structural engineers and architects to begin as early as the concept phase.

The contribution made by structural engineers in this first phase is bivalent. They should not discard ideas that, at first glance, seem to be unusual and absurd – not simply because of their differentness. Rather, such tasks require an openness to new, perhaps daring, and initially unrealistic-appearing ideas. Experience shows that within these is often the core of an outstanding problem-solving approach. This core must be recognized.

In the process, it must be ensured that still-unresolved points are also solvable at a later point in time. From experience, such a selective but hierarchical, priority-setting approach causes structural engineers considerable worries. Due to their training and their structured thought, they are not much inclined to allow project-relevant constructional issues to remain unresolved. It is, however, important to realize that design processes are dynamic and that, for reasons of time, all ideas cannot be examined down to the last detail right from the beginning.

In this connection, the currently applicable standards play an essential role. Unfortunately, they are too extensive and too complex to serve as a real aid in the design process. Each new generation of standards inevitably has more detailed knowledge – but its relevance is limited to a small portion of the buildings that are addressed. Likewise, security concepts are continuously refined with the goal of increasing the accuracy of the calculation models. Instead, standards should incorporate straightforward assumptions that describe, in simplifying terms, the very complex interrelationships among the load assumptions, the building material properties, and the modeling assumptions. In other words: the complexity of the standards means that the vulnerability to error increases considerably – a fact that should be food for thought. An approach's level of detail should therefore always be weighed against the practical applicability.

The relationships described here can be readily visualized by three projects: the Elbphilharmonie Hamburg, the CaixaForum Madrid, and the renovation of the Lido Ice Skating Arena in Rapperswil.

Elbphilharmonie Hamburg, Hamburg 2003–2016

The Elbphilharmonie consists of an existing quayside warehouse and new construction set atop it. Architectural considerations dictated that the new building should appear to be lifted above the existing warehouse. The resulting observation deck in the intervening space, known as the Plaza, thus forms a caesura, which is accentuated by setting back, or rather omitting the facade columns. The concert hall, which is centrally placed in the new building, extends transversely across the entire width of the building, thus abutting the north and south facades. Beneath the concert hall is a circulation area and foyer, whose amorphous spatial sequence develops upward along the concert hall, much like a cascade in reverse. On the east and west sides, the concert hall directly adjoins the peripheral uses: to the east, a fifteen-story hotel, and to the west, sixteen residential floors. The new building is covered by an undulating roofscape. From the peak of the residential section that rises one hundred and ten meters high, this roofscape gently slopes downward along the entire length of the building. The existing building is penetrated by three cores and an escalator leading upward. Along with the vertical circulation for the new building, the old quayside warehouse contains a parking garage with 521 parking spaces, a wellness area, the hotel's kitchen with numerous storage rooms and ancillary spaces, and a chamber music hall.

The additional loads of the new building are essentially supported by the existing foundation of 1,111 driven piles. The foundation was originally laid out for the substantial live loads of the warehouse. By converting the warehouse into parking areas, the live load of two tons is reduced to two hundred to three hundred kilograms per square meter.

The essential concept for the Elbphilharmonie Hamburg was to use the resulting load carrying capacity for the new added building. Accordingly, the dimensions and weight of the new extension are determined by the freed load capacity (fig. 1).

The visual separation of old and new buildings was an important architectural premise. To yield the desired separation of the added building form, the old and new buildings are linked together only toward the building's center – continuous facade columns would have interfered with this visual disassociation. The structural concept therefore specifies that the vertical forces of the facade plane above the Plaza are transferred to the row of columns behind them via inclined columns. In so doing, the facade columns on the plaza level are relieved and can be omitted. The horizontal forces originating from the inclined columns are brought together in the slabs and the differential forces are then directed into the cores. The inclined columns in the two to three stories above the Plaza have an impact on the sequence of spaces and the use of the areas along the facades. Since the corner areas should be kept open – diagonals cutting through the space were unwelcome here – the building corners are suspended. It goes without saying that the package of measures

1

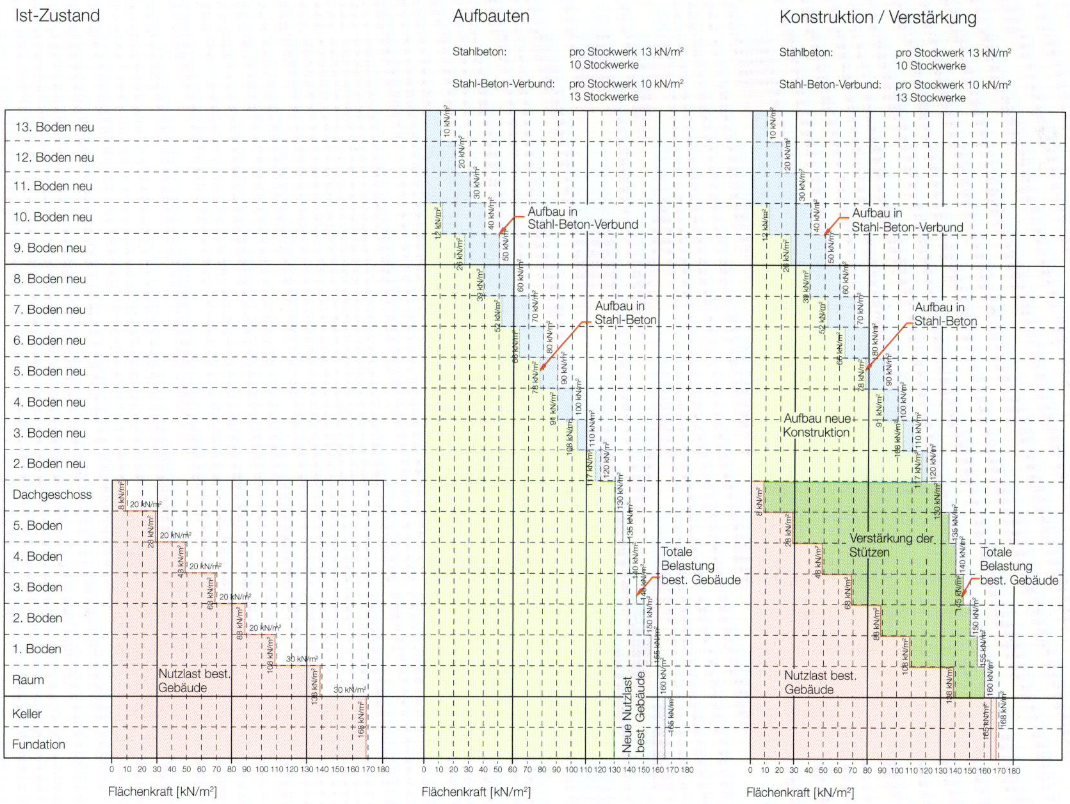

2

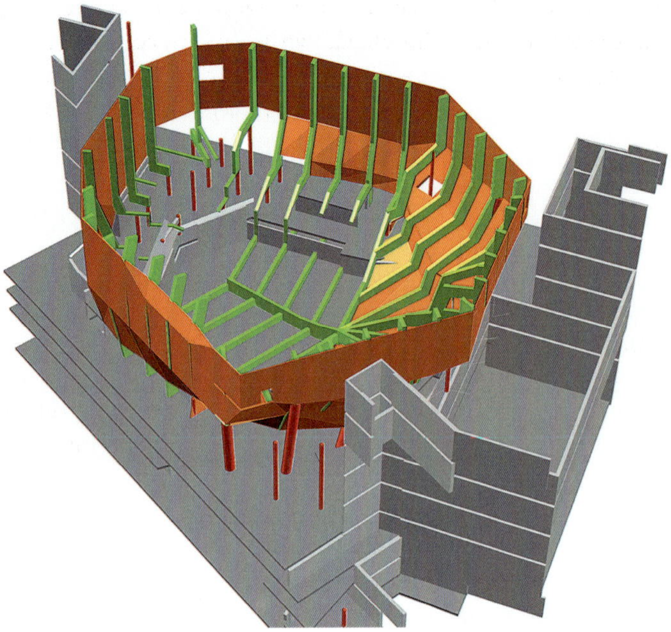

1 Diagram of the load reserves and possible superstructures in steel or concrete for the Elbphilharmonie Hamburg
2 Outer shell of the Grand Hall of the Elbphilharmonie Hamburg (computer model)
3 Analogy for the ribbed structure of the Grand Hall

3

specified for the structure of a building with a height of over one hundred meters entails a certain amount of effort.

Due to its construction and geometry, the actual concert hall itself likely represents the biggest challenge. On the one hand, the design must be in accordance with perfect concert hall acoustics and offer optimum viewing conditions for each one of the more than two thousand concertgoers. On the other hand, from a structural engineering perspective, it is like an egg set in the middle of the floor slabs, with a column-free interior of more than fifty meters that must be spanned by the concert hall's roof structure (fig. 2). Furthermore, the concert hall is surrounded by residential and hotel uses, with consequent problems regarding acoustics; the lovers of classical music may not to be disturbed by a ship's horn, and the residential and hotel area may not be impaired by unwanted noise. Such ambitious sound shielding requires appropriate measures: the walls of the concert hall were designed as a double skin (box-in-box construction), and the inner shell rests upon springs. From a structural engineering perspective, the hall is like a ship in dry dock. Analogous to a ship's hull with its ribbed frame, the Grand Hall's outer shell has ribs at regular intervals to collect the concentrated loads from the spring assemblies and convey them to the columns. These ribs run over the existing columns and gather together below in the "keel," which also rests on columns (fig. 3).

CaixaForum Madrid, Madrid 2008

The landmarked "Central Eléctrica del Mediodía" – Madrid's first coal-fired electric power plant – is situated directly on the Paseo del Prado, nestled between the Museo Nacional del Prado, the Museo Reina Sofia, and the Thyssen Museum. Despite the turbulent history of the monolithic brick building, which is closely linked with the industrial and political development of the country, the building has survived into the twenty-first century, and its external form was to become part of the new museum building.

The basic urban and architectural idea was to establish a pedestrian axis parallel to the street axis of the heavily traveled Paseo del Prado, as a link between the world-class museums. The space under the museum is part of this connecting axis, which had the conceptual and logical consequence of separating the granite plinth, the stone foundation, from the brick superstructure: The massive brick form is a giant block that floats over the plaza, supported by only three "legs" that contain the vertical circulation – the entire above-ground building structure stands on these three legs, and the plaza is covered by the floating "monolith."

Originally, a granite plinth up to two meters high formed a base around the whole building. During the construction of the basement levels, this rigid plinth made of large individual stones was strapped together with a steel band and supported by trestles on micropiles. After an initial excavation, the foundation pits for the three circulation cores were bored and the cores were built. These were used during the construction of the basement levels as stable points of support for the building pit bracing. This enabled us to build the exterior walls directly under the brick walls right up to the property line.

The primary load-bearing structure consists of two main supporting elements: the three circulation core and a perimeter wall that wraps around these cores to join everything together. The three concrete cores transmit all the vertical and horizontal loads into the subsoil. The perimeter wall attached to the cores surrounds the load-bearing structure of the building like a corset. It carries the loads from the facade and the building as well as the added, two-story steel construction that crowns the existing brick building. Together with two additional inner walls running parallel to each other, it forms a hollow, prestressed concrete box that acts as a macroscopic supporting structure to transfer all the building loads onto the aforementioned cores. At the same time, this concrete supporting structure and the wide-spanning inner walls enable a spatial division that make large-scale exhibition spaces possible. The reinforced concrete corset is prestressed. With the deviation forces from the prestressing, large concentrated loads are collected and released in a targeted manner to the lines of support via a negative curvature. The prestressing forces also pull together the old brick walls and counteract the tensile stresses in the brick that result from the load redistribution.

The existing building was completely gutted. Since the historical masonry facade is firmly connected with the newly created perimeter wall on the inside, the old brick exterior walls are fully integrated into the new building. Individual reinforced concrete studs that project from the load-bearing walls and are distributed over their entire height carry the existing masonry at numerous points. Through this interlocking, the vertical loads are transmitted continuously to the corset and a horizontal shear connection between the two different walls and materials is achieved. The closely spaced studs at the base of the existing masonry walls form linear bearings. The lowest row of bricks is held by a continuous steel plate that is connected to the corset. After slitting the plate into the joint between the granite plinth and the brick wall and then completing the corset, the granite plinth could be removed and the building was set down on the cores.

As the primary means of access, a spiral staircase leads from the plaza up to the first floor. From there, visitors can enter the exhibition spaces and reach the offices on the floors above as well as the auditoriums located in the basement levels. A suspended floor structure enables construction of the required ceiling height, ensures freedom in the plan layout, and guarantees a column-free space on the plaza beneath the museum. The floor of the entrance level is hung from the plate girder of the composite slab above the first floor, via delicate suspended columns at designated points. For reasons of weight, the floor plate is made purely of steel, and together with the suspended columns, it can be understood as a cluster of multiple inverted umbrellas: each of the freely arranged hangers represents a central umbrella post. The heights of the radial beams increase in relation to the hangers, thus forming the highly faceted ceiling over the plaza.

Lido Ice Skating Arena, Rapperswil-Jona, Switzerland, 2006

The indoor ice rink in Rapperswil from 1987 no longer met the present-day and future requirements for a modern and attractive sports facility. The operating body responsible for the indoor skating rink, the Lido Skating Rink Association, wanted to renovate, expand, and improve the overall quality of the hall at the current location. The maximum spectator capacity of the renovated hall was to remain unchanged at around six thousand people, but with less standing room capacity and more seats.

The basic structural concept for the extension to the grandstand and the roof was to completely reuse the existing steel structure. The necessary longer span was achieved with a two-sided extension and reinforcement of the main trusses. To accomplish this, the roof structure was raised to a higher position by using tension strands and prestressing jacks, and then set down on new foundations outside the existing hall, in alignment with the new facade. The procedure of lifting the existing roof structure was carried out in a compact sequence of work: extend the existing trusses and steel columns – raise the roof to the desired height with hydraulic presses and tie rods – relocate the extended roof onto the new columns. Lastly, the old columns were removed and the new tension-rod trussing was attached.

Due to the larger span length of 63.25 meters – before the renovation it was 50.5 meters – the trusses had to be reinforced. The buckling strength of the upper chord was increased in places, and on the tension side of the truss, simply installed polygonal tension-rod trussing was attached. With the induced deviation forces, the deflections could be easily inhibited. With the chosen cable geometry, the shear forces remained the same or even decreased slightly, so that reinforcement of the diagonals was not necessary.

The existing roof structure was completely integrated into the new roof structure, whereby the roof loads are sustained by a new pile foundation outside of the existing building. Thus, the existing piles were relieved and as a consequence, they could bear the additional loads of the enlarged grandstand. Moreover, that simplified and shortened the construction process – making the new foundations and carrying out all the preparatory work for the extension of the trusses could be done outside the building during ongoing operations.

The cantilevered extension (VIP level and technical floor) is supported by steel trusses at a spacing of six meters. The spacing of these trusses is aligned with the axes of the existing load-bearing structure. The trusses are on the technical floor and each rest on three columns. The tensile forces resulting from the cantilever are overridden by the loads from the roof structure. Horizontal bracing is provided by the floor slabs and through four vertical steel trusses.

Elbphilharmonie Hamburg
Hamburg 2003–2016

←← Concrete shell of the roof above the Grand Hall
↑ Steel structure of the inner shell and the rostrum trusses
← Steel structure of the roof peak above the Grand Hall
→ Interior space beneath the roof structure of the Grand Hall, with scaffolding and temporary columns (construction site photo)

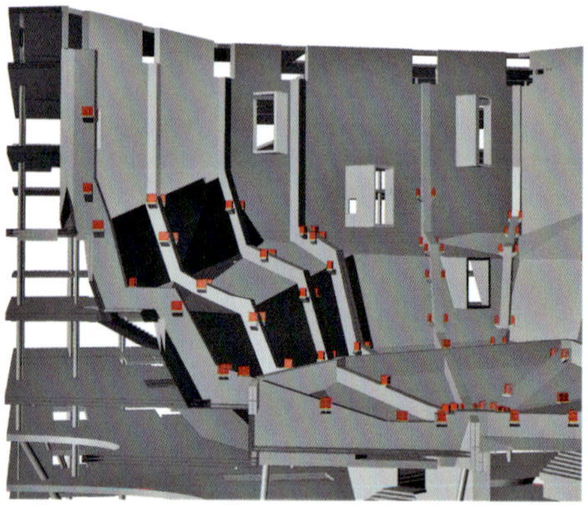

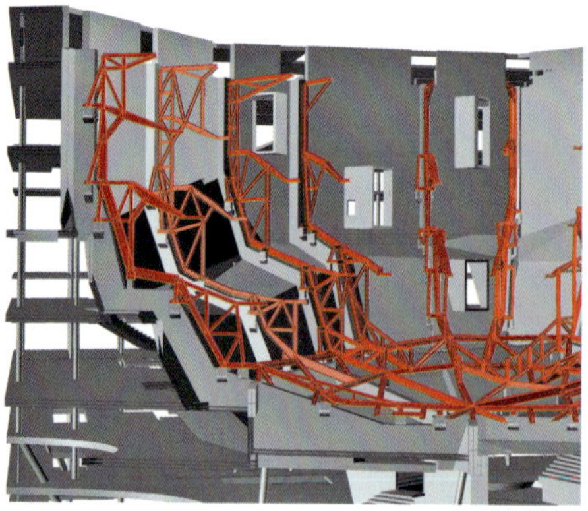

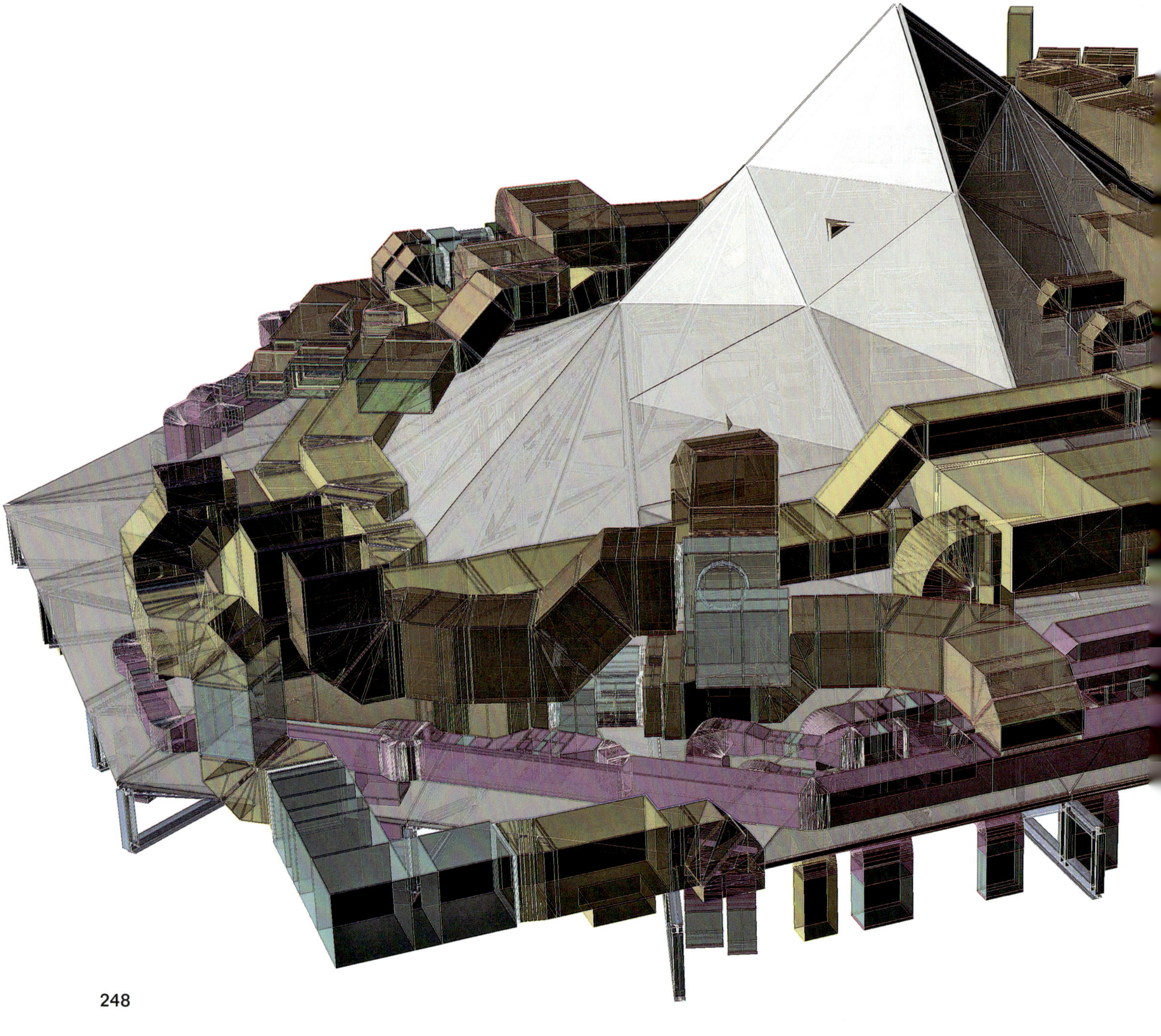

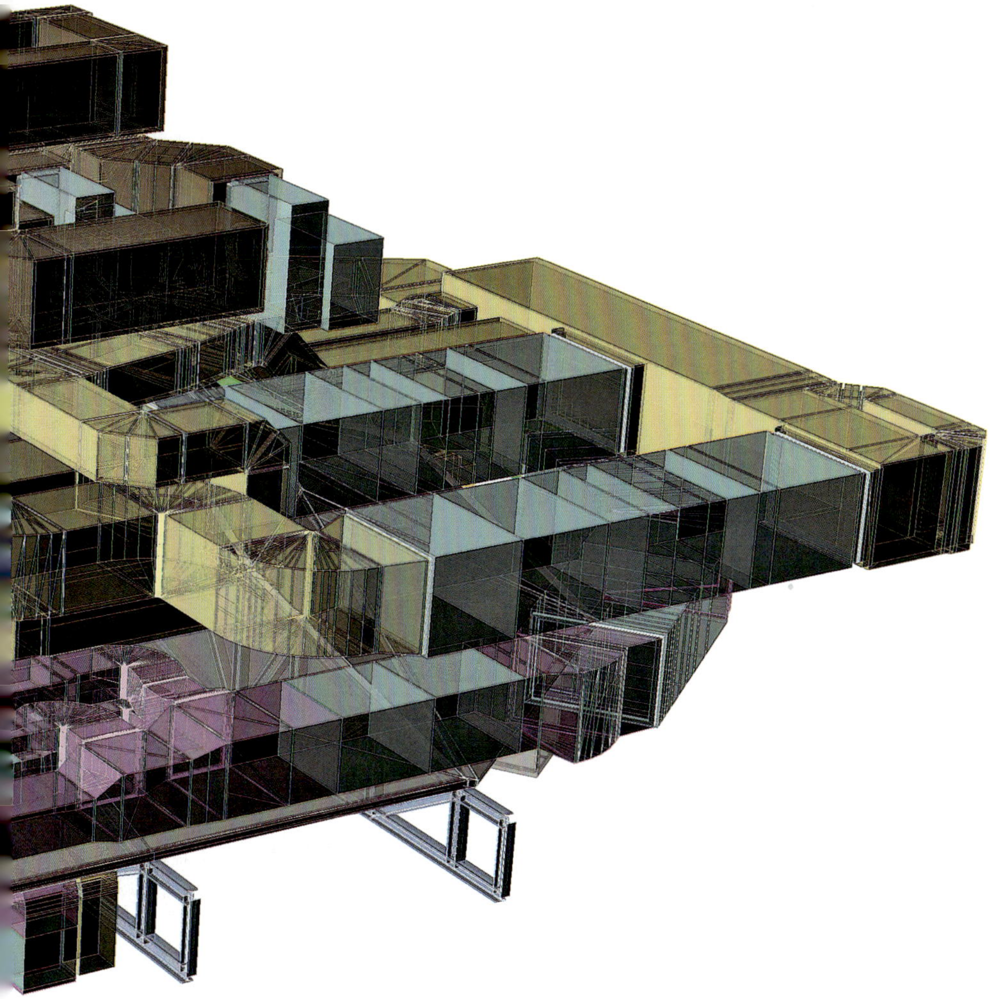

↑ Composition of the Grand Hall and construction process (computer model, with key elements of each stage highlighted in red): outer shell with the spring assemblies; steel structure of the inner shell; steel structure of the rostrum trusses; steel structure of the hall's roof. spring assemblies on the steel structure of the hall's roof; suspended acoustic ceiling
← Roof structure of the Grand Hall with the ventilation ducts and HVAC equipment situated above it (computer model)

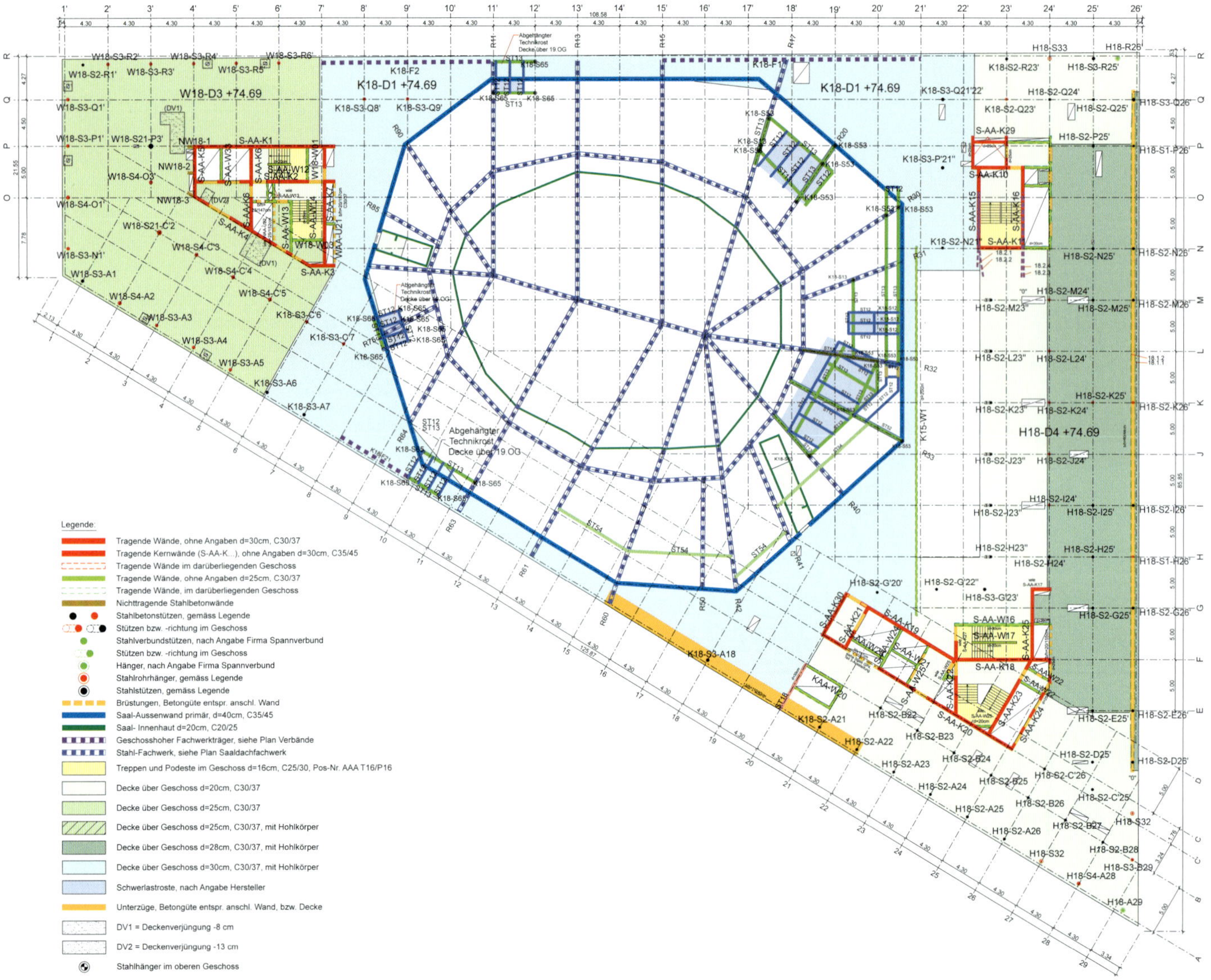

Positionierung der Stahlverbundstützen / Hänger:

H18-A29
- Achse
- Geschoss in dem die Stütze beginnt
- Nutzungsbereich

●	Stahlverbundstütze Ø nach Angabe FIRMA Spannverbund (=externe Statik)
◉	Hänger Ø nach Angabe FIRMA Spannverbund

Positionierung der Stahlbetonstützen:

H18-S2-A22
- Achse
- Durchmesser (Zuordnung siehe Tabelle)
- Geschoss in dem die Stütze beginnt
- Nutzungsbereich

Bereich	Durchmesser [mm]	Betongüte	Betonstahl
H = Hotel K = Konzert W = Wohnen	-S1 = Ø200 -S2 = Ø250 -S3 = Ø300 -S4 = Ø350 -S5 = Ø500 -S6 = Ø550 -S7 = Ø700 -S8 = Ø800 -S9 = Ø950 -S20 = Ø400 -S21 = Ø450 -S22 = Ø600 -S23 = Ø1050	● C80/95 ● C45/55	BSt500S

Positionierung der Stahlrohrhänger:

H18-S32
- Querschnitt, bzw. Tragfähigkeit
- Geschoss
- Nutzungsbereich

Bereich	D x t [mm]	Stahlgüte
H = Hotel K = Konzert W = Wohnen	-S30 = Ø244.5 x 20 -S31 = Ø193.7 x 16 -S32 = Ø168.3 x 10 -S33 = Ø139.7 x 8 -S34 = Ø323.9 x 25	S235JR

Positionierung der Stahlstützen

K18-S53
- Querschnitt
- Geschoss
- Nutzungsbereich

Bereich	Profil	Stahlgüte	Bem.
K = Konzert	-S53 = HEA280	S235...	-
	-S65 = HEA100	S235...	-
	-S13 = HEB240	S235...	-
	-S12 = HEB160	S235...	-

Positionierung der Vorspannkabel

18.1.1
- Vorspannkabel-Nummer
- Vorspannkabel-Gruppe
- Geschoss

Vorspannkabel 3/15,p f =1770N/mm□
Vorspannkabel 5/15,p f =1770N/mm□
"0" Nullpunkt der vertikalen Kabelgeometrie (gemäss Plan Vorspannkabel, vertikale Geometrie)

Positionierung der Stahlträger

Farbe	Profil	Stahlgüte	Bem.
blau	-ST12 = HEB160	S235...	-
ocker	-ST73 = UNP180	S235...	-
grün	-ST13 = HEB240	S235...	Anschlussdetail gemäss
rosa	-ST18 = HEM260	S235...	Positionspläne Details
grün	-ST52 = HEB240	S355...	-
hellgrün	-ST54 = HEB260	S235...	-
dunkelgrün	-ST57 = HEB280	S355...	

Gütegruppe der Baustähle	
S235JR	t ≤ 150mm
S355JR S355J0 S355J2 S355J2H	t ≤ 40mm 40 < t ≤ 120 120 < t ≤ 250 Hohlprofile

↑ Structural layout drawings for the slab above the eighteenth floor
→ View of the north and east facades

HOCHTIEF

CaixaForum Madrid
Madrid 2008

STRUCTURAL CONCEPT

Roof Loads / Facade Loads

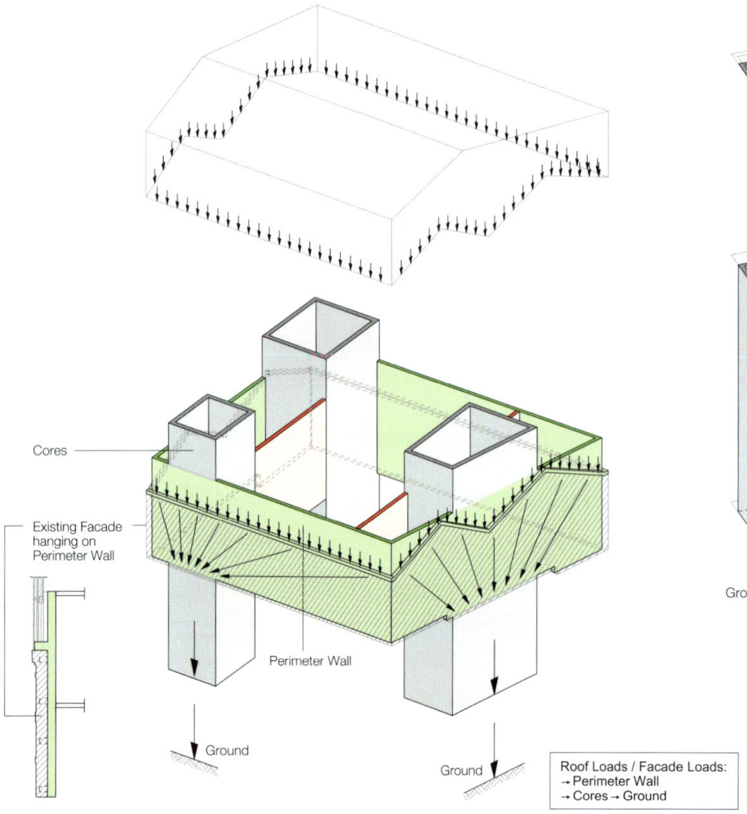

Slab Loads

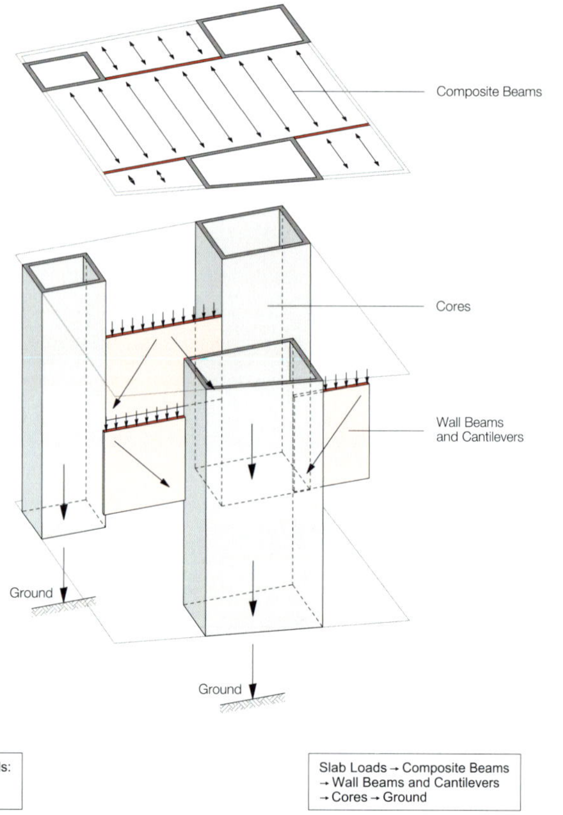

PERIMETER WALL

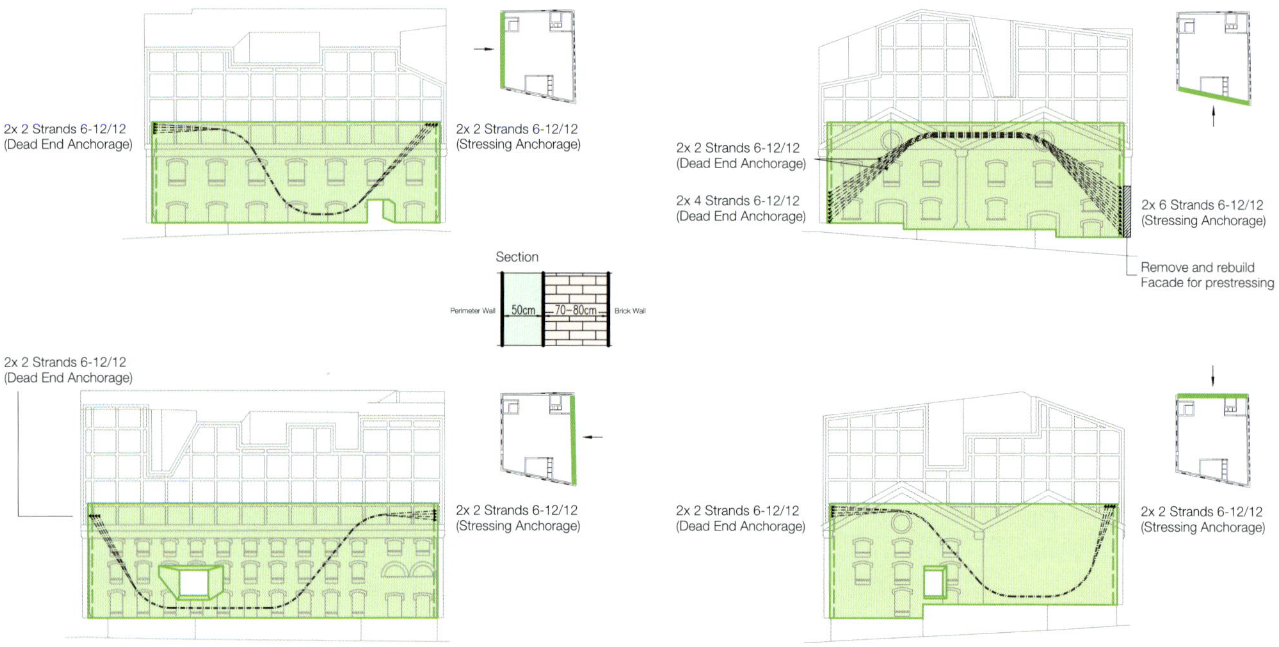

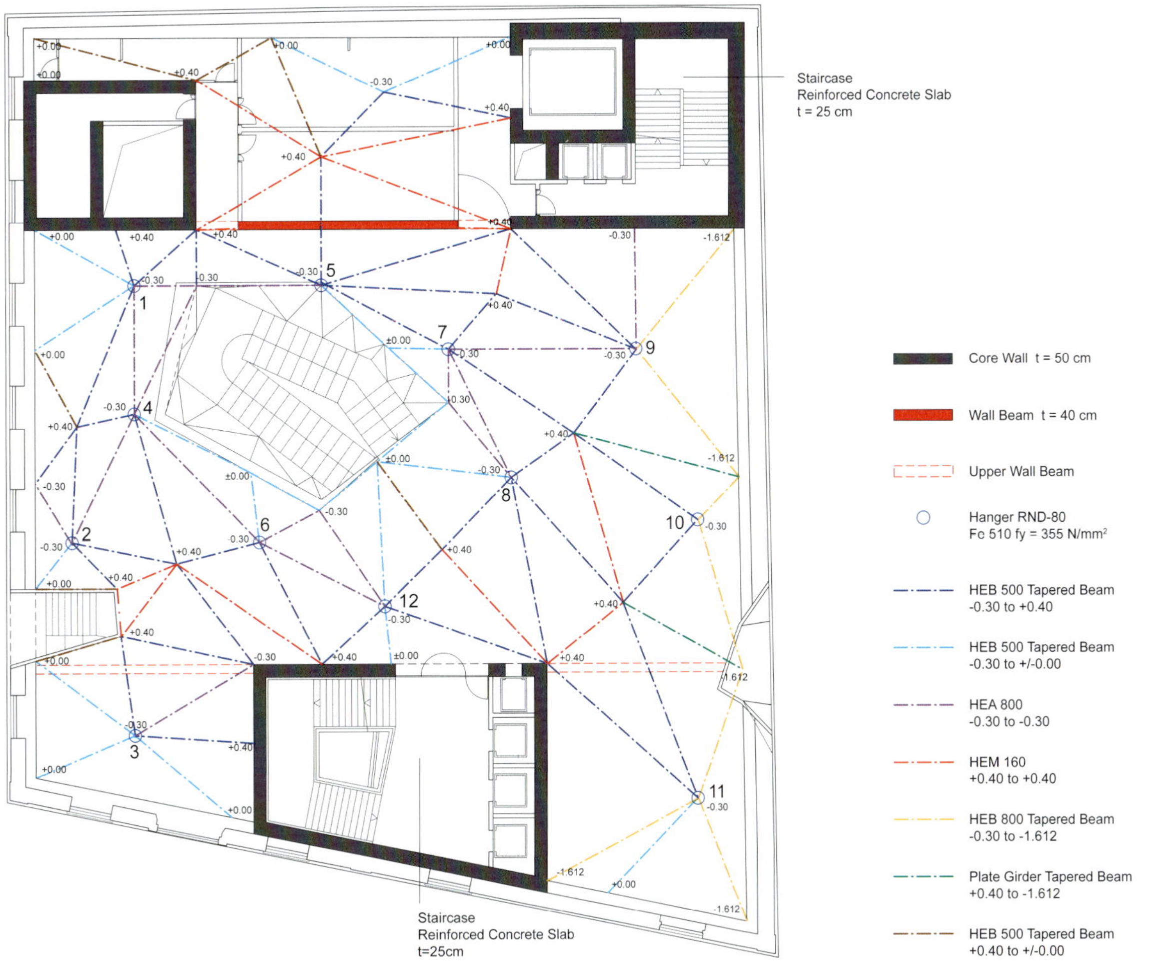

←← Freestanding facades during the excavation of both basement levels (construction site photo)
↖ Structural concept comprising three cores, a perimeter wall (green) that holds the existing walls, and two main wall beams (red) connected to the cores. This structure supports the steel structure of the addition. Slabs span between the main wall beams and the perimeter walls.
← Prestressing in the perimeter facade walls for prevention of cracks due to deformations in the existing facade walls.
↑ Geometry of the suspended "umbrella ceiling"
↗ Beam geometry and details for fabrication of the "umbrella elements"
→ Concept of the "umbrella" with a hanger at the center

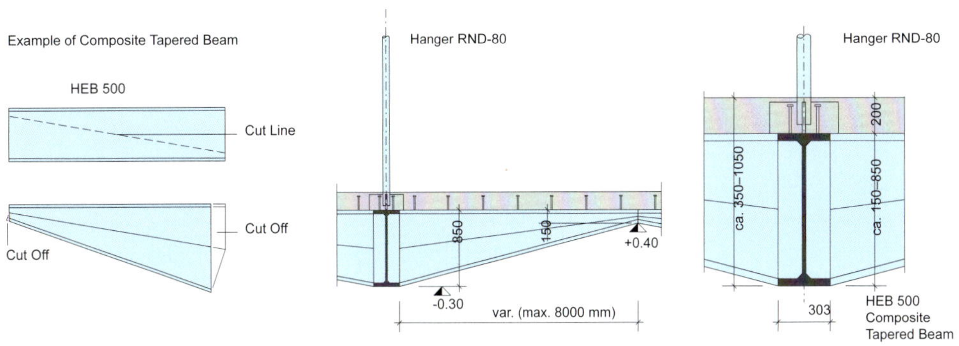

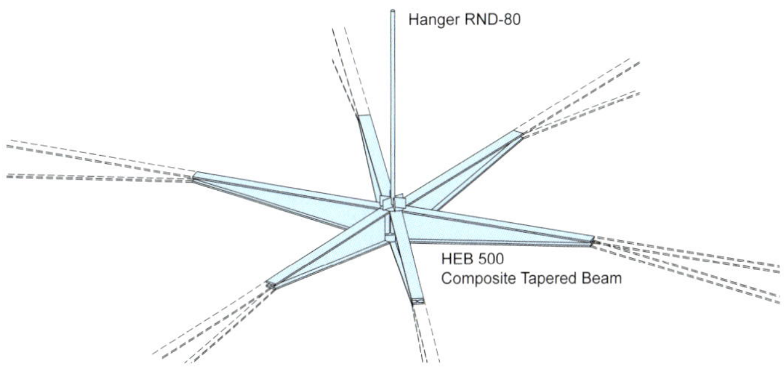

↑ Construction site photo of the foundation slab with core formwork (left side) and the temporary bracing (right side)
→ Aerial photo of the construction site taken from the crane basket
↗ The finished building. The steel structure atop the existing building is clad with solid steel sheets.
→→ Space under the building with stairway

CaixaForum

Lido Ice Skating Arena
Rapperswil-Jona, Switzerland, 2006

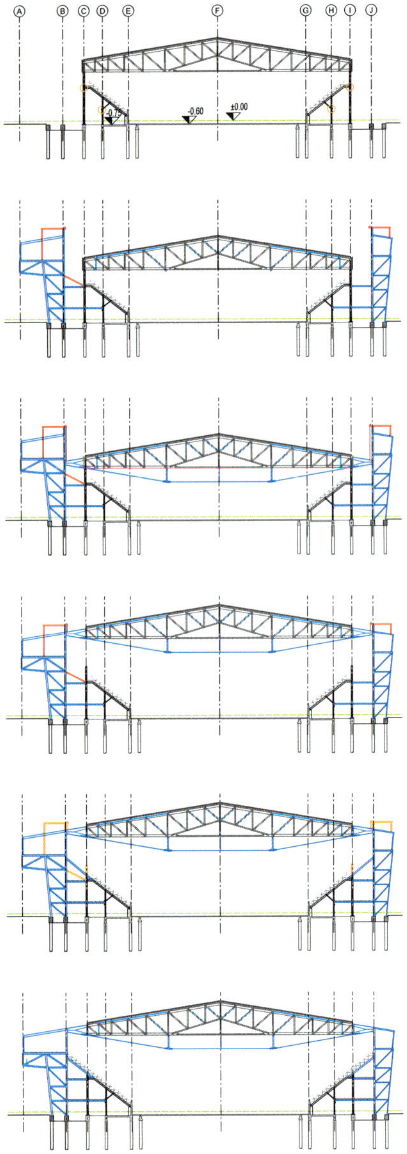

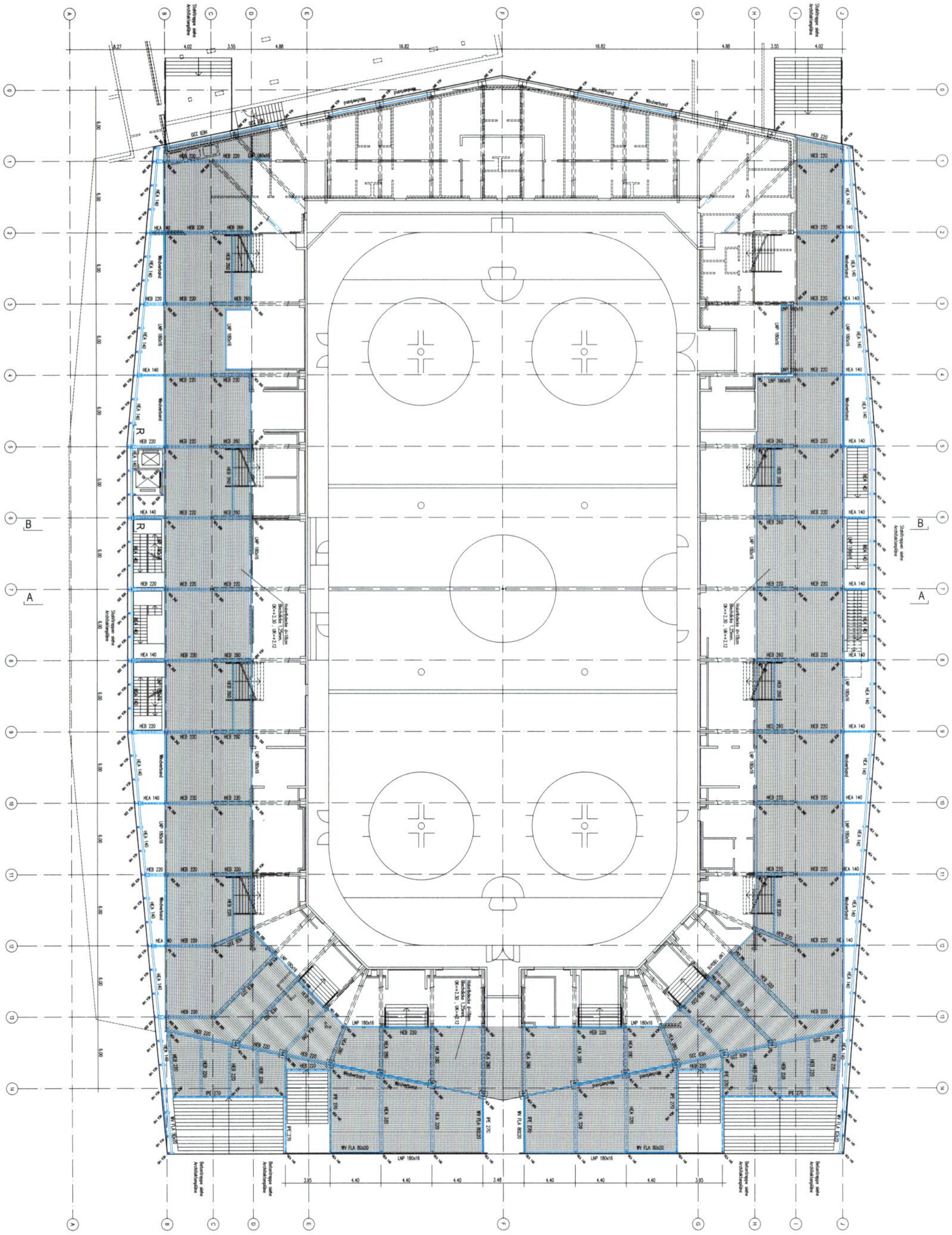

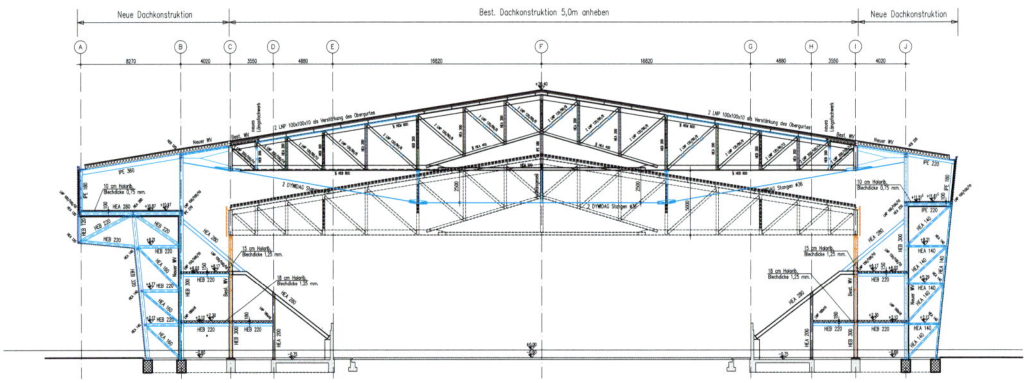

← The arena prior to addition of the facade (construction site photo)
↑ Diagrams of the construction process (from top): establish the new pile foundations; erect the new grandstand structure, with equipment for the lifting procedure; extend the trusses and install the tension-rod trussing; lift the roof structure; dismantle the equipment used for the lifting procedure; complete the grandstands

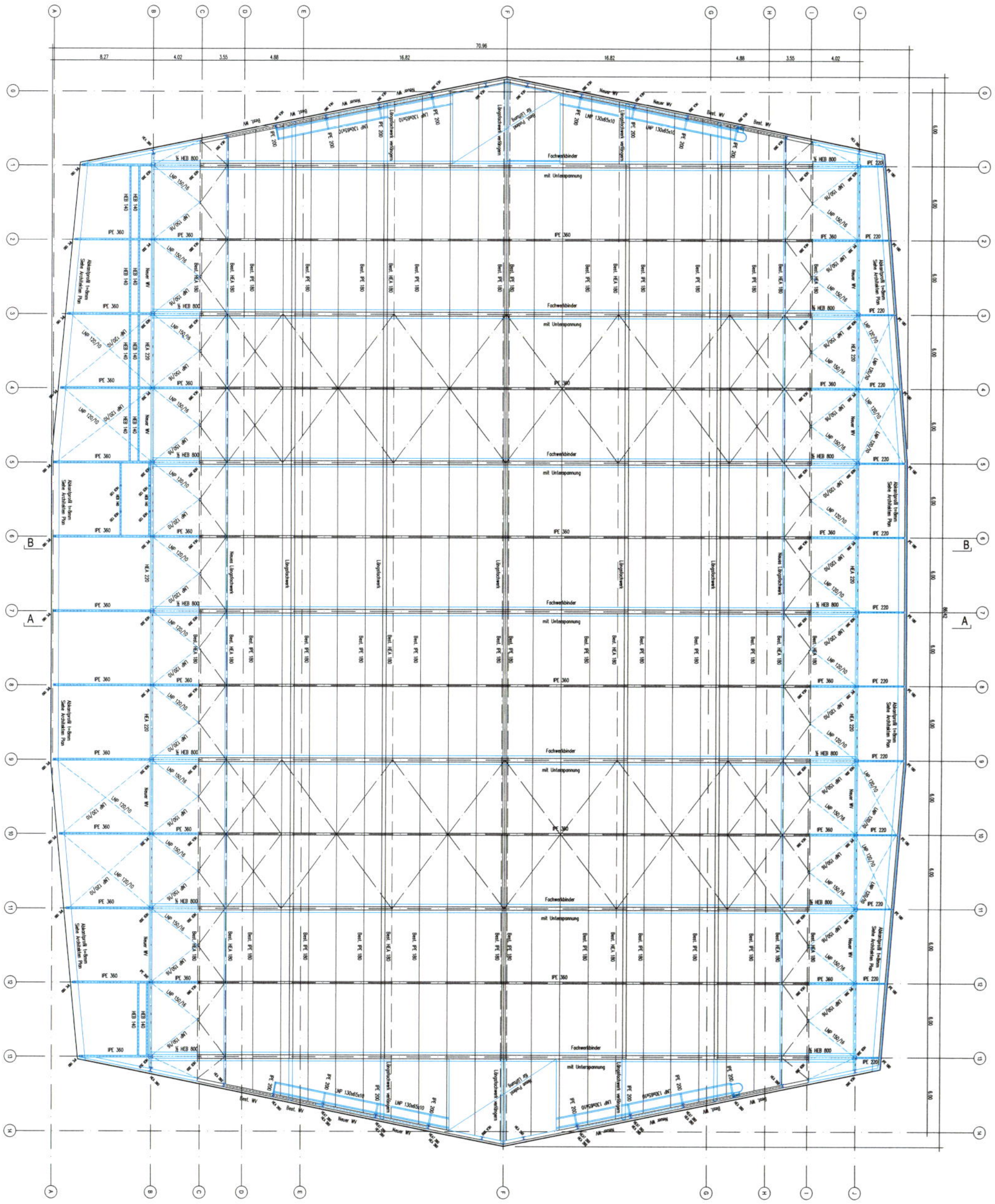

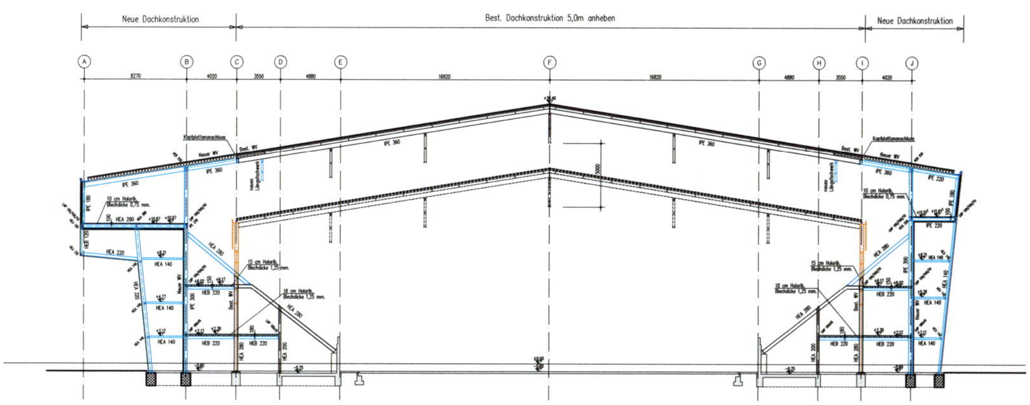

↖↖ Arena floor plan with the expansion of the grandstands (blue)
←← Section through the arena with the expansion of the grandstands, the extension of the trusses, and the added tension-rod trussing (blue)
↑ Roof framing plan with the truss reinforcement and the roof extension (blue)
← Section through the arena with expansion of the grandstands and roof extension

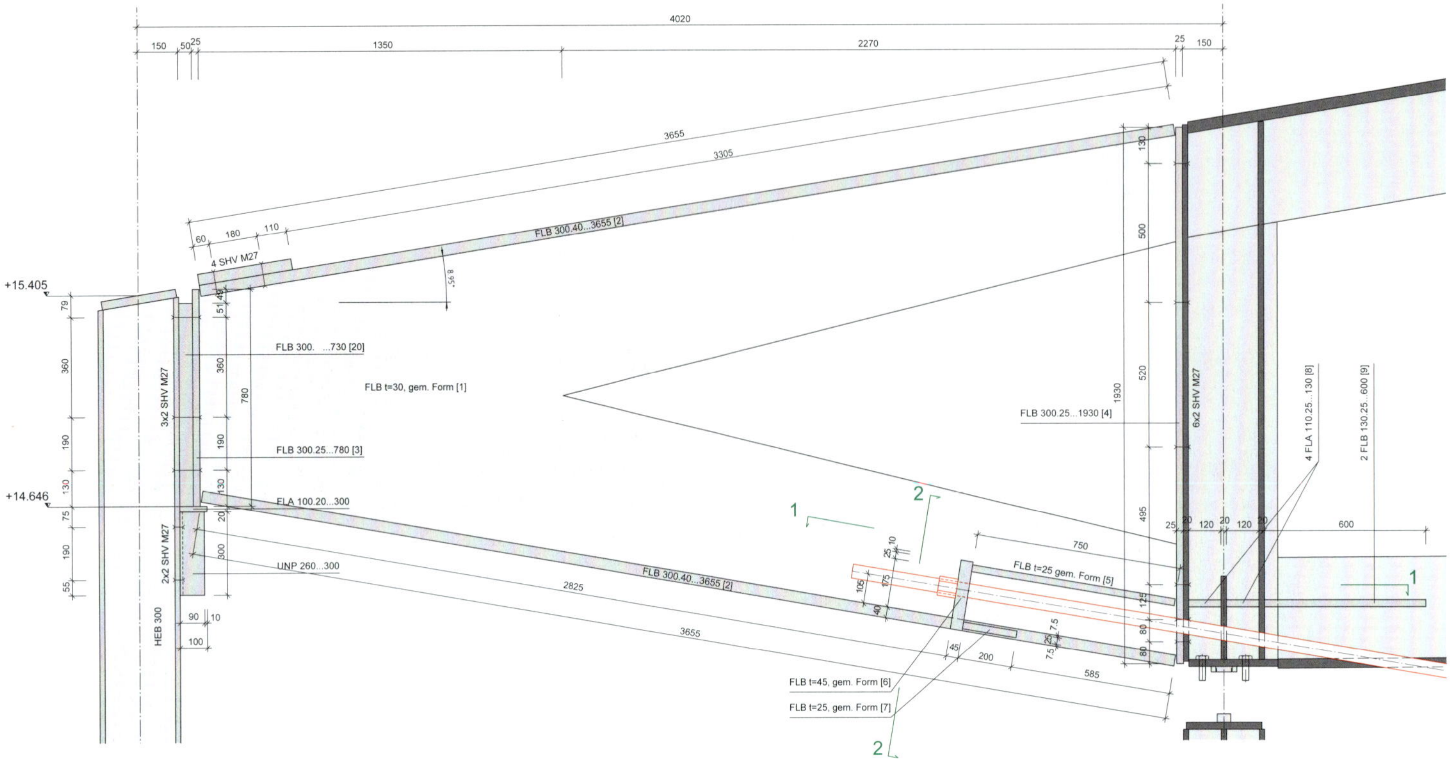

Ansicht 2 - 2

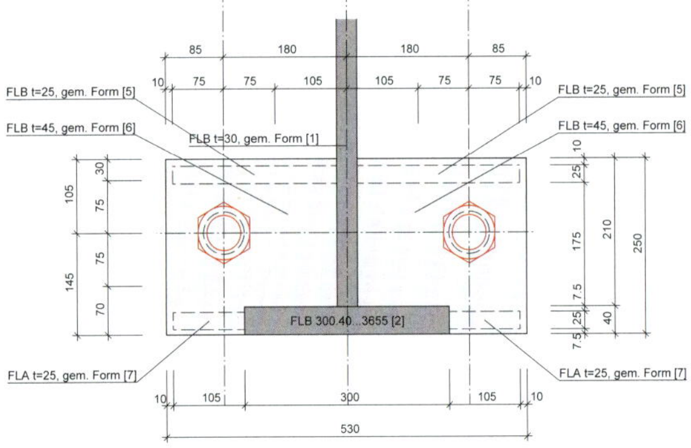

Ansicht 1-1

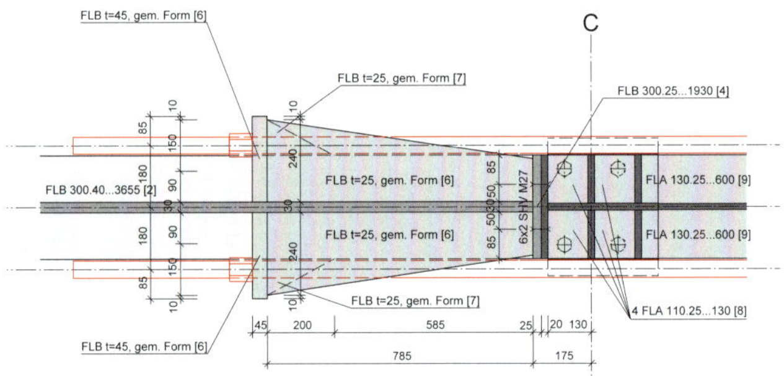

↖ Extension of the trusses, with bearing detail and anchor detail for the tension-rod trussing
← Expanded grandstand prior to lifting the roof structure (construction site photo)
↑ Interior view with roof trusses and tension-rod trussing

8 Bridges

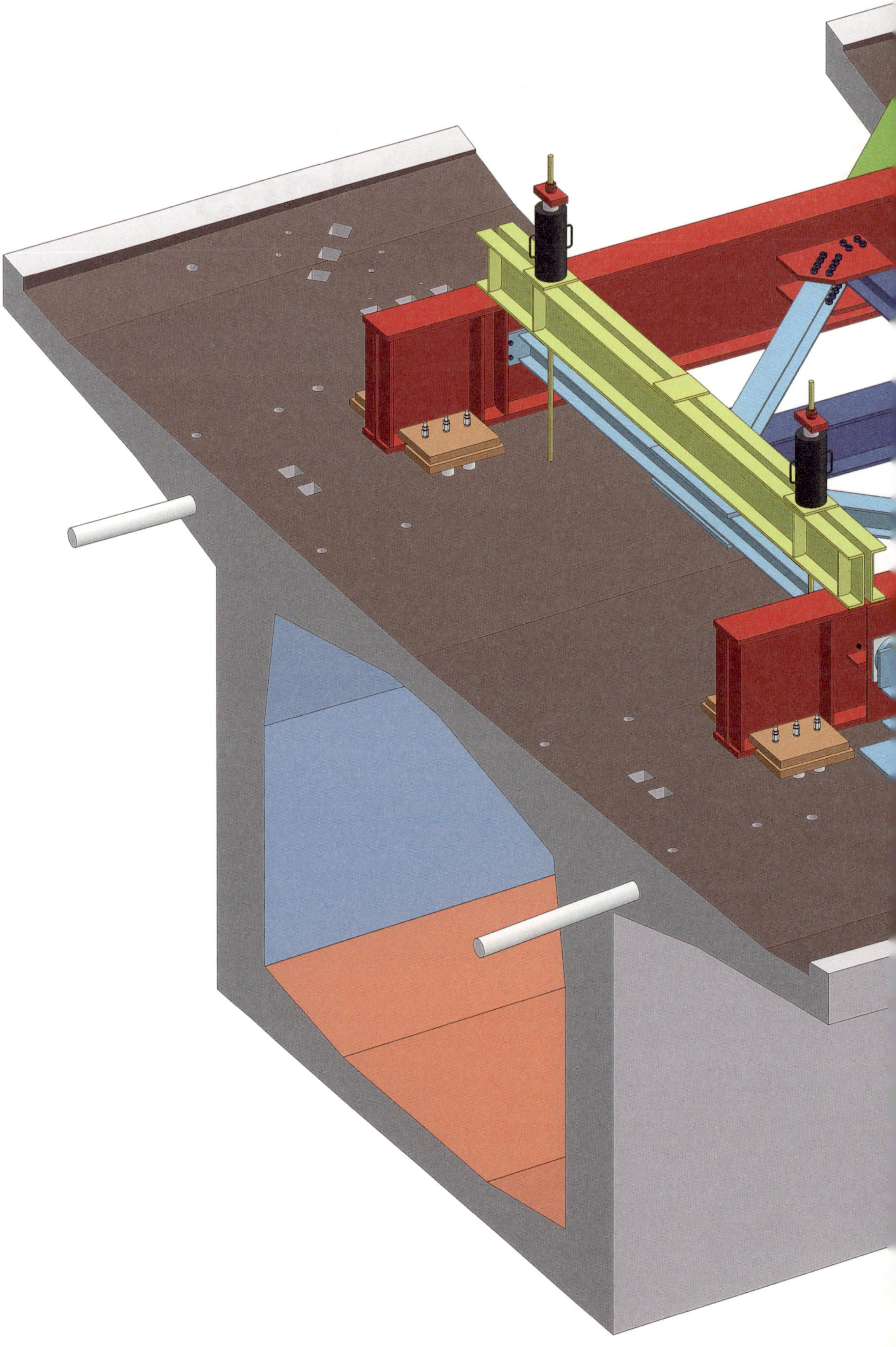

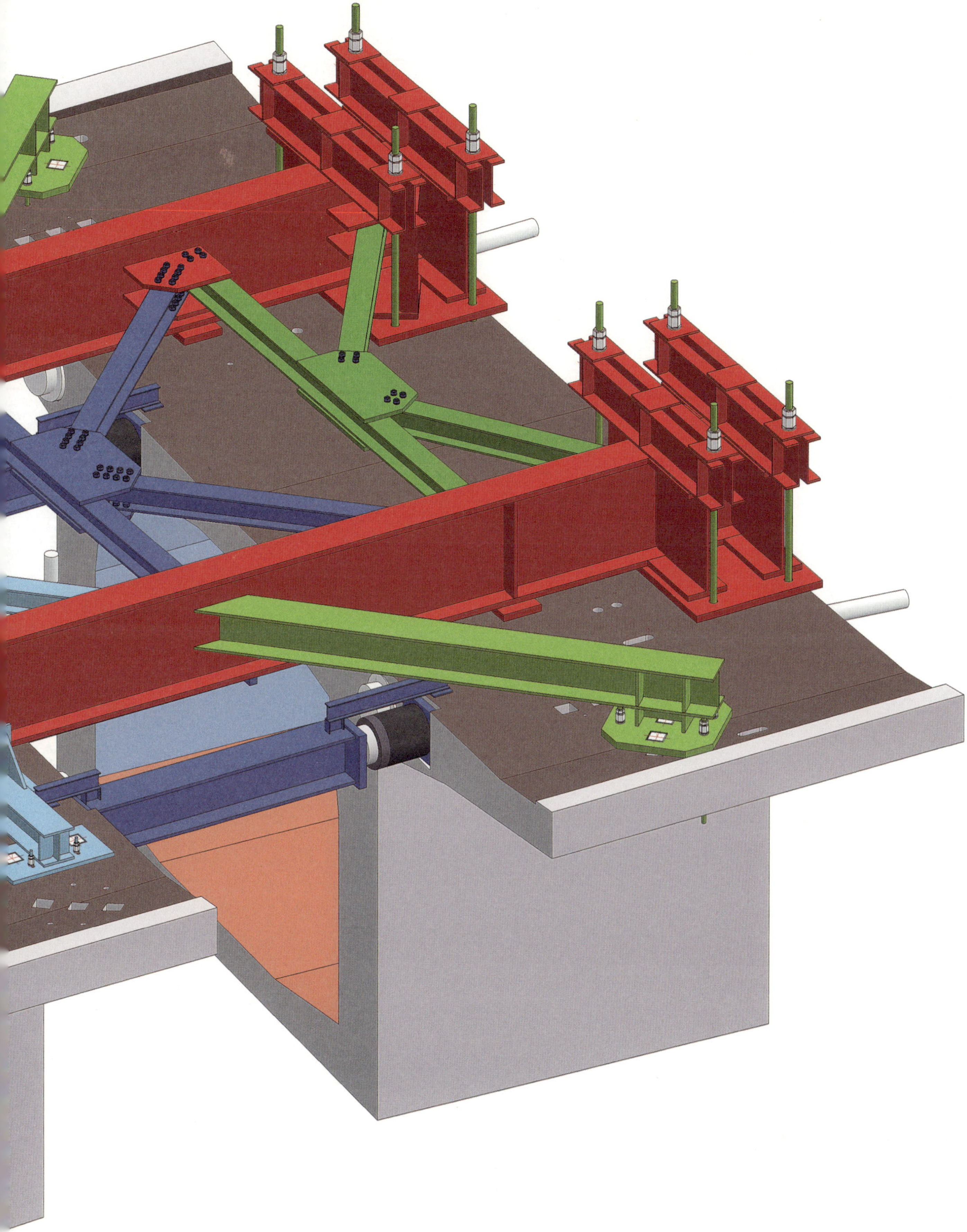

← Steel construction with hydraulic jacks for tensioning and aligning while closing the joints of the Second SBB Railway Bridge over the Rhine (computer model)

Conversations VIII

AF In his interview with David P. Billington, Jörg Schlaich, and Christian Menn in the film *Maillart's Bridges,* Heinz Emigholz poses the lapidary question: *What is a bridge, essentially?* What is a Bridge?

SB If we restrict ourselves to built bridges, they are primarily connections in a path, an easier connection in a path to reach a destination faster. With immense efforts, a structure is erected to make it easier to get from A to B.

TP The efficient connection is one aspect, the bridging of an obstacle is another.

HS The obstacle in the case of a bridge is a ravine, a stream, or a lake. When the obstacle is a mountain, the instrument for bypassing it would be a tunnel ... A bridge makes it possible to overcome an obstacle with economic and intelligent means. This must always be seen against the background of its historical origins and the corresponding materials and construction methods available at that time. In the fourteenth century, there were wooden bridges on stone piers; in general, wooden bridge construction was very advanced in Switzerland. Stone bridges appeared later, such as those along the Rhaetian Railway, then Maillart's reinforced concrete bridges, and finally prestressed concrete bridges and steel bridges.

AF You've just been citing very sober, functional aspects. But bridges are also moments of identification, if not to speak of lighthouses, of landmarks in the landscape!

TP Horizontal lighthouses! In any case, they are built structures with symbolic value. For us engineers, bridges are, of course, the challenge par excellence.

HS With larger span lengths, the weight of the building material becomes key. Reducing weight requires an intelligent construction. The large spans and the accompanying constructions always make a bridge, even a modest one, a very striking built structure.

AF I am interested in the relationship of a bridge to the terrain, to the landscape. How do you view the relationship between integration of a built structure into the landscape and its autonomy?

TP An example of a very successful integration is Maillart's Salginatobel Bridge. An arch resembling a strutted frame connects the two sides of the valley. In this case, the scenic context clearly determines the bridge form. Along a riverbank where, for example, there are no valleys or mountain slopes, and where, at most, a minimum clearance must be observed for the space beneath the roadway - supporting structures located above the roadway tend to be more prevalent. Even in an urban context, a bridge over a railway line, for example, has very specific conditions of its own. For the bridge's form, the perception of the bridge from the perspective of the driver or the pedestrian is always important, too.

HS The bridge over the Firth of Forth in Scotland breaks every scale. Still, we have become accustomed to its size and are fascinated by it. Not everything has to be "integral." A newer tendency currently points to the tailoring of bridges to a specific location. The winning project in the competition for Lucerne's Seebrücke, which features a lake side and an old-town side, serves as a good example. The bridge reacts with different pier spacing toward the old town and the lake. I find such hyper-contextual tendencies rather dubious, since they sometimes lead to absurd, forced solutions. I suspect an "architect solution" is behind the aforementioned example ...

AF The relationship between contextual reference and autonomy of the supporting structure is, in other words, a balancing act.

HS Yes. There are, of course, important urban issues that I would not want to dispute. The Pont du Poya in Fribourg, for instance, raises a question that would be similar for a new high-rise in the city: How high should – and may – the pylon be in relation to the cathedral?

TP You yourself just noted that people can become accustomed to the large scale of bridges. With the Golden Gate Bridge in San Francisco, the height of the church steeple in Sausalito was certainly not taken into consideration. Today, the bridge is an icon; it is a landmark for the entire city. The same applies to the Verrazano-Narrows Bridge in New York or the Sydney Harbour Bridge.

HS Those are, in my opinion, two very different urban contexts. The Golden Gate Bridge was, at the time of its creation, completely surrounded by small wood houses. The cathedral in Fribourg, on the other hand, has existed there for a thousand years.

TP In any case, bridges can potentially make important contributions in terms of urban design improvements. Bridges almost always elicit strong emotions and fuel the debate about a city's image. In this respect, they help define the city's image and shape the city itself. The city of Basel, as we know it today, actually emerged with the construction of the Mittlere Brücke in the thirteenth century, which established the connection between Kleinbasel and Grossbasel.

AF Since bridges naturally have relevance for urban design, one might question whether a new bridge solely belongs to the engineers. To this day, it is deemed as the supreme discipline within civil engineering – nowhere else are engineering achievements so apparent and directly tangible as in the icons of heroic bridge structures. In the recent past, the interest of architects has increasingly turned its attention to bridge construction. A large number of bridge competitions in Switzerland today are issued with a mandate to create a team. Opinions vary greatly, however, as to how productive collaborations between architects and engineers are for bridge designs, but are very disparate.

Bridges – The Engineer's Ultimate Challenge, or: Collaboration with Architects?

HS For bridge designs, too, I enjoy the interdisciplinary work with architects, because urbanistic components are indeed very important today. But engineers should keep the lead, because they are simply more competent concerning structural issues, construction sequencing, and the production process, and these aspects are a priority in bridge construction. When looking at bridges where the engineer had surrendered the lead to the architect, it's usually quite evident.

AF Regarding your bridge projects in Bhutan, I am interested in two aspects: on the one hand, the value of engineered structures for society, i.e. the social responsibility of the engineer in the creation of infrastructure and, on the other hand, the specific local starting point that compels taking local conditions and available materials into account when establishing the construction principles.

Bridge Construction in Developing Countries

HS When building in developing countries, you realize that many things date back a hundred years in comparison. For a start, a bridge profoundly enhances the quality of life of those in the area: suddenly, 10,000 people can sell their products on the market, thus improving their prosperity. It's also fascinating how a whole village emerges within a short time around a newly-opened bridge construction site! In terms of construction, too, we are much further. A pier cannot simply be set down in a river – we must consider using construction techniques that were actually current in Europe a hundred years ago. For an engineer, that's very exciting – going back to the roots of civil engineering, as it were.

TP In these cases, it is also consistent with our social responsibility to build the bridges in such a way that they can be maintained without outside help. To this end, we are also involved in training local engineers – so they have the appropriate knowledge to continue to build on their own.

AF How did your involvement in these countries come about?

HS We succeeded in winning one of the bridge competitions for Bhutan that was sponsored by Helvetas, the Swiss development organization. We were also able to realize the project and gained valuable experience in doing so. Now we are active in Nepal with another project, in which four thousand bridges are to be built over the next ten years. That's where we are involved with the training of local engineers. We must, as I said, consider construction techniques that make it possible to create a foundation in a river.

AF Are the planning-related possibilities, i.e. CAD, etc., commensurate with our standards? Are the limitations solely related to construction?

HS All the CAD programs are available, and all the planning resources are available, but not necessarily the methodology for dealing with them. The engineers need guidance in thinking simply. One of the central tasks of the training is to teach the modes of action of the building materials that are available and the constructions that are feasible – guidance in developing their gut instincts, so to speak.

Conceptual and Structural Creativity

Christian Menn

Bridge Construction – Conceptual Creativity

In the design of bridges, functionality (structural safety, durability and suitability for use) is both imperative and normalized. The costs (for construction, operation, maintenance, refurbishment or demolition) and the appearance of the bridge (structural shape and visual balance with its immediate and broader surroundings) cannot be normalized, but in view of maintaining a cultivated building culture, they are the goal of "conceptual creativity."

Until the mid-twentieth century, almost all of the important bridges were designed by just a few outstanding engineers, such as Roebling, Eiffel, Hennebique, Freyssinet, Dischinger, Leonhardt, Maillart, Ammann, etc. Due to the development of roads for the rapidly growing motorized traffic countless bridges were built after 1950, and many of these were designed by less experienced engineers. Thus, for the purposes of cost estimation and planning the construction process, the advice of contractors was often required; and when the appearance of prominent bridges was of concern, the involvement of architects was imposed. Relatively banal, prestressed girder bridges became ubiquitous, but the longer the span, the less satisfying they were. Corresponding to this development, building directors and chief engineers of municipal administrations expected special public recognition for their achievements in building culture and landscape conservation when they demanded the participation of architects in the design process and as jury members for competitions.

Architects, however, do not see it as their task to visualize structural efficiency, but mainly to pay heed to decorative and artistic aspects. In the design work, the personal influence of the engineer and the architect naturally has a huge impact on the appearance of modern bridges. If the engineer dominates, architects have little freedom; they must be content with minor artistic extras. On the other hand, if the engineer is concerned almost exclusively with structural and economic matters, the architect can have a decisive impact on the concept. The best situation is undoubtedly the middle ground, in which the engineer, due to his technical knowledge, suggests suitable concepts and the architect acts as a consultant – and when designing the details, the architect takes the lead and the engineer assumes the advisory role of a critic.

Of course, even today engineers design bridges without the involvement of an architect. That is perfectly okay, but it can be a delicate situation if the engineer has little feeling for forms but nevertheless feels the need to refine the design to be more artistic and original. It is equally tricky, however, if the architect designs the bridge autonomously and the engineer – lacking any absolutely compelling reasons – has nothing to say about it and is merely expected to make sure the architectural design can be built. Hence bridges have become "more colorful" in recent decades, but usually also more expensive.

Example: Seebrücke Competition, Lucerne

The competition entry by the engineering firm WGG Schnetzer Puskas for the new Seebrücke bridge in Lucerne was the result of a very smooth and productive collaboration between engineer and architect.

The bridge is located at the point where the Reuss river leaves Lake Lucerne, an exceedingly sensitive location in the urban fabric. The first decision taken by the engineer-architect team consisted of establishing general specifications for the level of the roadway and the design of the bridge abutments in order to protect, as well as possible, the bridge's valuable surroundings – including the square in front of the train station on the left bank, the Alpenquai on the right bank and the nearby Chapel Bridge – from impairment by the new bridge, irrespective of the future structural

1

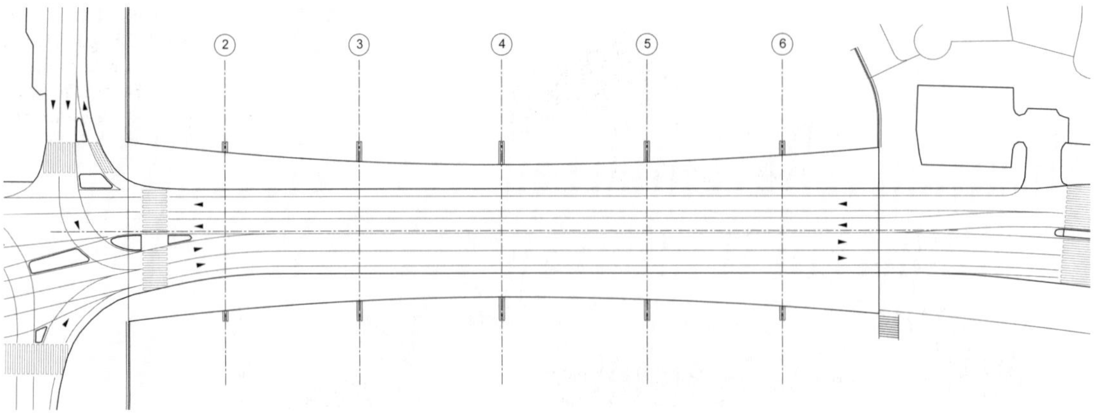

2

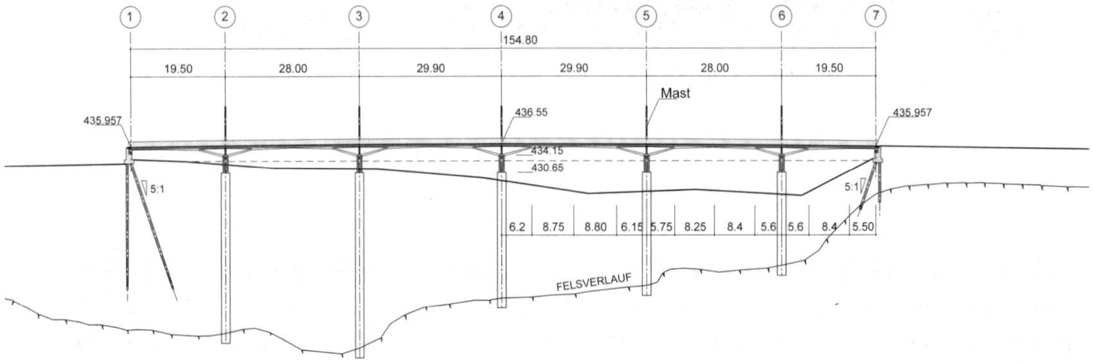

3

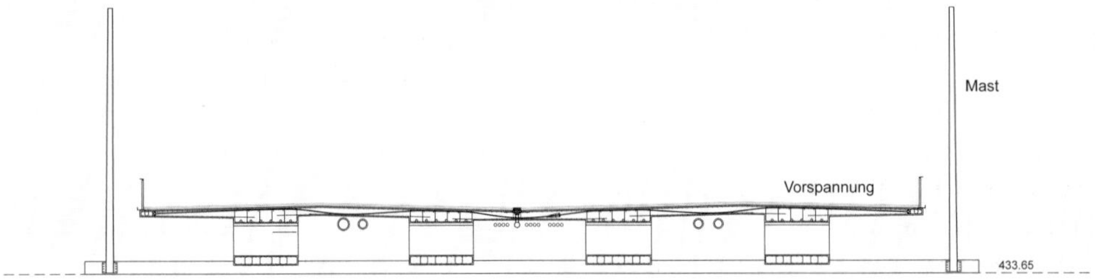

concept. The bridge abutments were to be optimally integrated with the shorelines and the existing transport network, and the bridge itself should be as flat as possible between the abutments. On this basis, the engineers developed an elegant, slender structure with six spans (instead of the previous seven) and a slab cross-section with four wide webs that extend only twenty centimeters beneath the slab. Above the piers, the slab would be reinforced in the longitudinal direction with shallow V-columns.

The relatively large, thoughtfully proportioned spans and the slenderness of the supporting structure were intended to mitigate the shadows cast from the twenty-six-meter-wide bridge slab as much as possible; with the V-columns, which are in principle load-bearing haunches that are part of the superstructure, it was possible to develop a compelling structural and formal solution despite the bridge's low height above the water. In addition, the doubled slabs above the pile caps marking the transition to the lake are a creative original detail.

Building Construction – Structural Creativity

For almost all buildings, the engineer can assist the architect through structural creativity. In doing so, it is not primarily about saving costs by

4

1 Site plan of the Seebrücke bridge in Lucerne
2 Longitudinal section of the Seebrücke bridge
3 Cross-section of the Seebrücke bridge directly in front of a pier
4 The cantilevered spatial forms of the Actelion Business Center

optimizing the span lengths of slabs. Much more important, for example, is creating useful, flexible spaces without precipitating significant extra costs.

More interesting, however, is working on the design of extraordinary buildings in collaboration with outstanding architects who see the building's artistic design and its interior space as having utmost priority – whereas its functionality (structural behavior, building physics, interior fit-out, suitability for use, etc.) and the costs are granted less importance. In such cases, the architect must entrust specialist engineers with many of the individual functional aspects, assuming responsibility primarily for optimizing and coordinating the developed proposals.

Throughout the design process, the structural engineer must – within many constraints and using suitable building materials and structural creativity – develop possible solutions that either present the architect with compelling design improvements or require design changes that are as few and as small as possible.

Example: Actelion Business Center, Allschwil, Switzerland

The Actelion Business Center is a highly complex, ring-shaped office building. The office spaces are stacked on top of each other in different directions. The architects were unwilling to adjust their extraordinary spatial concept to meet any wishes of the engineer. So the engineers had to make use of all the potential subsystems – these are generally steel space frames with complex node connections – to develop solutions for the orderly transfer of loads and for stabilizing the building. Finding paths for the flow of forces was not unlike navigating through a labyrinth: to ensure that the flow of forces is structurally efficient, for each confined area, the engineer will ascertain the center of gravity of the force and moment vectors and compare these, in terms of location, direction and magnitude, with the foundation reactions.

It is quite obvious that with these extraordinary structures, substantial costs can be saved through structural creativity.

Two Capricci: A Wooden Bridge and a Suspension Bridge

Marcel Baumgartner

Two exciting designs for small bridges, both created in short succession for public competitions, present two diametrically opposed solutions for comparable situations and functions. Upon closer inspection, the two designs demonstrate a striking affinity in the regularity and consistency of the structural and constructive concepts characteristic of the work of Schnetzer Puskas Engineers.

The projects were developed in close collaboration between the engineers and this author and, in the case of the wooden bridge, together with a resourceful contractor. The analytical approach of the engineer, the spirited pragmatism of the contractor and the endeavors of the architect to read things spatially and articulate them accordingly in his usual manner – even though, as in this case, the objects of concern were bridges – all came together during the design process in a creative way. Common ground was established by an unbiased attitude open to various results and the joy of constructive thought. Although the collaboration proceeded in an objective and sober way, it was continually driven by vibrant curiosity and tenacity.

The Wooden Bridge

In light of the significant tradition of wood construction in the Emmental region, the call for an innovative wooden structure was an important premise for the design of the Emme Bridge.[1] The response that pervades the entire design, motivated in part by the spontaneous enthusiasm of the carpentry contractor, was the idea of making the bridge girders from massive tree trunks of silver fir: the wood was to be taken directly from the forest, and joined – without glued or steel connections – in a manner based on old production techniques, but making use of current methods of machining.

The structural concept is based on a continuous beam extending across three fields with a total length of 76.3 meters and individual spans of 16.3, 45.3 and 14.7 meters – the slightly asymmetrical position of the foreshore piers is prescribed by lot boundaries. The two main girders flanking the bikeway are stacked together from 40 x 50 cm cross-sections of rough-sawn and dried solid wood. The stack and the resulting girder form are designed in such a way that no tensile splices are required between the beam ends; any tensile forces that occur are taken up by the beam above or below. Oak wedges that are carefully incorporated between the girder's individual beams serve to transmit the tensile and shear forces. The entire stack is held together with vertical tie rods. The oak wedges allow not only the individual beams to be prestressed against each other, but also the girder as a whole. This makes it possible to compensate for shrinkage deformations and control the deformation of the bridge. The tie rods share these traits, too; they can be subsequently retightened, thus compensating for changes resulting from the drying process of the wood.

The entire structure is protected by a large roof, which is built as a light wooden structure and supported by the main girders. The bridge's spatial concept derives directly from the structure that has been developed. With the powerful girders and the projecting roof, the area along the roadway essentially becomes an interior space with an emphasis on its longitudinal extent. Near the densely wooded foreshore, this space is strongly delimited and compressed by the maximum height of the girders at the supports. Toward the middle of the bridge, the height of the girders decreases, leaving a large opening in the transverse direction that opens up a surprising view along the axis of the tightly contained space along the river.

The seemingly archaic, solid construction of the main girders forms a stable, trough-like vessel. The light and airy, delicate roof establishes an exciting contrast to the body of the bridge and awakens memories of pavilion-like architectures that invite you to linger.

The Suspension Bridge

The starting points for the design of the Linthsteg footbridge[2] were the sensitive landscape scenery where Lake Walen flows into the Linth river and the idea of giving the bridge the appearance of a nearly floating structure that is as light as a feather by using a flat-lying, delicate cable structure.

An abandoned military bunker, covered by a primitive forest belt, serves as the abutment on the south shore. As a counterpart to this, a new, solid concrete foundation was developed on the north bank. Between these, three pairs of steel cables (support, walkway and railing) are spanned over two slightly inclined hinged columns at the edge of the foreshore. The subdivision into thirds of the span's 82.5-meter total length is determined by the position of the column pair, resulting in a primary span of 55 meters above the channel and a secondary span of 27.5 meters over the foreshore. The abutments serve as supports for the vertical loads, but they mainly transmit the horizontal cable forces into the ground. Due to its considerable size and weight, the existing bunker is able to impart the cable forces directly into the subsoil. In contrast, the new abutment must be anchored with driven piles.

Fully encapsulated steel wire ropes, like those also used for aerial cable cars, are employed for the bridge. Instead of the hanger cables that are characteristic for this bridge type, nets that transmit the loads and simultaneously serve as fall protection are spanned between the support, walkway and railing cables. The diagonal orientation of the individual wires of the net enables it to contribute to the horizontal stiffening of the structure.

The structural system has a very low dead load and can therefore be kept extremely flat, with a small camber of just two meters. The gently curving support and railing cables and the soft arch of the walkway, whose curvature follows levels defined by flooding and navigation, form a slender, recumbent figure with an elegant silhouette.

The nets between the cables have two different mesh sizes, which envelops the form of the web-like bridge in varying shades and thus makes it appear lighter. From a distance, the structure seems like a fine band stretched taut to connect the two shorelines. The transparent design offers pedestrians on the bridge an unobstructed view of the lake, its outlet and the impressive panoramic landscape.

Concept

As different as the two bridge designs are, they share an underlying interest in following – as consistently as possible – a logic stemming from the initial selection of material and construction method. The chosen systems are both surprisingly direct: the rudimentary stacking together of a bridge girder from crude tree trunks in once case and the spanning of a few steel cables that are interwoven with fine wires to form a band.

The selection of material and construction method is not made a priori from technical constraints. In one case, the programmatic circumstances of the competition were decisive, while in the other it was a free association with the existing context: The course of history and the engineers involved have caused the surroundings of the Linthsteg to undergo a radical reshaping of nature and landscape. To this day, the canalization of the Linth and the series of World War II bunkers along the northern front of the Swiss National Redoubt set the mood and inspire similarly unsentimental structures – technical and cool.

On the one hand, the designs represent didactic rigor in deriving the structural system, the formation of space and the expression from the material. On the other hand, they represent a willful exaggeration of the purely pragmatic chains of decisions asserted at the outset. This has the intention of ultimately creating a "cultivated" and characterful built structure that goes beyond the intrinsic logic of the technology. In exemplary manner, the massiveness and solidity of the wood structure is seen alongside the radical transparency of the cable structure. Both are intrinsically spatially powerful and tangible concepts that have been developed from both their logic of construction and the search for a coherent and sensual whole. In one case, the resulting expression plays with historical and quite romantic images of traditional wooden bridges, and in the other case with more ordinary memories of provisional or military infrastructures.

Moreover, the common denominator of the projects is their assiduous economy. Both designs are intentionally conceived as cost-effective structures and feature estimated construction costs of comparable magnitudes (each considerably under one million Swiss franc). In the consistency of their structural parameters, both the wooden bridge and the suspension bridge are each reduced to just that which is absolutely necessary for their construction, making the most of the respective properties of their materials. In such processes of rationalization and self-imposed economy of means, form-finding is not ideological, but results instead from the interplay between structural development and spatial sensitivity.

1 Public design-build competition for the Emme Bridge between Burgdorf and Kirchberg, 2011, 2nd place; architect: Marcel Baumgartner, Zurich; carpentry contractor: Brawand Zimmerei AG, Grindelwald (see List of Works p. 351).

2 Open competition for the Linthsteg between Weesen and Glarus Nord, 2012, honorable mention; architect: Marcel Baumgartner, Zurich (see List of Works p. 351).

Bridges

Heinrich Schnetzer

Along with creative and technical challenges, working on a competition always involves substantial economic factors for the those involved in the design and planning. The ultimate goal of genuine and unaffected collaboration within a team is therefore the design of an economical system. Because only when everyone involved – from the client all the way to the contractor – wants to support the project, move it forward, and ultimately realize it, will the project actually get built; and only then will the upfront investment of planning services be remunerated. Close cooperation among all those involved, especially between the architect and the structural engineer, is naturally self-evident and may be required of each individual planner and consultant on the basis of the shared overarching objective.

The team's organization and its composition depend solely on the task at hand and can be objectively substantiated; claims to power by individual participants are inappropriate and senseless. In the case of investor competitions involving great financial risk, the lender or investor who bears the financial consequences of the construction project must take on leadership of the team. In the case of urban design competitions, and for tasks that revolve primarily around design in general, the architect is best suited to lead the team. The strengths of structural engineers, on the other hand, lie in the design of efficient structural systems for all building phases, and in their proficiency with construction, especially pertaining to civil engineering and foundations: the development and optimization of load-bearing structures with due regard to construction procedures and geotechnical conditions is the core competence of the structural engineer. Hence, for complex tasks in these fields, the engineer is the one who must take the lead. This is particularly true for bridges, since only a structural engineer understands how the supporting structure depends on the flow of forces, loading, and the construction process. Because of larger span lengths and a much lower ratio of live to dead load, these correlations have a far more substantial impact in the design and construction of bridges than they do for buildings – they are of key importance to the economics of a structural design.

The economic production of load-bearing structures is, however, strongly subject to social change. It relates directly to the ratio of material to labor costs. These days, wages on construction sites represent the decisive cost factor. The material – if it can be installed easily, that is, without any unusually great expenditure of time – plays a much smaller role in the economics of buildings than it did fifty or more years ago. This shift also affects the design of bridges, because it pulls the central design tool out of the hands of the engineer, who thinks first of all about material economy. Today, material-intensive built structures are less expensive than structural systems that require little material, but which can only be constructed on-site with much expended effort.

We studied the correlations between material and labor costs and the economical production of a bridge as a whole in the context of a competition for a new bridge in Bhutan. Starting with these considerations, we sought to implement an economical bridge construction project.

Puna Tsang Chhu Bridge,
Wangdue Phodrang, Bhutan, 2002

In order to build the bridge with the limited local resources available, an adequate structural concept was needed. Based on the local structural possibilities, new insights on bridge construction were to be adapted and economic constraints were to be taken into account.

The designated location for the bridge had been selected more than two hundred years earlier for a traditional wooden bridge. The geotechnical conditions at this site and the lack of possibilities

1

1 The Puna Tsang Chhu Bridge upon completion, with the old pier of the previous wooden bridge and the Wanghdi Dsong (monastic castle) on the hilltop

for building a pier foundation in the river, however, necessitated a bridge structure with a span of about one hundred meters – a span that would be no small feat even in technically advanced Switzerland. An arch bridge with abutments in the bedrock of the valley walls best fulfilled the geotechnical, geometric, and economic conditions. Our inspiration for this structural system were the arch structures built in the first half of the last century in Switzerland, such as the Salginatobel Bridge by Robert Maillart from 1930. With a high ratio of material to labor costs, these are very economical in terms of material. This link between the Bhutan of today and the Switzerland of yesterday made sense in that the current economic conditions, or rather the ratio of labor costs to material costs in Bhutan are comparable to those in Switzerland from about eighty years ago, in the heyday of reinforced concrete arch bridge construction.

Due to the recurring monsoon rains in summer and the subsequent higher water level, construction in the riverbed was limited to the winter months. Thus, the arch construction – including falsework on five steel supports – and the foundations all had to be built in the winter half of one year. In order to deal with the short time available for construction, the concept for building the arch was based on old, well-known methods that were supplemented with modern construction procedures that are also feasible in Bhutan: The arch slab and the roadway deck were partially prefabricated of concrete elements in a small production facility close to the construction site and, after being placed in position, they served as formwork for the concrete topping. The arch ribs were cast of in-situ concrete. In contrast, the steel beams for the falsework were produced in Calcutta and transported from there to the construction site. To compensate for the high transport costs for the steel beams, they were made in such a way that they could be removed after finishing the concrete arch and reused as composite beams for the roadway deck.

A cable crane was installed along the bridge axis and used to move the concrete elements for the bridge and the parts of the auxiliary construction, such as stones enclosed in nets of wire mesh. These were submerged in the water and functioned as temporary foundations for the falsework towers set up mid-river. Once the falsework was in place, the arch segments were set in position and the two-centimeter-thick joints were then closed with mortar. After setting the last arch segment – the so-called keystone – all the elements already began working together as a compressive arch, thus relieving the falsework. As a result, limit load reserves were freed for additional loads and the concrete topping on the arch could be poured, and the arch ribs that stabilize the thin arch could be completed. The stable condition that resulted was reached at the end of the seventh month. Only then were the piers and abutments constructed and the falsework dismantled in order to install the pieces again as roadway beams and lay the precast roadway elements. The steel edging for connecting the larch walkway along the side was already been cast into the concrete topping of the roadway deck as a pour stop. The hand-carved wood posts in combination with the railings reflect Bhutan tradition.

Second SBB Railway Bridge over the Rhine, Basel 2013

The commission for the railway bridge over the Rhine in Basel was also the result of a competition (1994), as in the previous project. Severe constraints were, in this case, the stipulated pier locations and the maximum construction height. On this basis, we developed a prestressed reinforced concrete bridge. The structural system is a cantilevered bridge extending over three fields, with spans of 59.7, 117.2 and 59.7 meters. The columns are each constrained in the superstructure and at the base and, as a result of the traffic loads, are subjected to bending. The frame system shrinks and creeps, as temperature changes create secondary stresses for the structure. The frame system, which was applied here for the first time to a large railway bridge, is advantageous for several reasons: Due to the restraint, the pier slabs have a much greater resistance to ship impact. Moreover, the frame system is stiffer, which makes a cantilevered bridge with limited height possible in the first place. In addition, no large, virtually non-replaceable bridge bearings are required.

A polygonal, longitudinally prestressed haunched girder with a cantilevered roadway deck slab on both sides is the principal supporting element of the bridge structure. The shipping clearance profile limits the depth of the main span, and the end spans are treated symmetrically with polygonal soffits. The cross-section is 7.3 meters high above the pillars and 3.4 at mid-span; all structural elements are kept below the railway tracks. Above the abutments and the piers in the river, cross girders conduct the forces into the bearings and supports. Along the bridge's longitudinal axis, additional cross girders direct the deflection forces – caused by the polygonal course of the lower compression plate – into the webs. The cross girders at the bridge apex absorb the deflection forces resulting from the tensile force in the lower slab of the box girder. The box girder is filled solid with concrete for a length of four meters near the abutments and due to its weight, it prevents the bearing from lifting off. As is common for cantilever construction, the girder's prestressing consists of cantilever prestressing and mid-span prestressing. Additionally, the bridge's box profile has continuity prestressing. It is externally prestressed and diverted over the cross girders, thereby favorably affecting the deformation behavior.

The 12.35-meter-high reinforced concrete piers stand in the river and are rotated by four degrees in the direction of the river's flow so as to align with the new highway bridge. Each is fixed in a pile cap that is 2.5 to 3 meters thick and measures roughly 9.2 by 13.3 meters, and which, in turn, transmits the loads into four foundation shafts. These have a diameter of three meters and a depth of about six meters, allowing them to anchor over a length of about five meters in the compact Cyrene marl. The pier foundations in the river were constructed within a sheet pile cofferdam that is driven into the marl. After construction was completed, the sheet piles were trimmed off at the top edge of the foundation slab, in such a way that the remaining portion acts as protection against scouring. The sheet pile cofferdams were not dimensioned to withstand ship impact, as the two piers were built at different times and it was therefore possible to redirect the maritime traffic to the opposite side for a short time while the foundation and pier work was carried out. Once the piers were completed, the sheet pile cofferdam was braced against the pier, so it would have resisted a ship impact. The actual bridge piers were built within the sheet pile cofferdams using conventional concrete construction. Immediately after completing the initial stage with its sixteen-meter cantilever, the construction pit was flooded. The cantilever construction then proceeded in a weekly pace in stages that were initially 3 and later 3.9 meters in length, whereby the work began from the Grossbasel side and subsequently continued from the Kleinbasel side. After the joint in the middle of the bridge was closed, the cantilever heads were cast in lengths of eight meters using movable work platforms starting from the abutments.

Competition Emmequerung
Neumatt Ey, Switzerland, 2011

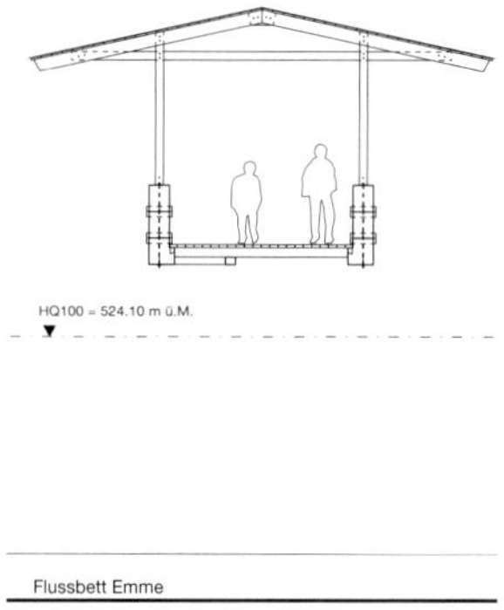

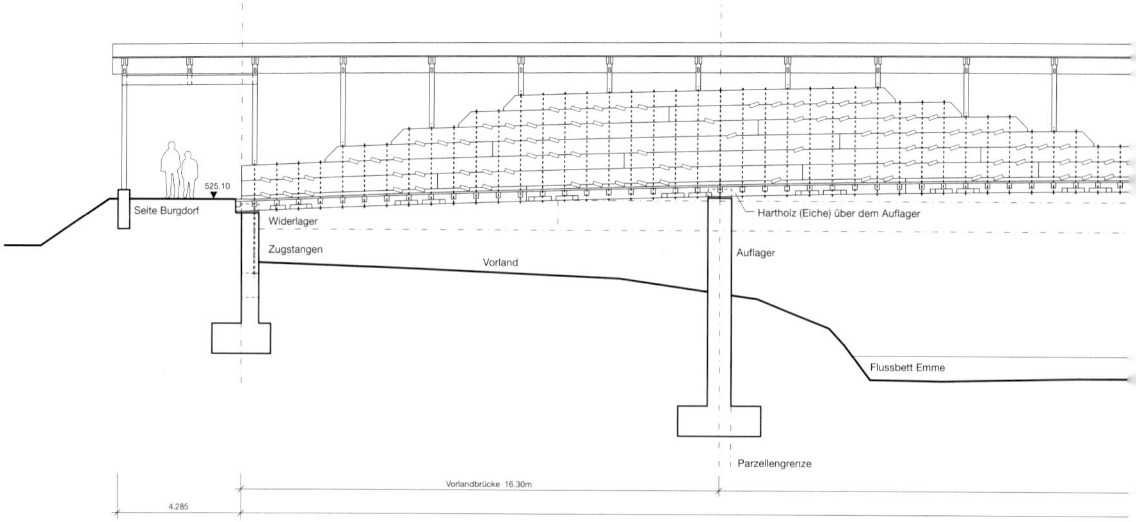

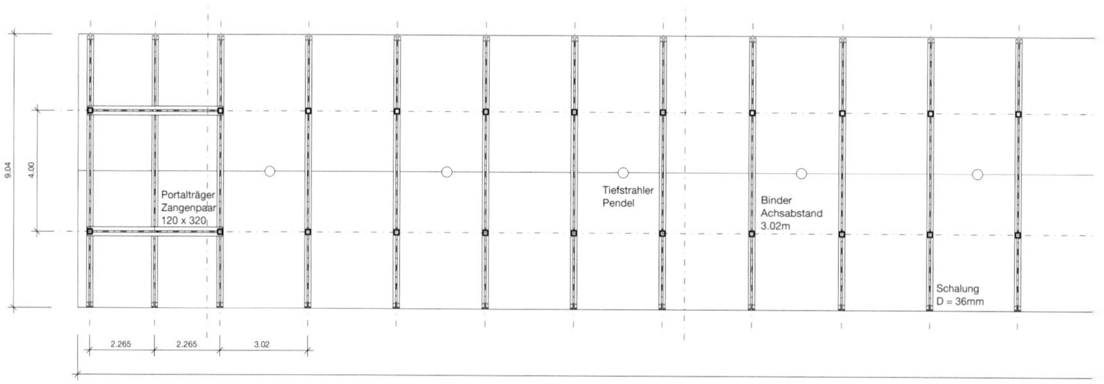

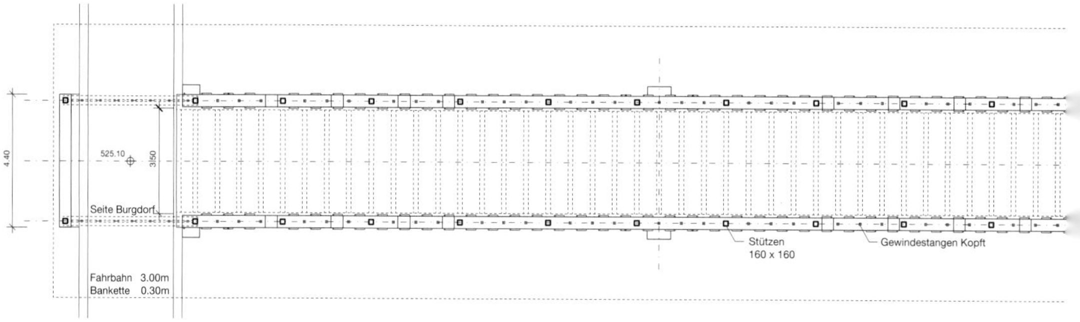

←← Rendering of the wooden bridge
↑ Longitudinal section
↗ Roof framing plan
→ Bikeway plan

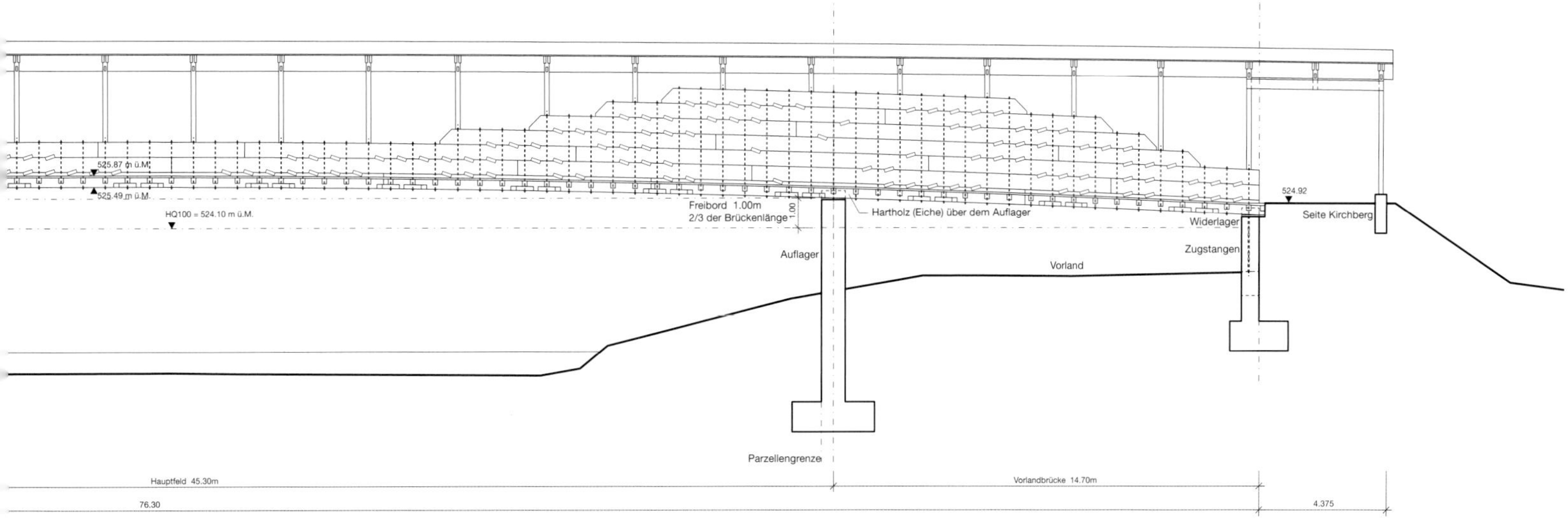
525.87 m ü.M.
525.49 m ü.M.
HQ100 = 524.10 m ü.M.
Freibord 1.00m
2/3 der Brückenlänge
1.00
Hartholz (Eiche) über dem Auflager
524.92
Seite Kirchberg
Widerlager
Zugstangen
Auflager
Vorland
Parzellengrenze
Hauptfeld 45.30m
Vorlandbrücke 14.70m
76.30
4.375

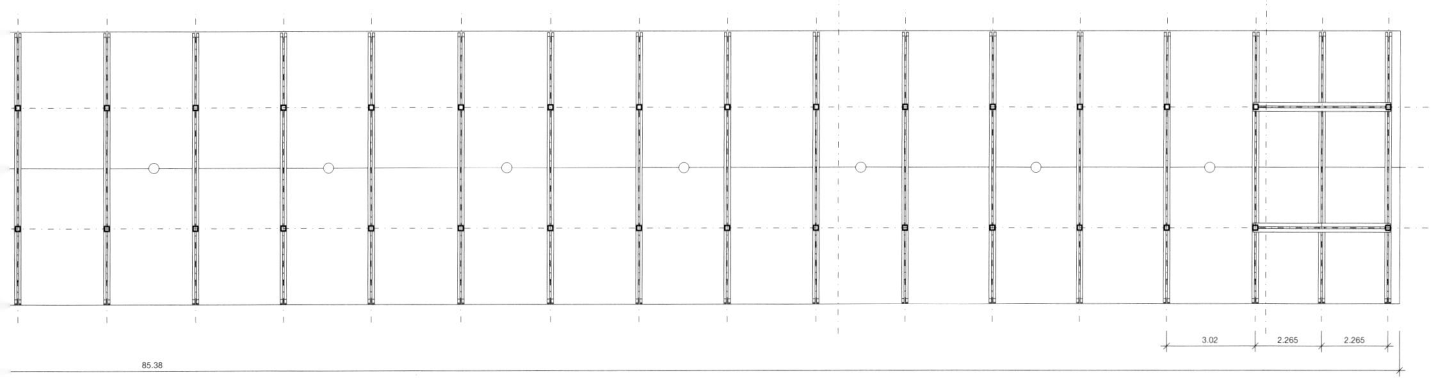
3.02
2.265
2.265
85.38

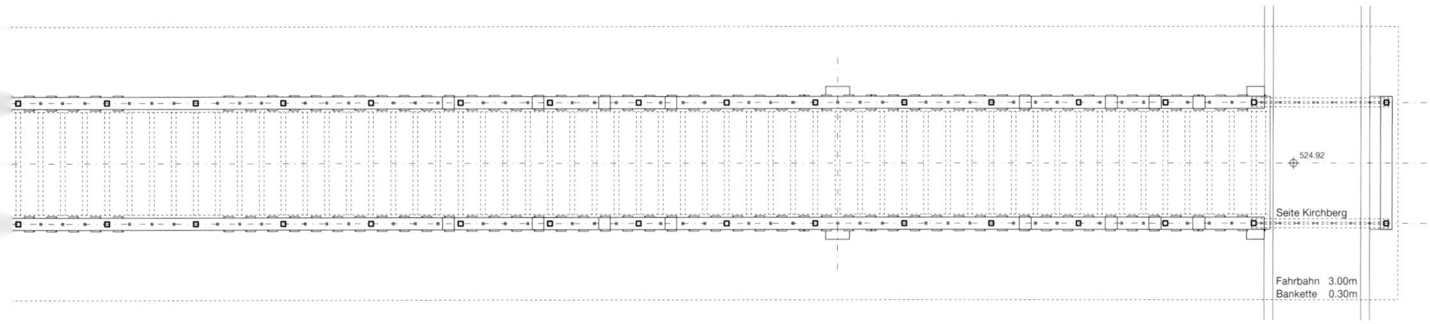
524.92
Seite Kirchberg
Fahrbahn 3.00m
Bankette 0.30m

Competition Linthsteg
Weesen, Switzerland, 2011

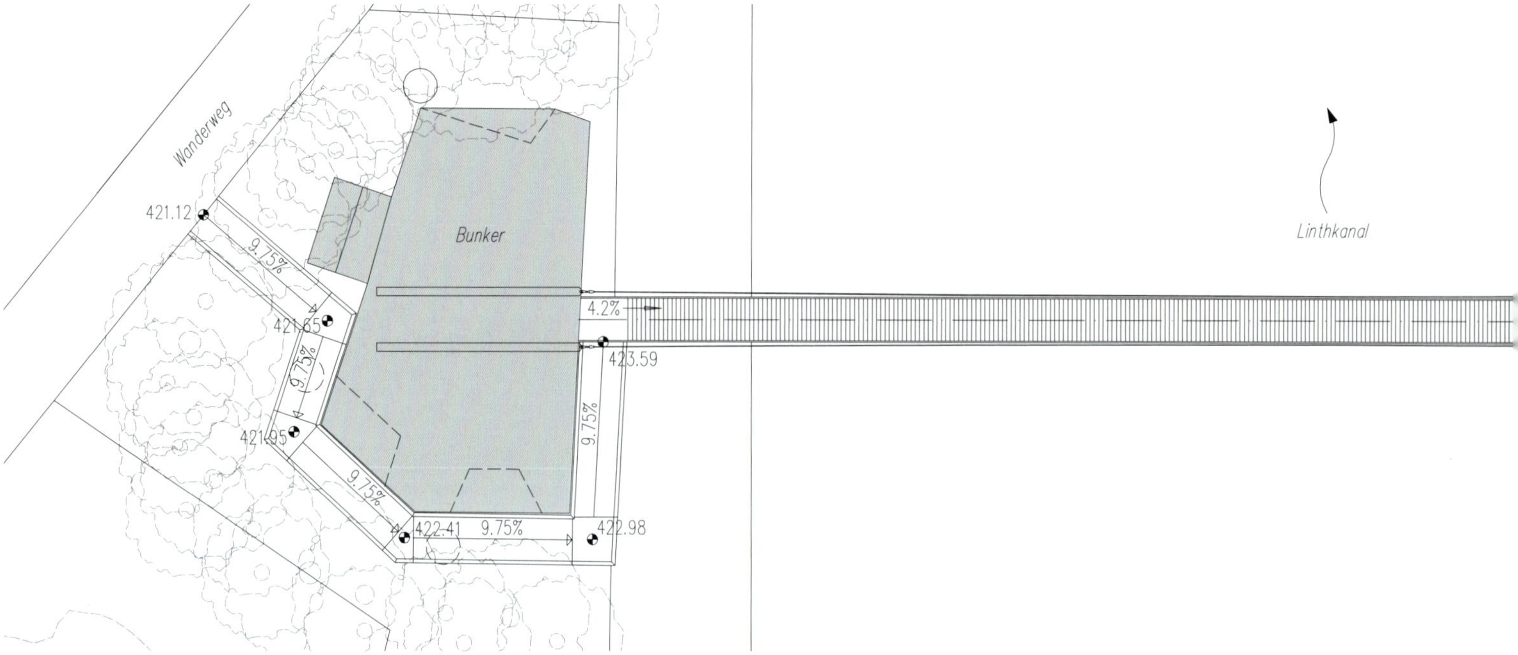

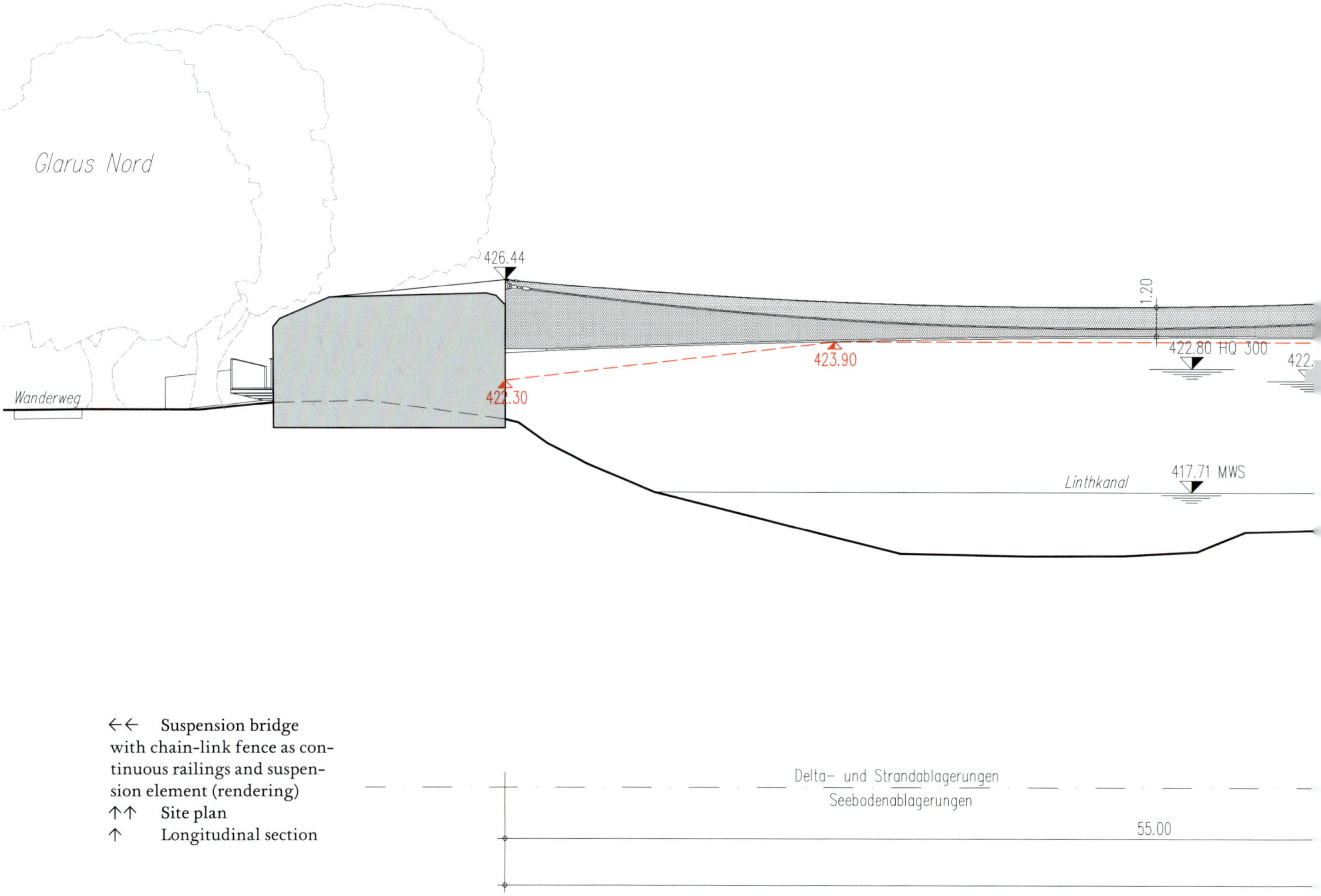

←← Suspension bridge with chain-link fence as continuous railings and suspension element (rendering)
↑↑ Site plan
↑ Longitudinal section

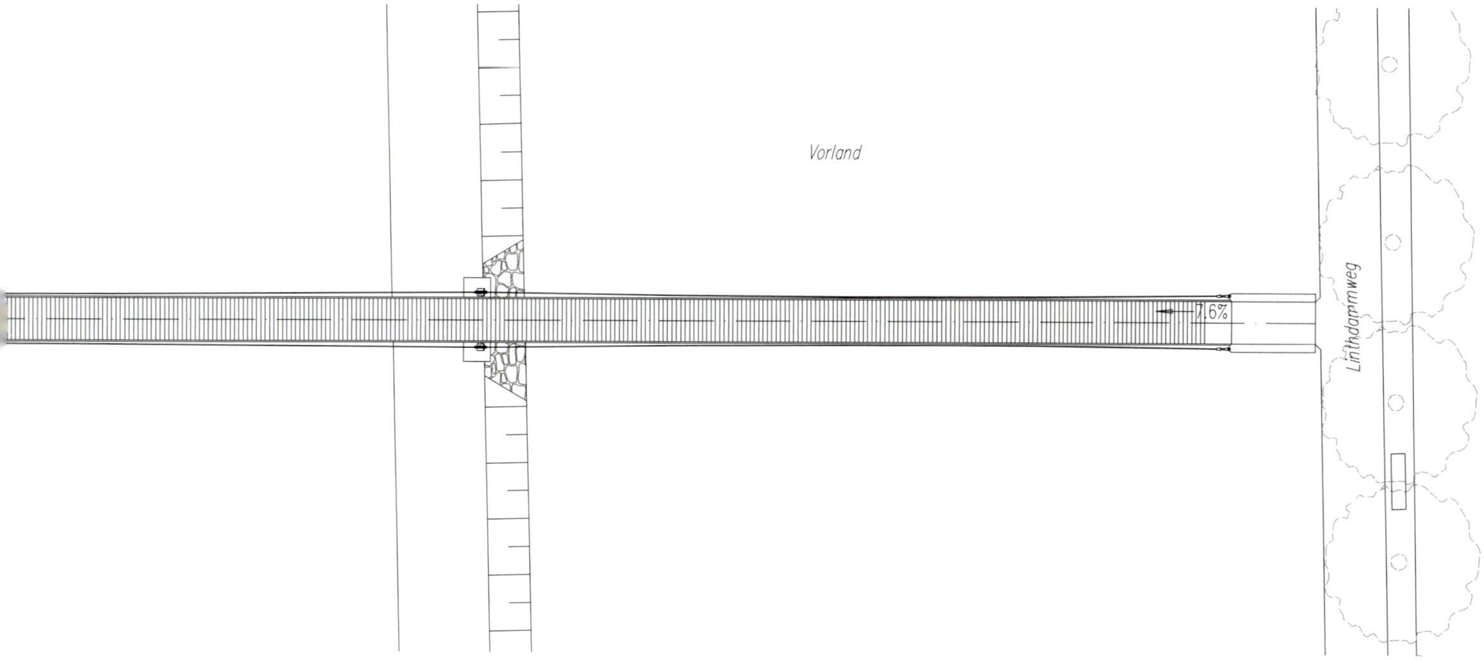
Vorland
7.6%
Linthdammweg

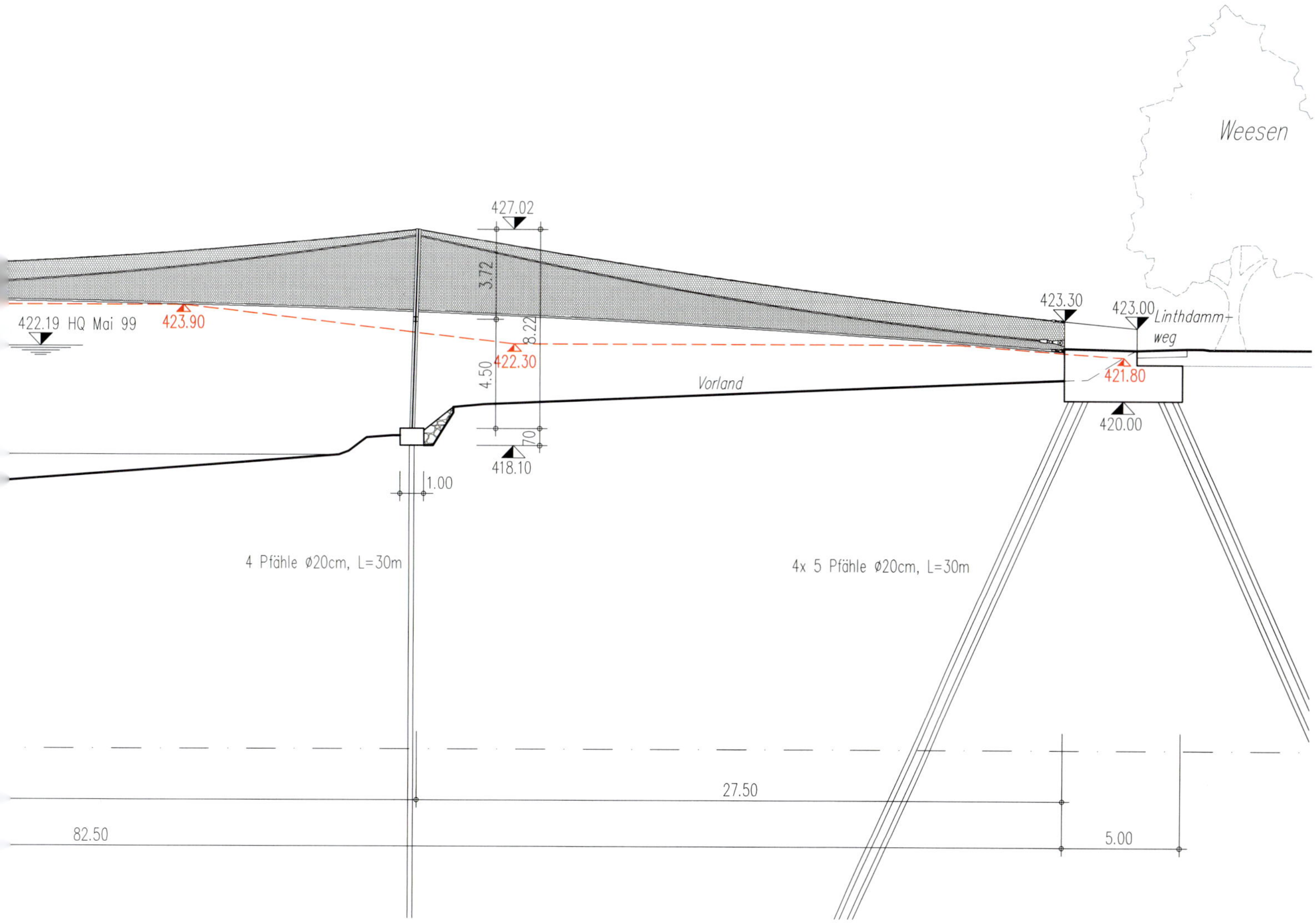
Weesen
427.02
3.72
8.22
4.50
70
418.10
423.30
423.00
Linthdamm-
weg
422.19 HQ Mai 99
423.90
422.30
421.80
420.00
Vorland
1.00
4 Pfähle ø20cm, L=30m
4x 5 Pfähle ø20cm, L=30m
27.50
82.50
5.00

Second SBB Railway Bridge over the Rhine
Basel 2013

Kontinuitäts- und Feldvorspannung, Längsschnitt

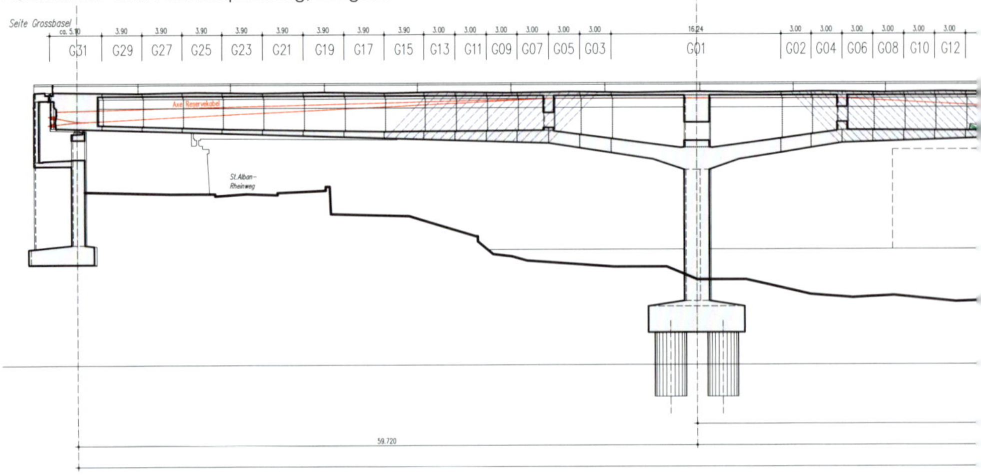

Schubvorspannung (Spannköpfe oben)

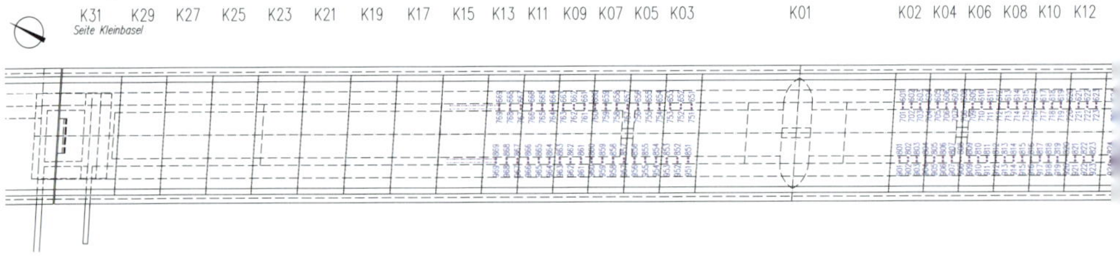

Kragarmvorspannung

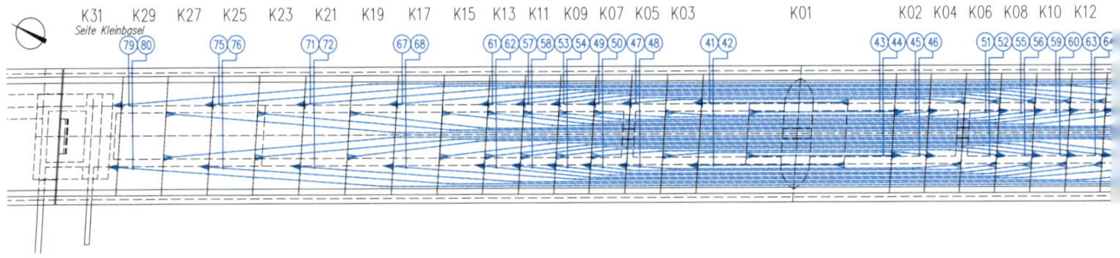

Feldvorspannung

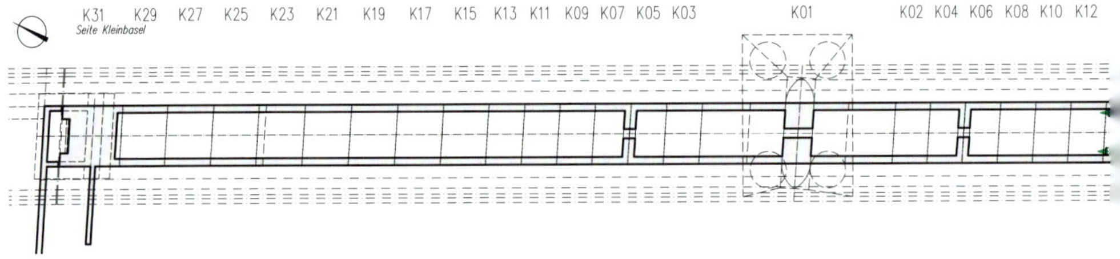

Kontinuitätsvorspannung

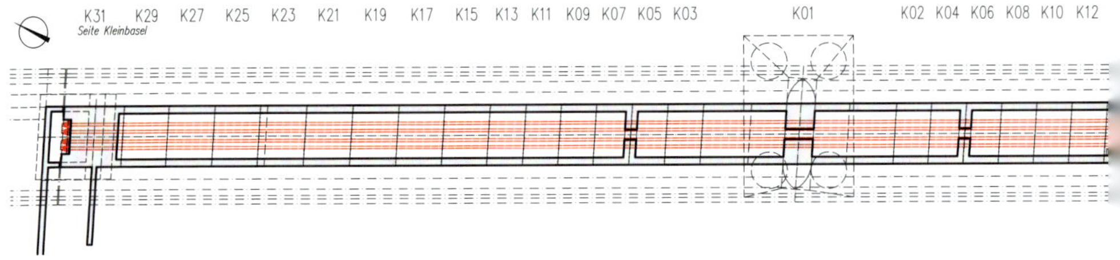

←← Pier reinforcing with the existing steel bridge and the Schwarzwald Bridge in the background (construction site photo)
↗↗ Longitudinal section with shear prestressing (violet), continuity prestressing (red), and division into cantilever construction stages
↗ Prestressing concept: cantilever prestressing (blue), mid-span prestressing (green), and continuity prestressing (red)
→ Plan with anchorage points for the shear prestressing (violet); plan of the tendon layout for the cantilever prestressing (blue)
Horizontal section through bridge girder with mid-span prestressing (green); horizontal section through the bridge girder with continuity prestressing (red)

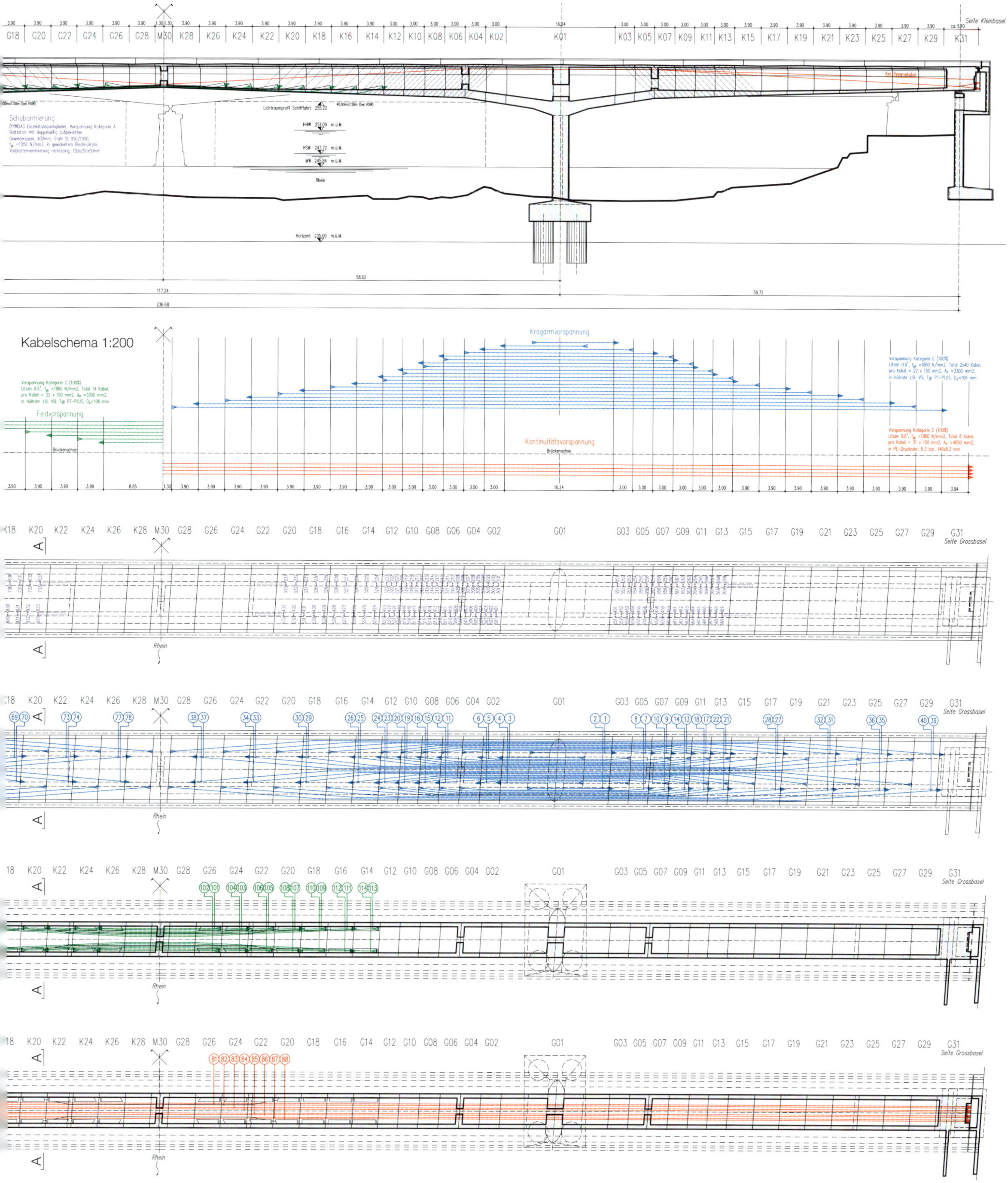
Seite Kleinbasel
Schubarmierung
Lichtraumprofil Schifffahrt
Rhein
Kabelschema 1:200
Kragarmvorspannung
Feldvorspannung
Kontinuitätsvorspannung
Brückenachse
Seite Grossbasel

Prinzip-Innenansicht Steg

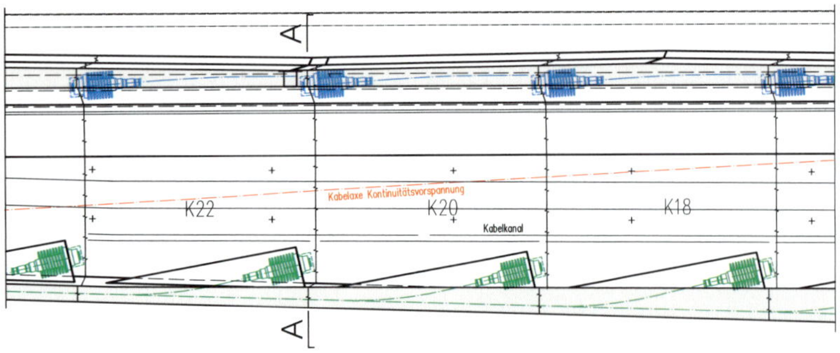

Querschnitt A-A

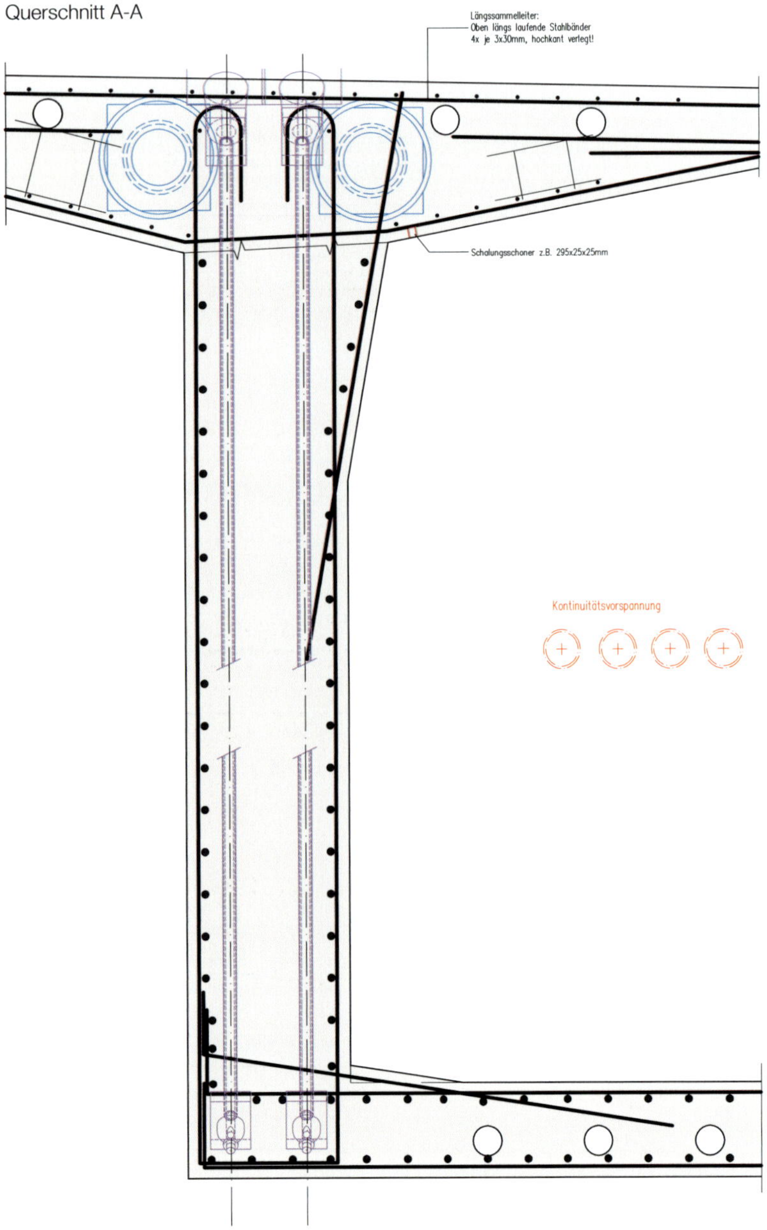

↖ Detail of the longitudinal section of the bridge, with anchorages for cantilever and mid-span prestressing
← Detail of the bridge cross-section, with shear prestressing and anchorages for cantilever and mid-span prestressing
↗ Horizontal section of the initial stage with tendon geometry and anchorages for the cantilever prestressing (blue)
→ Longitudinal section of the initial stage
→→ The finished bridge

Untersicht Platte oben

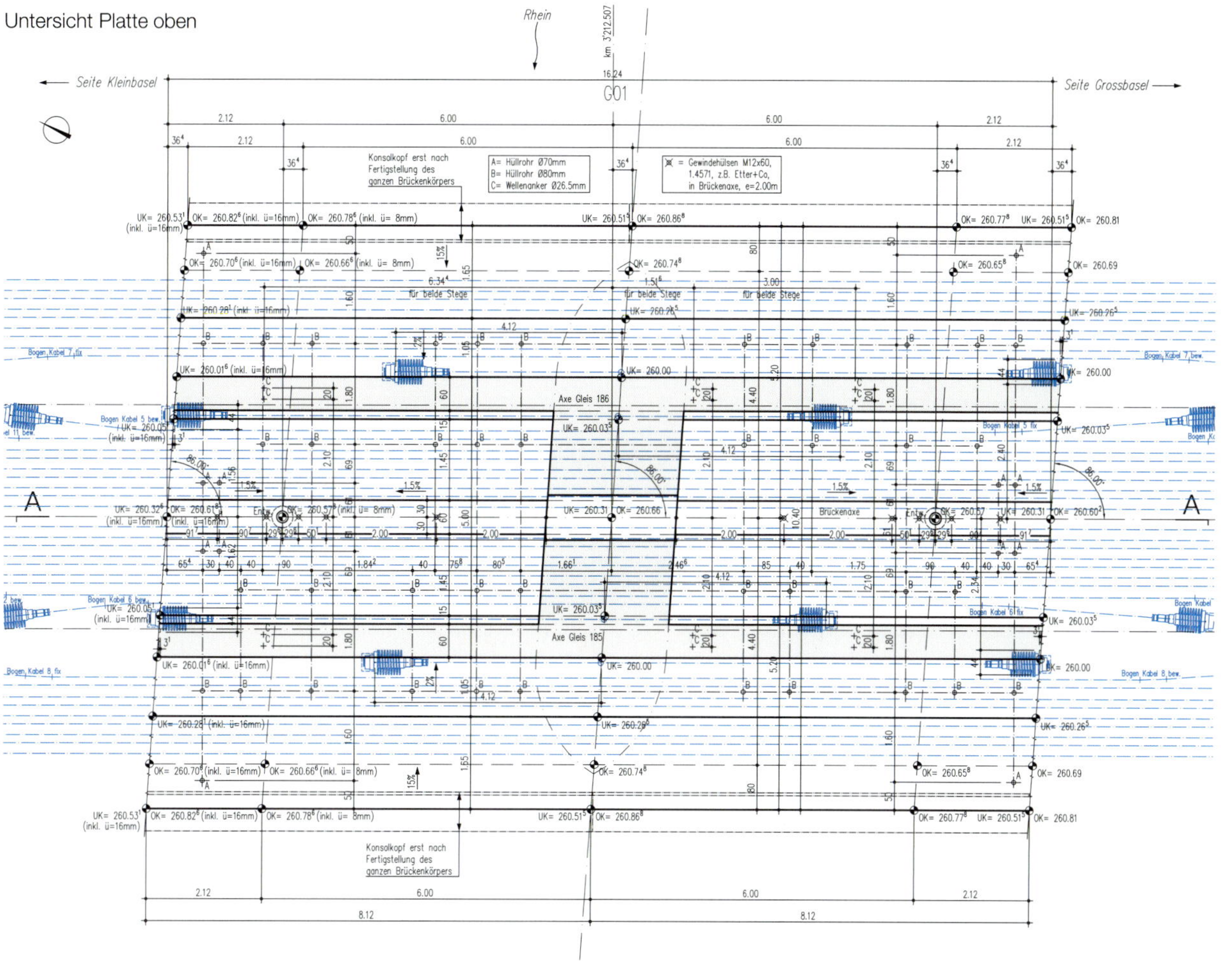

Schnitt A-A (in Brückenlängsaxe, in Fliessrichtung betrachtet)

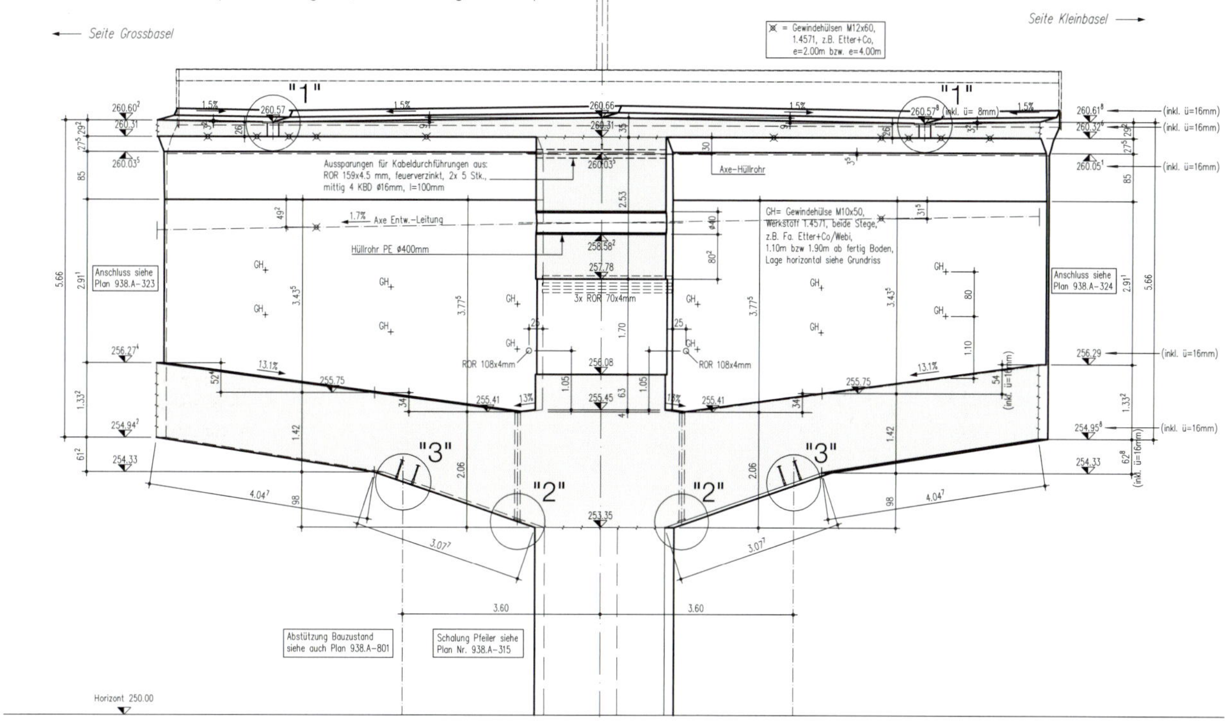

Puna Tsang Chhu Bridge
Wangdue Phodrang, Bhutan, 2002

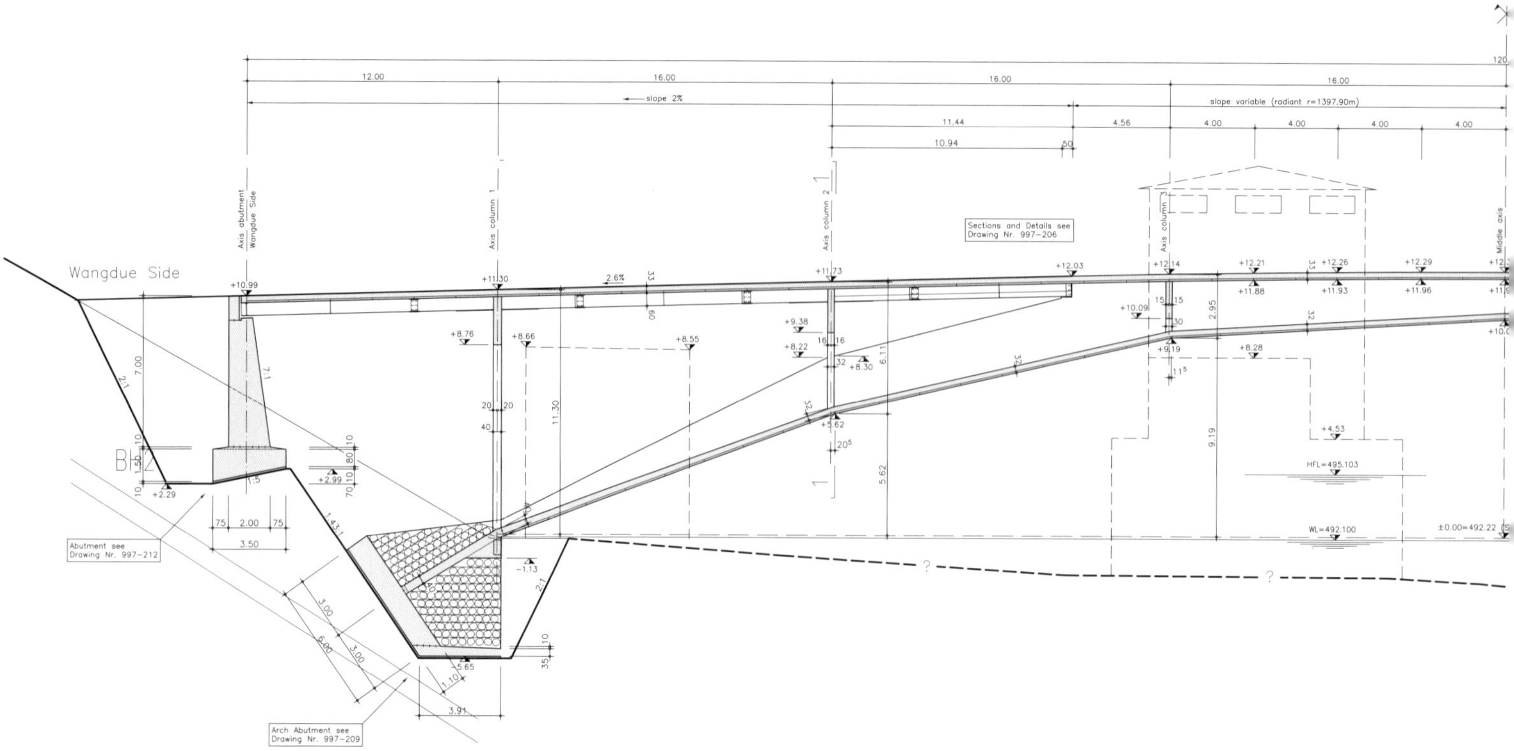

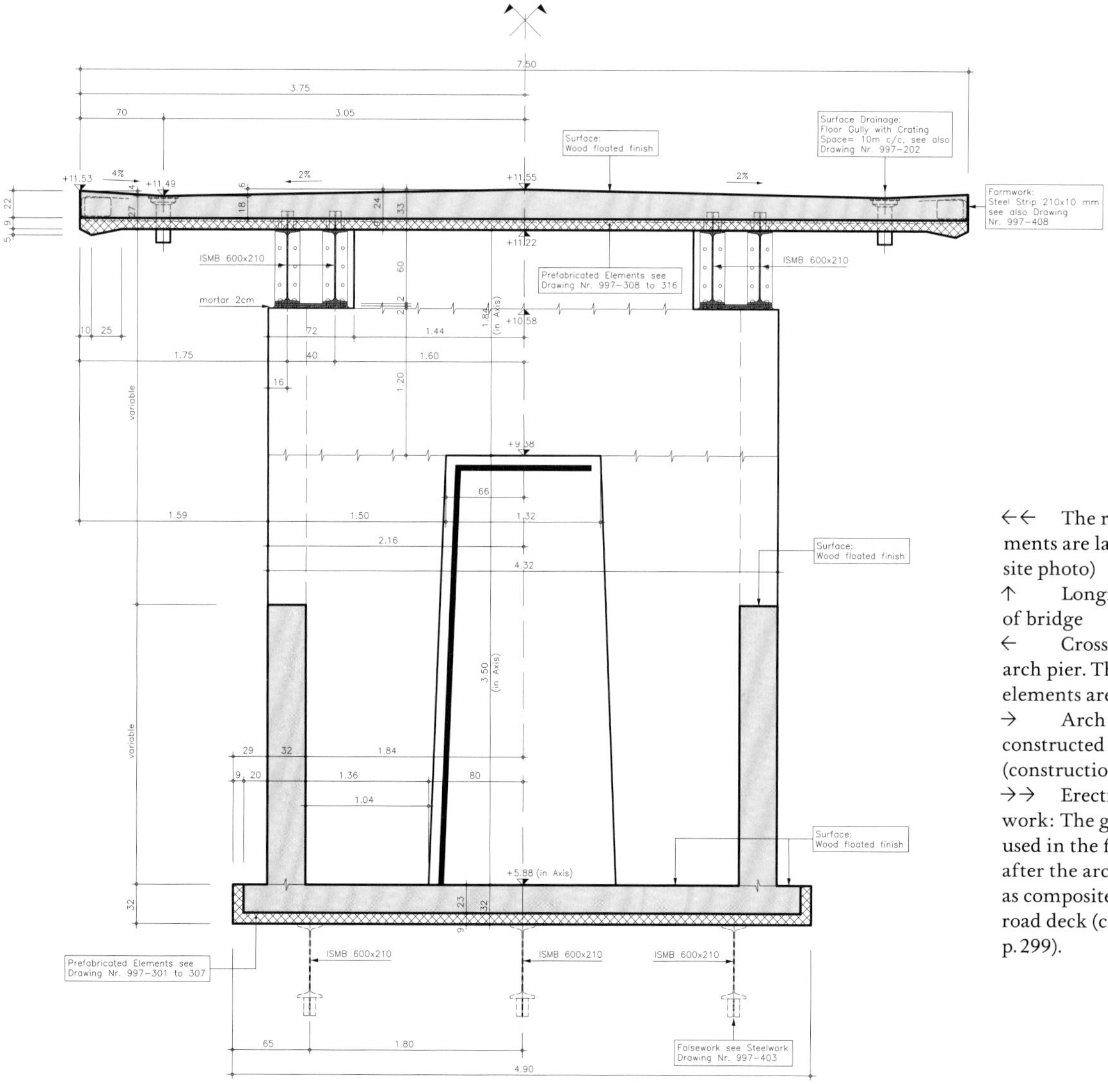

←← The road deck elements are laid (construction site photo)
↑ Longitudinal section of bridge
← Cross-section at first arch pier. The prefabricated elements are hatched.
→ Arch abutment being constructed of concrete. (construction site photo)
→→ Erection of the falsework: The girders will be used in the finished state – after the arch is constructed – as composite beams for the road deck (cf. illustration p. 299).

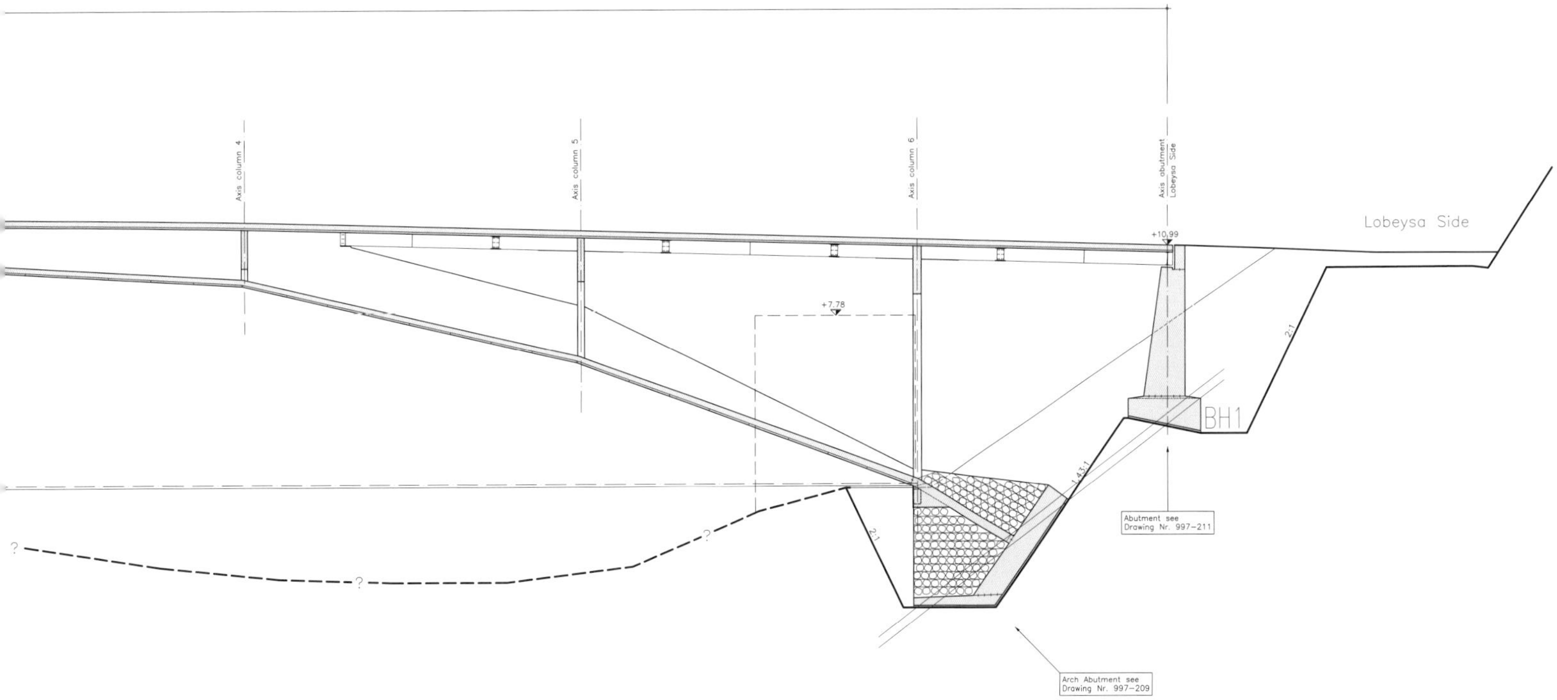

Axis column 4
Axis column 5
Axis column 6
Axis abutment Lobeysa Side
Lobeysa Side
+10.99
+7.78
2:1
BH1
1.43:1
Abutment see Drawing Nr. 997-211
Arch Abutment see Drawing Nr. 997-209

↑ The finished bridge as seen from the water, with the Wanghdi monastic castle, the pier of the old Basam (traditional wooden bridge), and the temporary Bailey bridge
→ The finished bridge with the Wanghdi Dsong (monastic castle; upper right)

9 Special Tasks

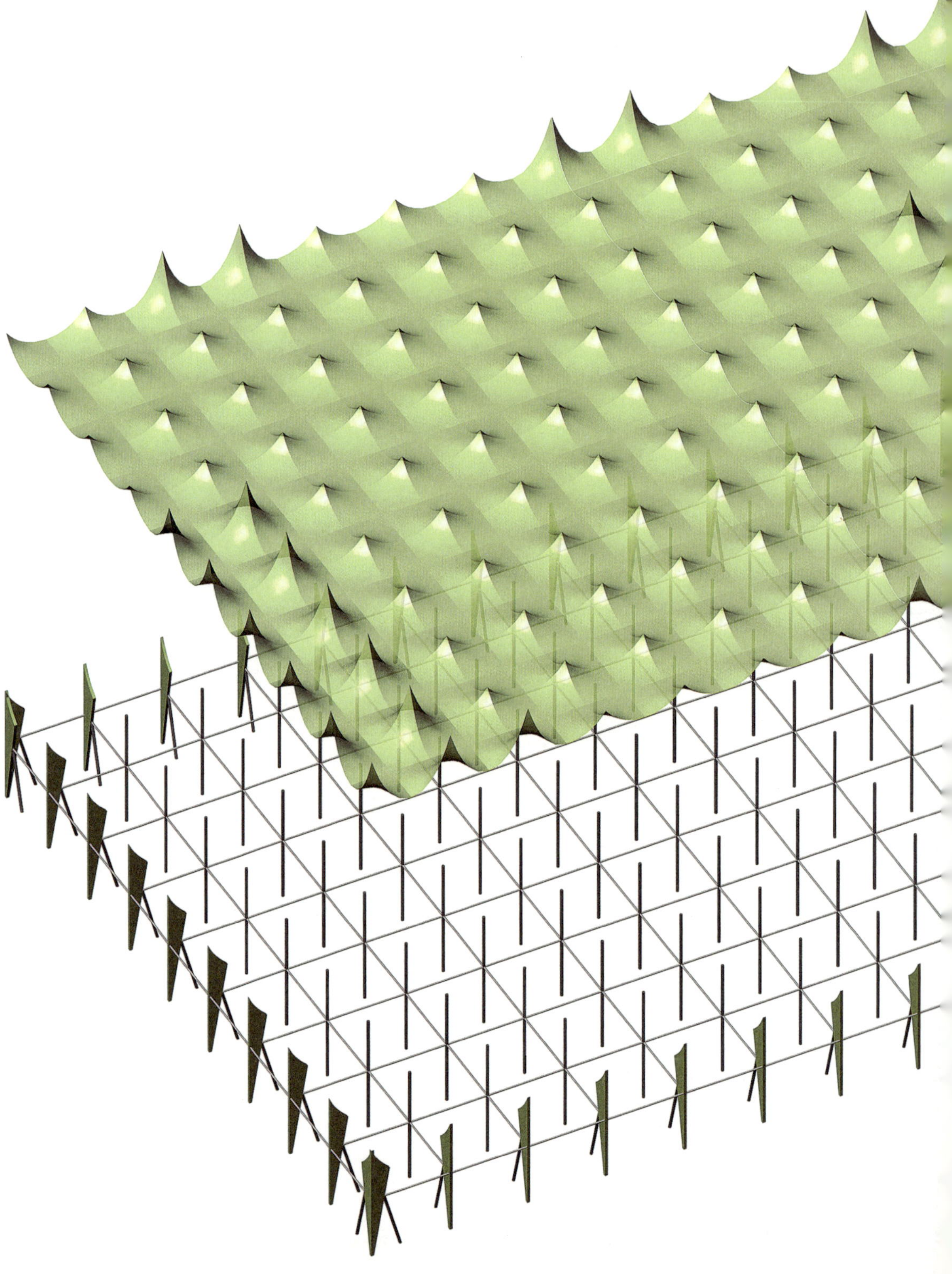

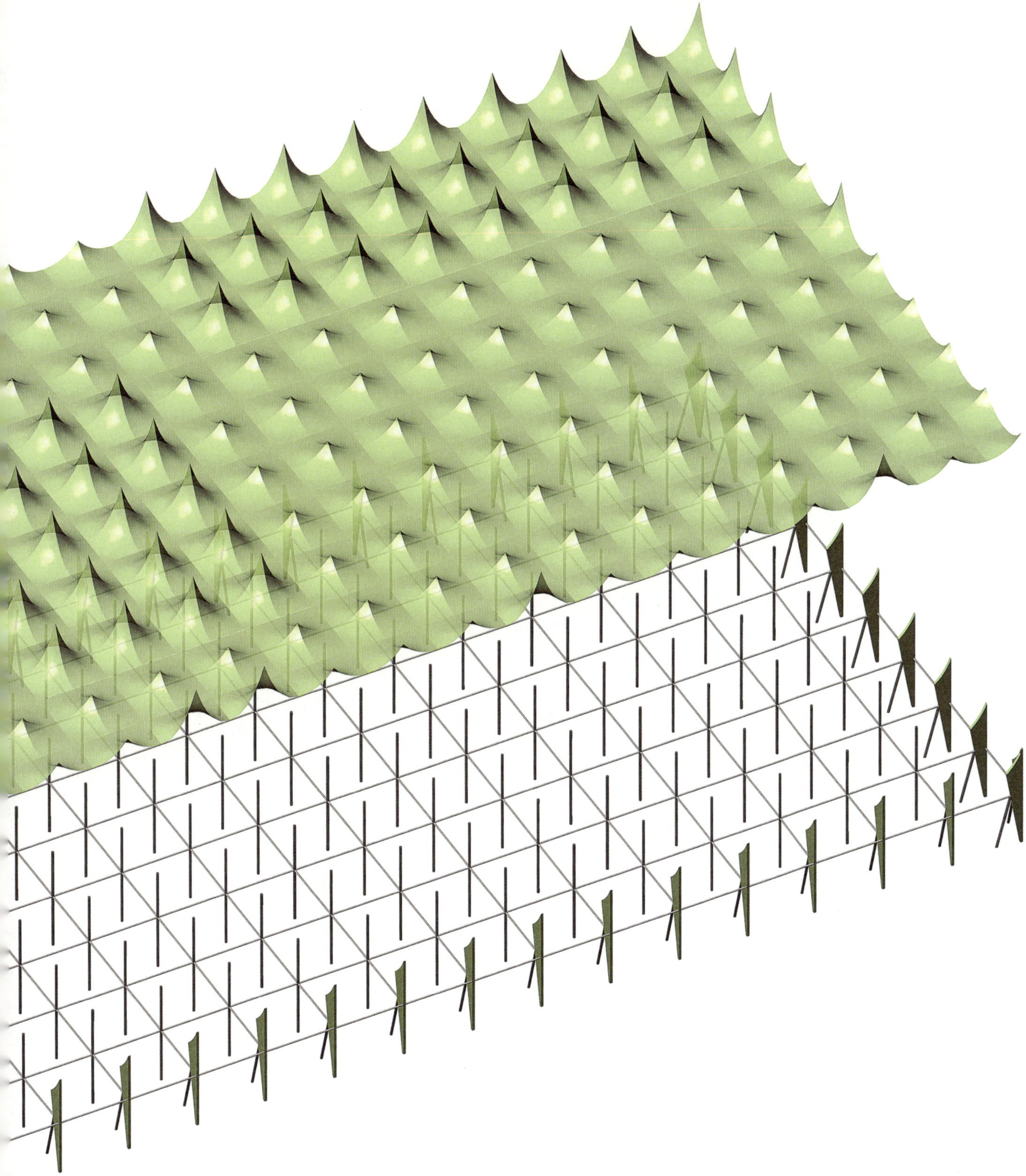

← Roof structure of the Prada Levanella, Warehouse and Distribution Center

Conversations IX

AF Let's talk about civil engineering: are infrastructure works built structures with no aesthetic standards, no design intentions? Infrastructure

TP That retaining walls and tunnel portals should be designed is not something we learned from Rino Tami. The entire realm of civil engineering is essential for the built environment. Nevertheless, a great deficit prevails: The problem begins with bidding procedures that are focused solely on price. In so-called fee competitions for infrastructure works, we have been the only ones to have also brought in architects. Unfortunately, public clients are not yet so advanced that they would attach higher value to the aesthetic design of civil engineering works.

HS Today, more areas for vehicular traffic are designed somewhat randomly by engineers applying standards than there is land with buildings. That can be visualized well on a 1:25000 scale map. My hair stands on end regrettably often when I see everything that is simply set into the earth, from noise barriers to guardrails and fall protection, etc. – everything you can find in a catalog.

AF Isn't it a political question?

HS Yes, but also one of sensitizing people: raising the awareness of the general population, and especially of the engineers themselves! Beginning with their early training, we must make clear to the engineers that they should actively consider where a retaining wall starts, where and how it stops, etc. That's completely lacking in our training.

AF Not entirely anymore...

HS Right, thanks to you, that's now also conveyed at the ETH!

AF That's right.

TP We must not allow competition to be reduced to a question of fees; it must also capture creative aspects.

AF A central problem is, in my opinion, that decisions affecting large areas – such as roadway alignments – are made by people with no spatial competence.

HS In the past, the alignment of a road was established so that the bridges wound up in prudent locations. To that end, a longer road would perhaps even be accepted. Today, the design speed is primarily an issue, for instance one hundred and twenty kilometers per hour, which means the bridge needs to be in a place that allows that speed.

Ephemeral Structures

AF Do you have a special interest in ephemeral structures, where the constructive more likely comes into play?

HS We are very interested in them. Our projects in Saint-Prex and Avenches belong to this group.

TP ... or exhibition pavilions. Together with a Japanese architect, we have just designed a very interesting stand for a well-known leather goods manufacturer: a double-curved wood facade made of wafer-thin slats that are attached with magnets, so there are no visible connections. A delicate lattice that is curved in two directions.

AF Magnetic wood?

TP Yes, the magnets are pressed into the wood. That means the construction can be quickly assembled and taken apart and is reusable.

Assessment

AF What criteria do you use to evaluate the success of a built structure? When you are satisfied with your designs?

TP If the structure and architecture merge symbiotically, we are very satisfied. Such cases have been known to occur.

HS With a successful structure, the various requirements come together in a natural and straightforward way. One might call it "simple unity." In retrospect, I also take a critical view of many of the buildings I've designed. What's nice is that we can always think of improvements!

AF Gaining valuable experience can be an important criterion for one's own assessment of a work.

SB After working on the planning for a long time, it's satisfying when the contractors not only can, but also want to implement the proposals as we have planned them.

AF ... when the digital and physical models correspond. Finally, I would be interested to know what questions I have forgotten – or could not ask because I'm not an engineer. What would you engineers ask yourselves?

HS For instance; why are we not really present in society's perception?

TP How something works is not valued so highly these days. It's often more about a cult of surfaces and less about content. We first get a chance in the event of a crisis. Until that point, all technology is taken for granted. In addition, the big role models are missing from the engineering profession today.

AF Like Heinrich Figi, the head of civil engineering structures at the civil engineering department in Graubünden, once aptly remarked in a lecture, civil engineering works, and engineered structures in general, are perceived fatefully by the youth – that is, as something non-designable; as given. For today's young people, the built environment seems incapable of being influenced – perhaps because the amount of emptiness that is critical for a conscious perception of the environment – and which we were still able to experience – is something they have never even known. At most, the glamor of becoming a star architect still entices and deludes as an ideal profession.

TP The cult of star architects is associated with fashion and design – that's chic.

AF It also raises the question of why architects, but also engineers, receive inadequate compensation when measured against the responsibility they carry. The remuneration for the services provided is entirely disproportionate to the salaries of bankers, managers, etc.

TP That's crazy, but those are also simply the signs of our time. Another time will come again. Besides, our profession has a high intrinsic value and that's why I'm happy to be engineer, even today!

AF To conclude, what would you give young engineers to take along on their way? My ears are still resounding with the words of Luigi Snozzi, who, at an event at the ETH Zurich, recently recommended in his characteristic way that the students leave school as quickly as possible.

SB I would recommend that they take with them as much as possible from their training. Even though all the aspects cannot be covered there, the school provides a valuable foundation of study – the horizons can also be subsequently expanded with aspects that may have received short shrift in school.

HS That reminds me of a saying by Willi Ritschard: What good is a tiger in the tank if an ass is driving?

Special Tasks

Stefan Bänziger

Not all the tasks that we deal with pertain to a classic branch of structural engineering. The tasks are sometimes very complex, interdisciplinary or, in fact, primarily concern other fields. The unique and unprecedented problems that our clients sometimes confront us with invariably hold a special allure. With our specialized knowledge of engineering and our construction experience, we contribute to the development of new structural concepts – often together with architects, landscape architects, and artists.

Balloon for the St. Prex Classics, St. Prex, Switzerland, 2012

On Lake Geneva in the Canton of Vaud, performances by dancers and musicians are held annually under the open sky in the medieval town of St. Prex. A covered venue for the event was installed for the first time in 2012. This roof consists of a hemispherical balloon with an inner ring at its bottom edge to absorb the deflection forces from the balloon. The balloon itself is fixed to the inner ring at forty-eight attachment points. The hemisphere is attached at six load transmission points to the truss-like stage ring at a height of about twelve meters, which corresponds to roughly half its diameter. The ring, in turn, rests on three groups of columns that are hinged to individual concrete foundations. Three load transmission points transmit only compressive forces, while the remaining three can be locked in place and also transmit tensile forces. The stage ring is tilted about eight degrees from the horizontal. The higher part covers the spectator seating and the lower part covers the stage.

When filled with helium, the balloon and its inner ring can float above the stage ring. The hemisphere will ascend into the sky while anchored by three pairs of cables, whose upper ends are fixed to the load transmission points of the balloon's inner ring. To be able to regulate the position of the balloon in space as desired, one cable from each pair is secured vertically and the other diagonally. This permits adjustments in all six degrees of freedom.

The design of the helium balloon and the precise constraints it mandated were developed in consultation with a firm that specializes in constructing large balloons.

"Tree Pot" Leutschenpark, Zurich 2008

Leutschenpark is a landscaped open space of approximately fifteen thousand square meters in Leutschenbach, a previously industrial and commercial district in northern Zurich. A conglomeration of park elements – including a lawn, a fountain, a playscape, a stage area and seating elements – centers around a feature that is designed like a huge "tree pot" to envelop a bank of earth that once served as the bullet trap for a shooting range. Due to its contamination with lead, this area needed to be kept inaccessible to park visitors. Instead of going through a labor-intensive and costly process of removing the contaminated mound of earth, it was enclosed with a sarcophagus-like concrete ring.

Approximately three meters high, this feature is built monolithically in a way reminiscent of a flowerpot, with a bench forming the pot's base. The ring is a rounded-off rectangle in plan with a length of sixty meters and a width of twenty meters, and has no expansion joints. Centric longitudinal prestressing ensures that the concrete structure nevertheless remains substantially free of cracks. Analogous to a cantilevered retaining wall, the pot's cross-section has a projecting heel and toe. This allows the structure to withstand the earth pressure from the backfill and enabled it to be poured in three stages. The surface is curved biaxially – in cross section and in plan – and therefore required complex formwork. The biaxially symmetrical plan, however, enabled a certain degree of repetition, which thus allowed the form-

1

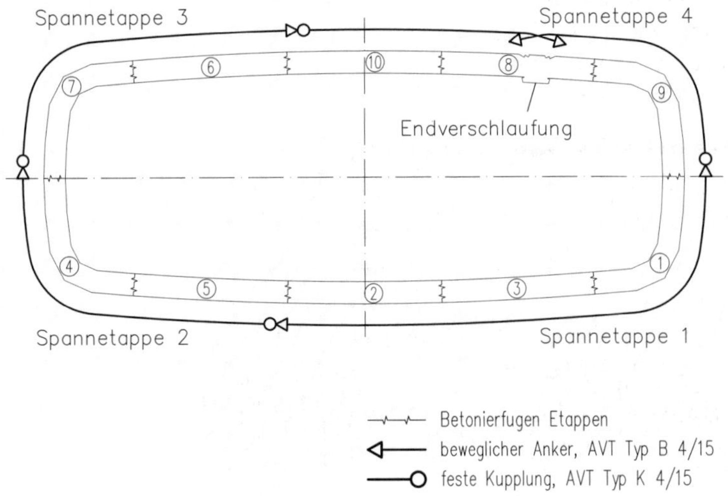

1 Diagram of staged prestressing for the "Tree Pot" Leutschenpark

work elements to be used repeatedly. Polypropylene oval inserts that were cast into the top of the wall allowed the lighting to be integrated elegantly into the cross-section.

The admixture of white cement resulted in a very bright and smooth exposed concrete surface. To prevent the formation of ugly rust streaks, galvanized reinforcing bars were used in the above-grade walls.

In plan, the concreting work was carried out in ten stages, whereby each stage was individually prestressed and then subsequently coupled to the following section. The prestressing is terminated by an end loop on the inside of the pot, where the tensioning cables are "tied together" like shoelaces, or rather tightened against one another.

Prada Levanella, Warehouse and Distribution Center, Montevarchi, Arezzo, Italy, 2004

We were able to work together with Herzog & de Meuron on the preliminary design for the "Polo logistico PRADA" project in Levanella – an innovative roof structure for an exhibition showroom. The budget for the proposed shell structure had been established based on a prefab industrial shed, so keeping to the budget required an efficient structural system with repetitive construction sequences. The structural system is extremely economical with respect to material consumption. It has a structurally effective height of roughly four meters and a span length of fourteen meters, which yields a thin shell that uses little steel.

But shell structures require stabilization – the horizontal forces must be anchored to the system's edges – which makes a decisive contribution to the total costs. A system of pipes stabilizes the columns at a height of about ten-and-a-half meters. It is capable of acting in tension and in compression and is supported by A-shaped columns along the building's edges. This results in a system that is stable in itself.

Stretched atop this system of struts is a cloth-like, doubly curved membrane composed of many individual concrete elements in tension. Combined with the relatively large shear stiffness of concrete structures, the elements' properties yield a high degree of rigidity. The prestressed concrete shells are designed and constructed to correspond to the path of forces from the dead and imposed loads. As a result, primarily just axial forces arise in the shells – especially in the middle section of the building. Thus, a single layer of reinforcement and a shell thickness of eighty millimeters are sufficient here. Along the edges, however, where the concrete shells are also subjected to considerable bending, there are two layers of reinforcement and the shells have a thickness of one hundred and twenty millimeters. Both reinforcement conditions provide the thirty-millimeter concrete cover needed to satisfy fire resistance requirements.

Suspended structures generally demonstrate a very "forgiving," tolerant behavior. Overloading leads to greater deformations and, as a result of the increasing static height, also to increased resistance – so in the case of overloading, the structures seek to return to a state of internal equilibrium by themselves, therefore making them exceedingly robust. In addition, because of the pure tensile loading, there are also no stability problems such as bulging and buckling, even with thin cross-sections.

2

2 Exhibition stand for Hermes at BaselWorld 2013
3 Node details of Hermes exhibition stand

3

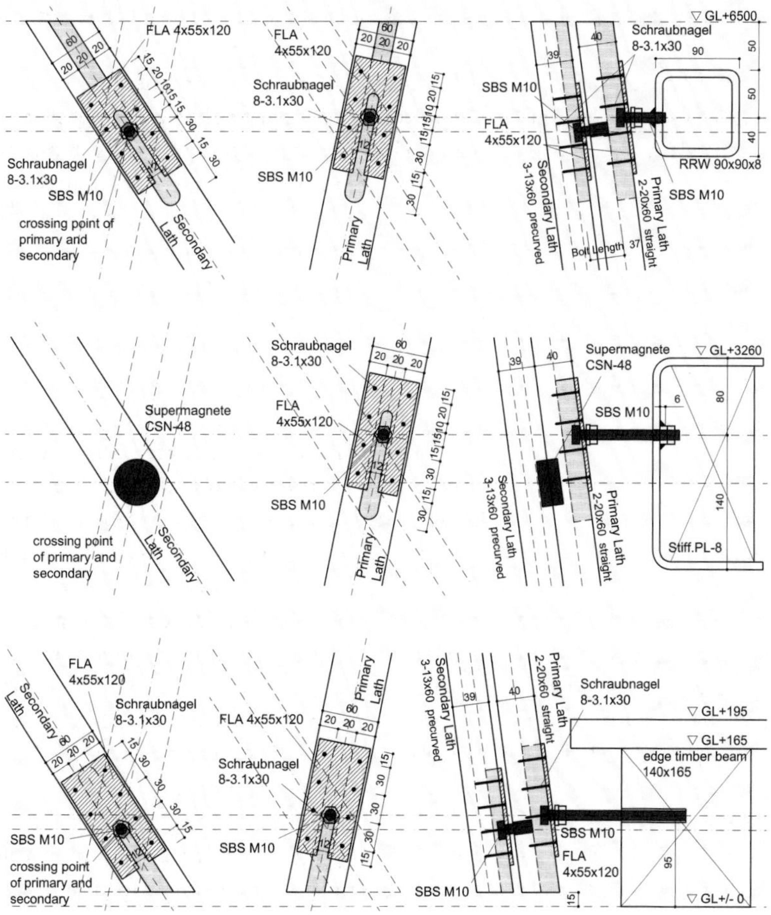

Prada Levanella, Warehouse and Distribution Center
Montevarchi, Arezzo, Italy, 2004

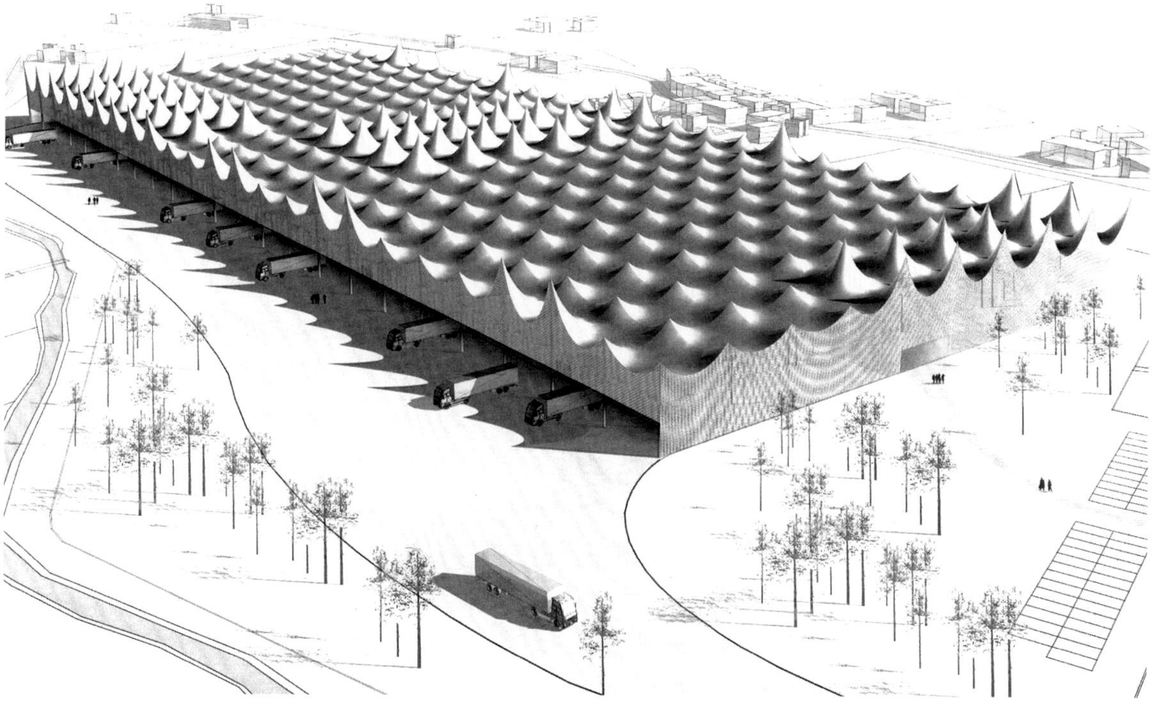

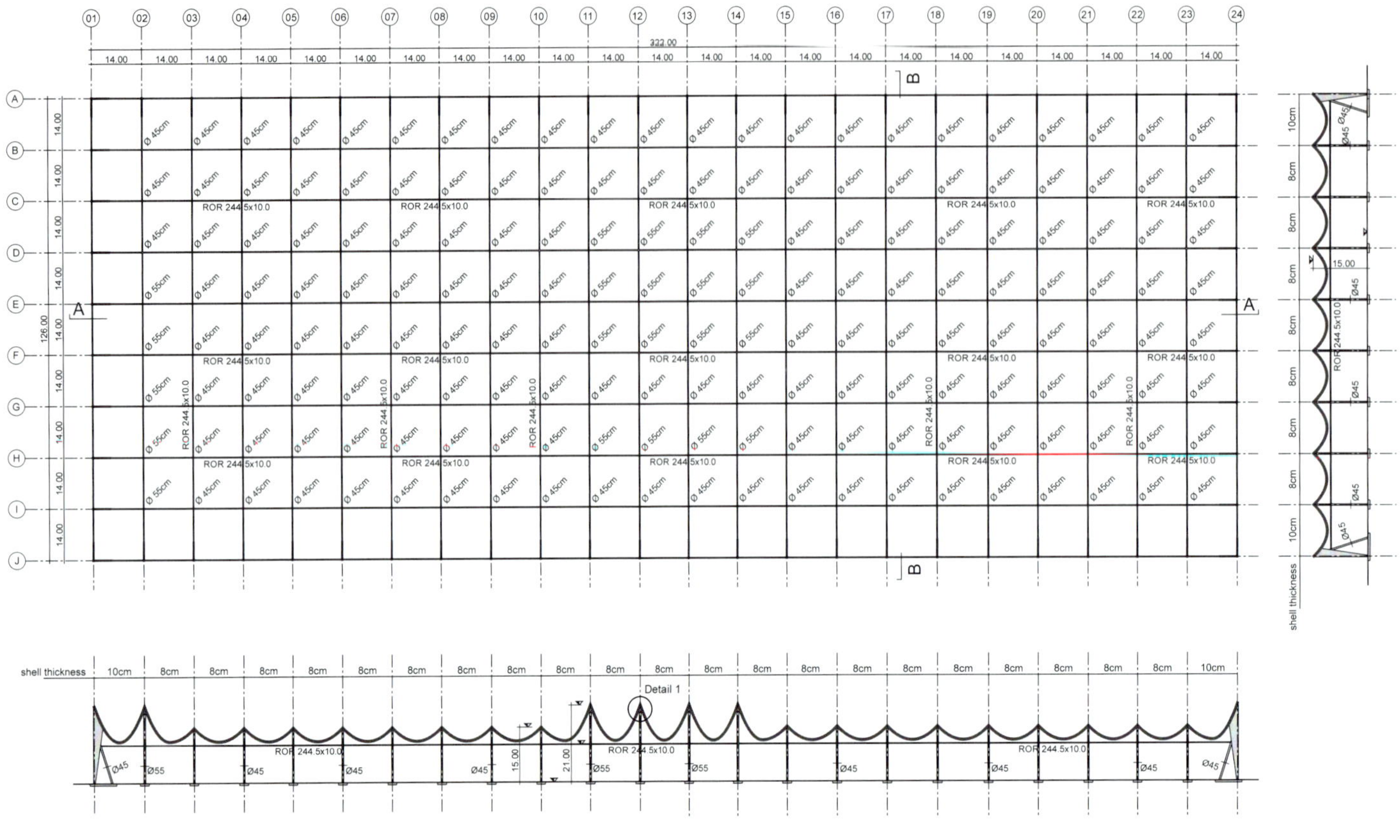

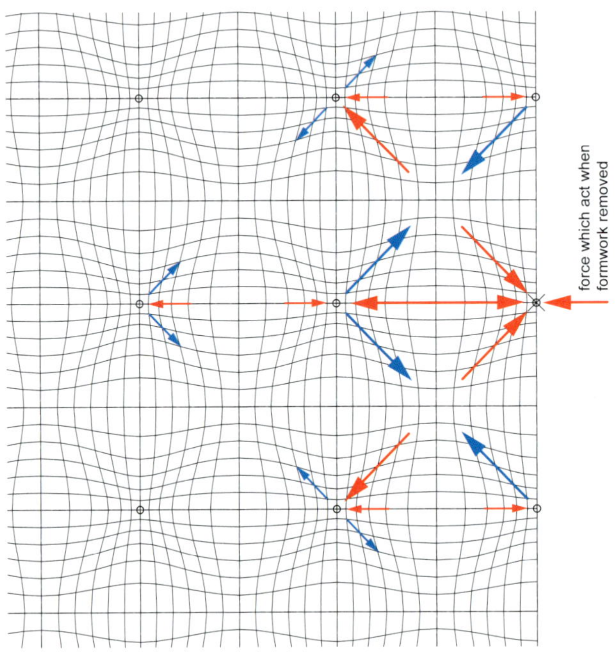

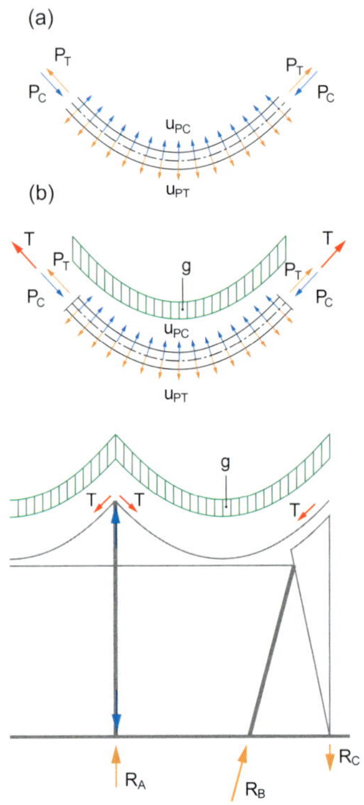

← Overall project (visualization)
↑ Equilibrium studies
→ Structural design of the concrete membrane
→→ Local equilibrium of forces in the tensile shell

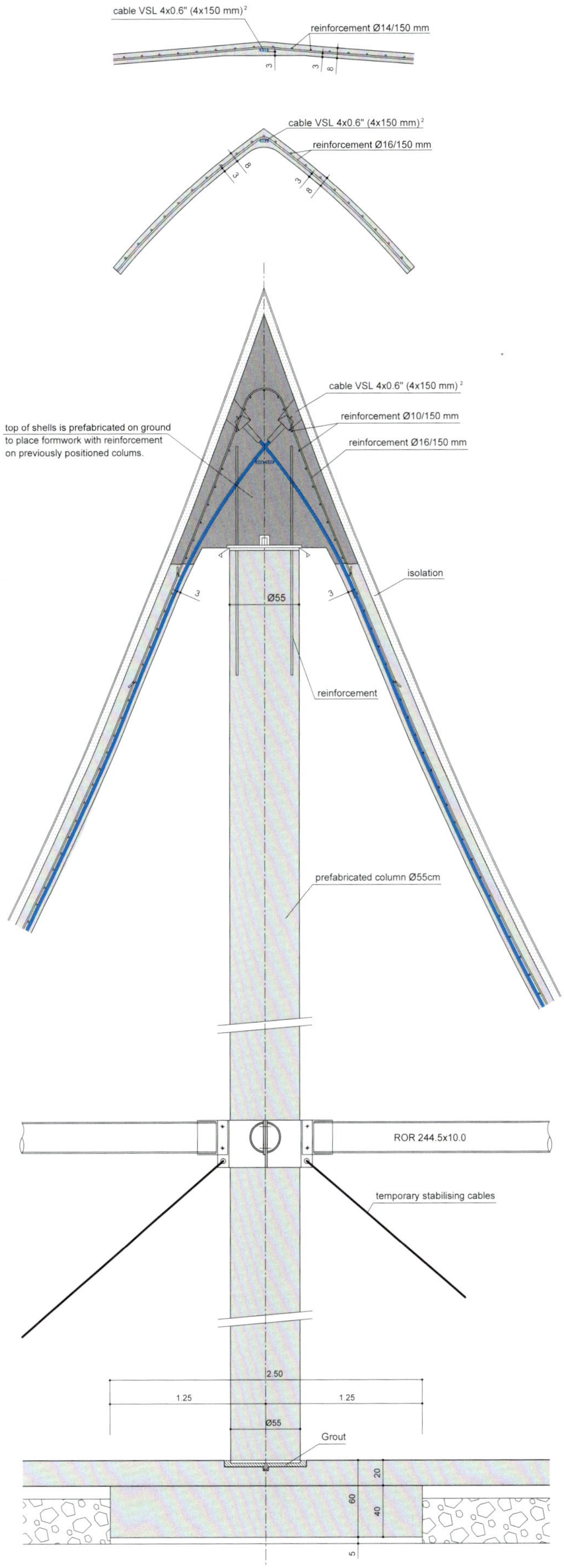

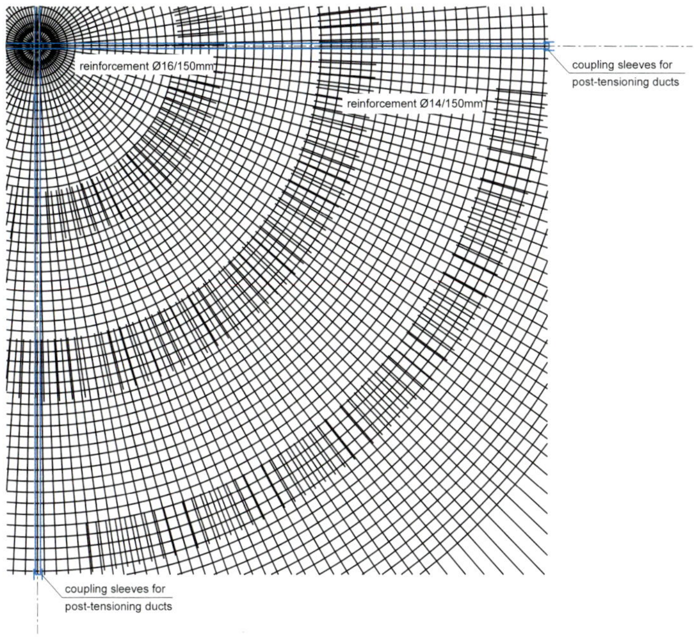

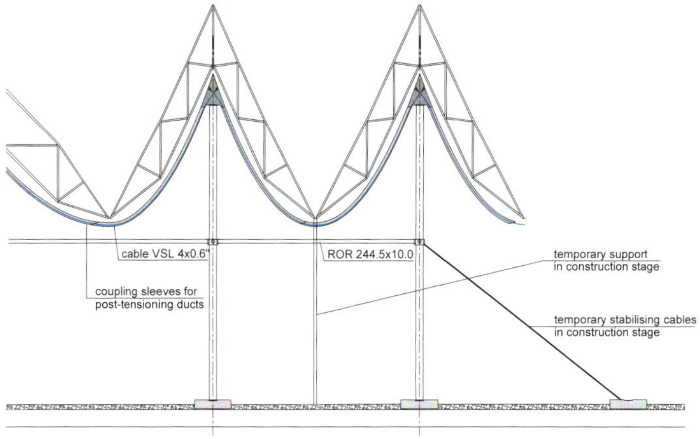

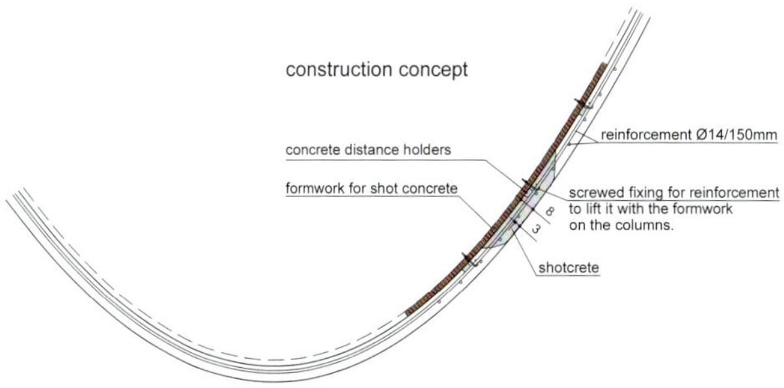

↑↑ Reinforcement pattern of the concrete shell
↑ Temporary structures during construction stage
← Construction detail at column

“Tree Pot” Leutschenpark Zurich 2008

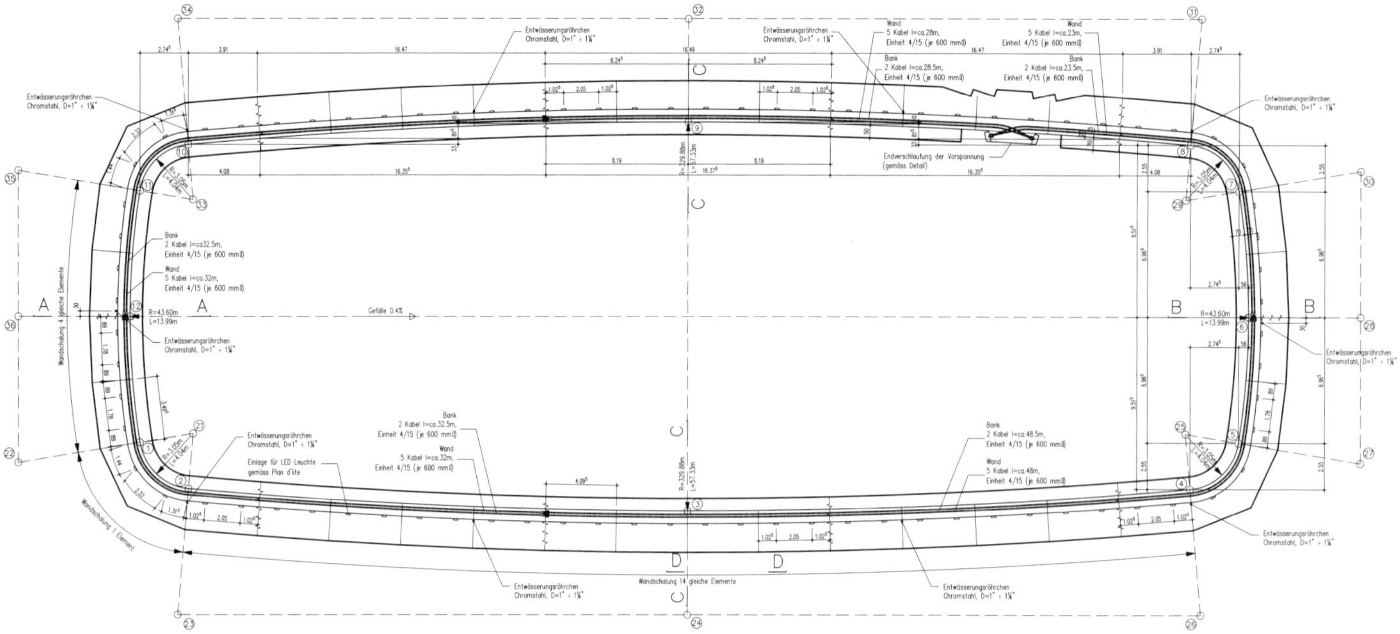

Entwässerungsröhrchen
Chromstahl, D=1" ÷ 1¼"
Gefälle 0.4%
Endverschlaufung der Vorspannung (gemäss Detail)
Einlage für LED Leuchte gemäss Plan d'ite
Wandschalung 14 gleiche Elemente

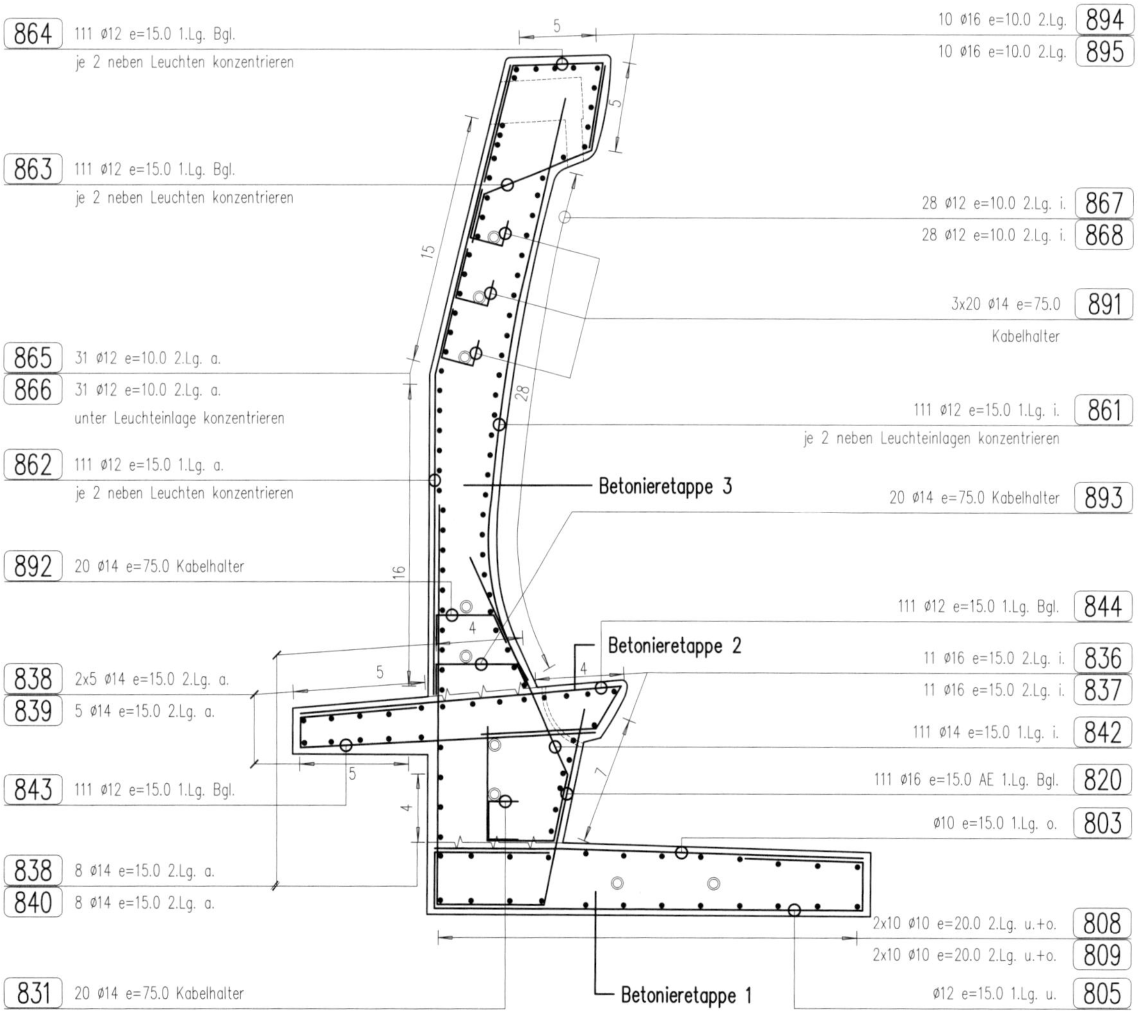

864 111 ø12 e=15.0 1.Lg. Bgl.
je 2 neben Leuchten konzentrieren
863 111 ø12 e=15.0 1.Lg. Bgl.
je 2 neben Leuchten konzentrieren
865 31 ø12 e=10.0 2.Lg. a.
866 31 ø12 e=10.0 2.Lg. a.
unter Leuchteinlage konzentrieren
862 111 ø12 e=15.0 1.Lg. a.
je 2 neben Leuchten konzentrieren
892 20 ø14 e=75.0 Kabelhalter
838 2x5 ø14 e=15.0 2.Lg. a.
839 5 ø14 e=15.0 2.Lg. a.
843 111 ø12 e=15.0 1.Lg. Bgl.
838 8 ø14 e=15.0 2.Lg. a.
840 8 ø14 e=15.0 2.Lg. a.
831 20 ø14 e=75.0 Kabelhalter
10 ø16 e=10.0 2.Lg. 894
10 ø16 e=10.0 2.Lg. 895
28 ø12 e=10.0 2.Lg. i. 867
28 ø12 e=10.0 2.Lg. i. 868
3x20 ø14 e=75.0 891
Kabelhalter
111 ø12 e=15.0 1.Lg. i. 861
je 2 neben Leuchteinlagen konzentrieren
Betonieretappe 3
20 ø14 e=75.0 Kabelhalter 893
111 ø12 e=15.0 1.Lg. Bgl. 844
Betonieretappe 2
11 ø16 e=15.0 2.Lg. i. 836
11 ø16 e=15.0 2.Lg. i. 837
111 ø14 e=15.0 1.Lg. i. 842
111 ø16 e=15.0 AE 1.Lg. Bgl. 820
ø10 e=15.0 1.Lg. o. 803
2x10 ø10 e=20.0 2.Lg. u.+o. 808
2x10 ø10 e=20.0 2.Lg. u.+o. 809
Betonieretappe 1
ø12 e=15.0 1.Lg. u. 805

←← End face of a concreting stage, with extended reinforcement and anchors for the post-tensioning cables
↖ Formwork drawing, plan
← Reinforcement drawing, cross-section
↑ Prepared formwork elements
→ Light fixtures are inset into the top of the concrete construction.

Balloon for the St. Prex Classics
St. Prex, Switzerland, 2012

←← Assembly of the column base anchorage
← Geometric definition
↙ Cross-section detail of inner ring and stage ring
↓ Detail of column base anchorage, cross-section (above), plan (below)
→ Visualization with rising balloon

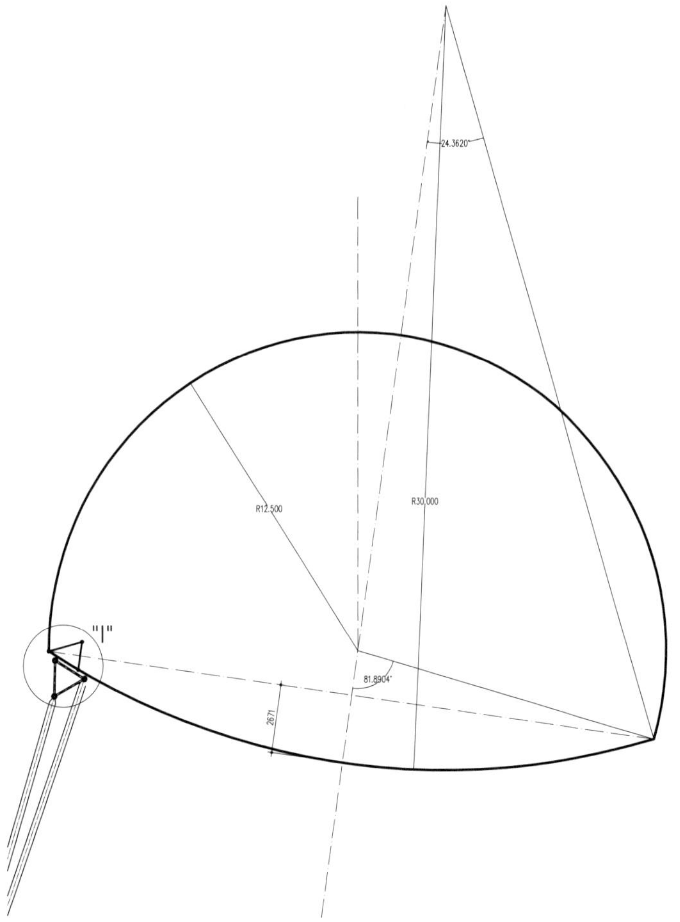

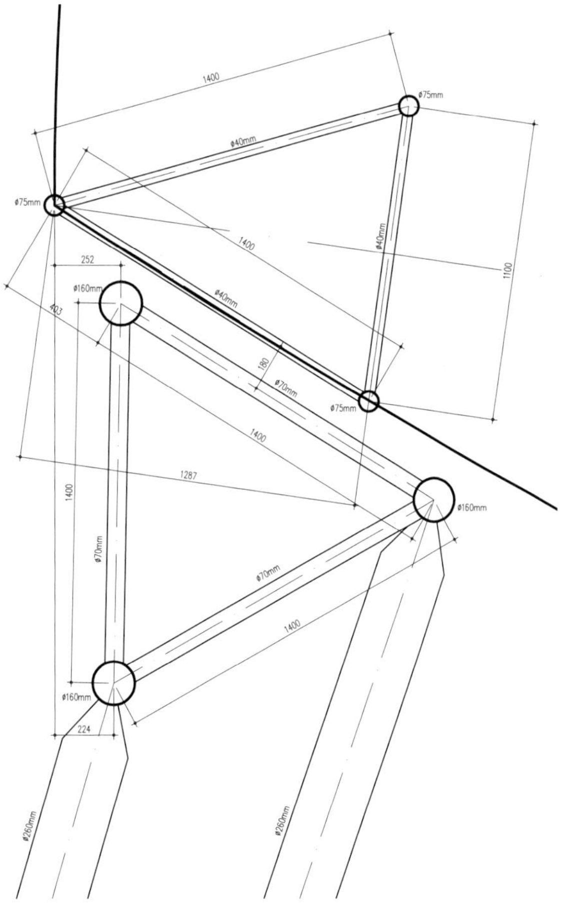

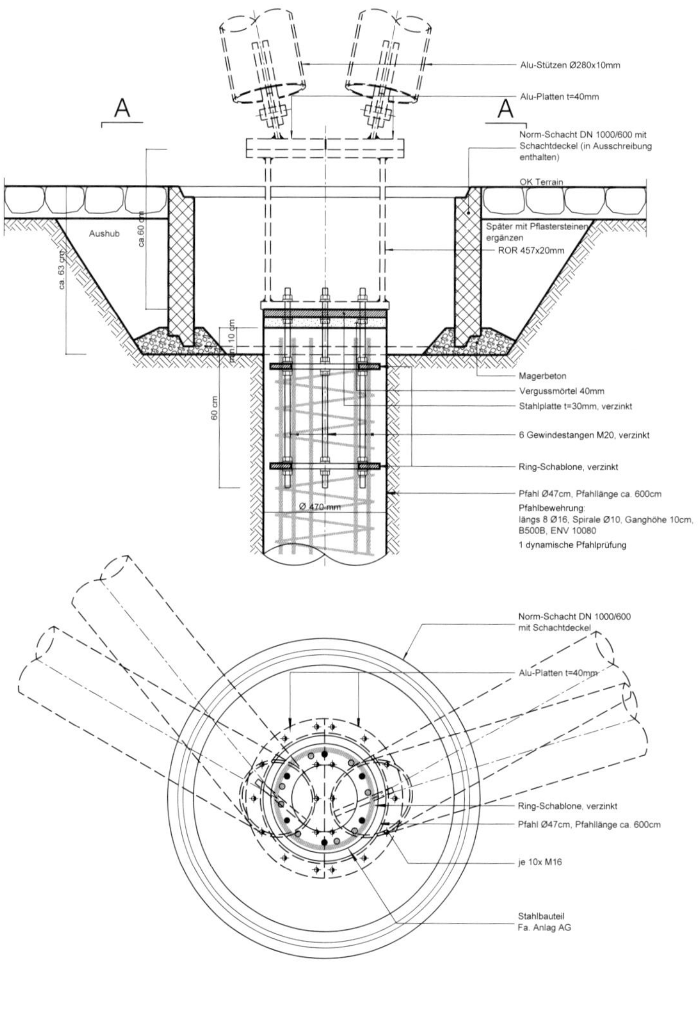

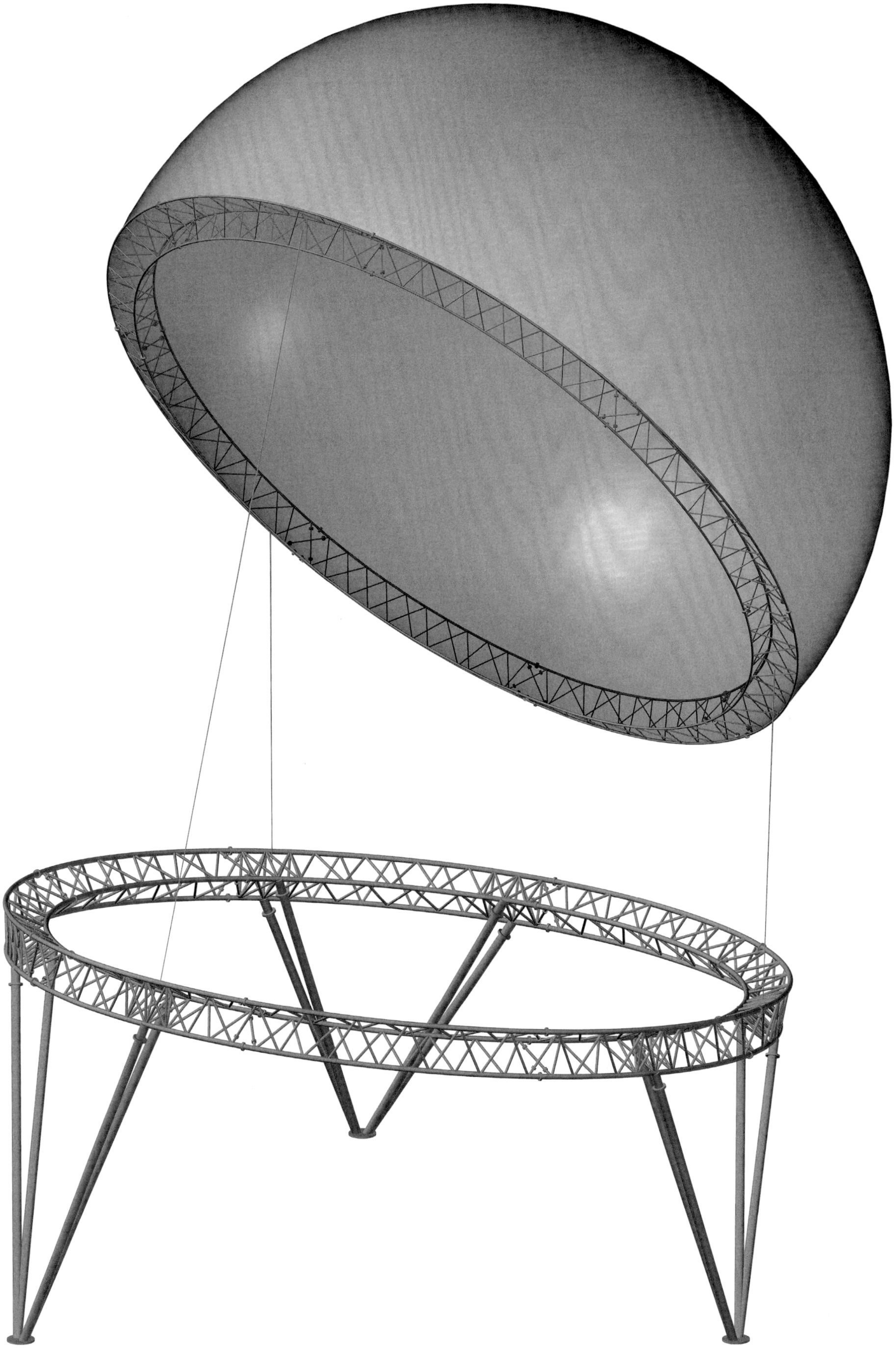

↑ Finished interior space
→ Impression of the luminescent balloon at night

List of Works

Second SBB Rhine Bridge Basel

Project no. 0938a (competition)
Engineering consortium WGG Schnetzer Puskas Ingenieure AG, Basel / CSD Ingenieure AG, Berne
Realization 2009–2013
Structural concept Heinrich Schnetzer, Prof. Dr. C. Menn
Project management Heinrich Schnetzer, Rémy Jabas (bridge), Mark Eitel (pedestrian footbridge), Thomas Gutzwiller (construction management)
Architecture Lorenz & Musso Architekten, Sion
Client SBB AG, Olten
Contractors Frutiger AG / Walo Bertschinger AG

Puna Tsang Chhu Bridge Wangdue Phodrang (BT)

Project no. 997 (competition)
Realization 2000–2002
Structural concept Heinrich Schnetzer
Project management Heinrich Schnetzer
Client Royal Government of Bhutan (BT); Helvetas Swiss Intercooperation, Thimphu (BT)

Renovation of Kunsthalle Basel

Project no. 0998u02
Realization 2003/04
Structural concept Tivadar Puskas
Project management Patrick Grieder
Architecture Miller & Maranta dipl. Architekten ETH BSA SIA, Basel
Client Basler Kunstverein Kunsthalle, Basel
Contractor Glanzmann AG, Basel

Singeisenhof Development Riehen (CH)

Project no. 1010 (competition)
Realization 1999–2001
Structural concept Tivadar Puskas
Project management Tivadar Puskas
Architecture Morger & Degelo Architekten AG, Basel
Client Municipality of Riehen
Contractor Marti Bauunternehmung AG, Basel

Aegerten Tunnel (Bahn 2000) Mattstetten–Rothrist (CH)

Project no. 1029
Realization 2000–2002
Engineering consortium Rapp AG, Basel / WGG Schnetzer Puskas Ingenieure AG, Basel
Structural concept Tivadar Puskas
Project management Stefan Bänziger
Client SBB AG, Berne
Contractors Frutiger AG / Marti AG / König Bauunternehmung AG

St. Jakob Park Stadium, Basel. Football Stadium, Commercial Center and Senior Citizens' Residential Home Basel

Project no. 1034
Realization 1999–2001
Engineering consortium Rothpletz Lienhard + Cie AG, Berne / Elektrowatt Infra AG, Zurich / WGG Schnetzer Puskas Ingenieure AG, Basel
Structural concept Walter Wiedmer (Rothpletz Lienhard + Cie AG, Berne), Willy Wüthrich (Elektrowatt Infra AG, Zurich) (stadium roof), Tivadar Puskas (structural engineering)
Architecture Herzog & de Meuron Architekten AG, Basel
Client Winterthur Lebensversicherung, Winterthur; Suva Schweiz, Lucerne; Basel government employees' pension fund, Basel
General contractor Marazzi Generalunternehmung AG, Berne

Extension of St. Jakob Park Stadium, Basel Basel

Project no. 1034z02
Realization 2005–2008
Structural concept Tivadar Puskas, Stefan Bänziger
Project management Jan Stebler
Engineering subcontractor: Rothpletz, Lienhard + Cie AG, Berne
Architecture Herzog & de Meuron Architekten AG, Basel
Client Genossenschaft Fussballstadion St. Jakob-Park, Basel
General contractor/contractor Batigroup AG, Basel

Images by project number: 0938a: Foto-Werk, Michael Fritschi, Basel; 0997: Hans Pfaffen, Chur; 0998u02: Ruedi Walti, Basel; 1010: Ruedi Walti, Basel; 1029: unknown; 1034: Ruedi Walti, Basel; 1034z02: Juri Weiss, Basel

Messeturm (Trade Fair Tower) Basel

Project no. 1058 (competition)
Realization 2001–2003
Structural concept Tivadar Puskas
Project management Tivadar Puskas, Reto von Salis, Stefan Bänziger
Architecture Morger & Degelo Architekten AG, Basel; Marques AG, Lucerne
Client Swiss Prime Site AG, Olten
General contractor Batigroup Generalunternehmung AG, Basel
Steel construction Josef Meyer Stahl und Metall AG, Emmen
Contractor Batigroup AG, Basel

School Building Extension in Brühl Erschwil (CH)

Project no. 1063
Realization 2000
Structural concept Tivadar Puskas
Project management Patrick Grieder
Architecture Jürg Stäuble, Solothurn; Christoph Wild Architekten BSA ETH SIA, Berne
Client Einwohnergemeinde Erschwil
Contractor Meier + Jäggi AG, Basel

Basel International School Reinach (CH)

Project no. 1073
Realization 2001
Structural concept Tivadar Puskas
Project management Stefan von Ah
Architecture Burckhardt & Partner AG, Basel
Client International School Basel, Reinach
General contractor HRS Hauser Rutishauser Suter AG, Basel

Refurbishment of the Kollegienhaus Building at the University of Basel Basel

Project no. 1086
Realization 2001–2003
Structural concept Stefan Bänziger
Project management Stefan Bänziger
Architecture Fiertz Architekten AG, Basel
Client Municipal Building and Planning Department, Basel-Stadt
Contractor Meier + Jäggi AG, Basel

School for Guide Dogs for the Blind Allschwil (CH)

Project no. 1092
Realization 2000–2003
Structural concept Heinrich Schnetzer
Project management Heinrich Schnetzer, Patrick Grieder
Architecture Stump & Schibli Architekten BSA, Basel
Client Stiftung Schweizerische Schule für Blindenführhunde, Allschwil
Contractor Gebrüder Heid AG, Lausen

Residential Development, Reinach Community Center Reinach (CH)

Project no. 1095 (competition)
Realization 2000–2002
Structural concept Tivadar Puskas
Project management Stefan von Ah
Architecture Morger & Degelo Architekten AG, Basel
Client Municipality of Reinach; Pensimo Management AG, Zurich
Contractor Meier + Jäggi AG, Basel

Prada Aoyama Tokyo (JP)

Project no. 1101
Realization 2001–2003
Structural concept Heinrich Schnetzer
Project management Heinrich Schnetzer
Architecture Herzog & de Meuron Architekten AG, Basel
Client Prada Japan Co. Ltd., Minami-Aoyama (JP)
General contractor Takenaka Corporation, Japan
Steel construction Kawada Industries, Japan

Apartment House, Giornicostrasse 223 Basel

Project no. 1106
Realization 1994–1996
Structural concept Tivadar Puskas
Project management Tivadar Puskas
Architecture Trinkler Engler Ferrara Architekten, Basel
Client private
Contractor Meier + Jäggi AG, Basel

1058: Ruedi Walti, Basel; 1063: Jürg Stäuble, Solothurn; 1073: unknown; 1086: Tom Bisig, Basel; 1092: Ruedi Walti, Basel; 1095: unknown; 1101: Herzog & de Meuron, Basel; 1106: unknown

St. Jakob Indoor Ice Rink Basel

Project no. 1108
Realization 2001/02
Structural concept Heinrich Schnetzer
Project management Stefan von Ah
Architecture Zwimpfer Partner / Krarup Furrer Architekten SIA AG, Basel / Berrel Architekten SIA BSA, Basel
Client Genossenschaft St. Jakob-Arena, Basel
General contractor Huber Straub AG, Basel + Spaini Bau AG, Basel

Apartment Building, Engelgasse Basel

Project no. 1122
Realization 2001/02
Structural concept Tivadar Puskas
Project management Patrick Grieder
Architecture B. Trinkler + H. R. Engler Dipl. Architekten ETH HTL SIA, Basel
Client Baukonsortium Engelgasse, Basel
Contractor Marti AG Bauunternehmung, Basel

Ameropa Apartment / Office Building Binningen (CH)

Project no. 1130
Realization 2001/02
Structural concept Heinrich Schnetzer
Project management Stefan von Ah
Architecture Herzog & de Meuron Architekten AG, Basel
Client Ameropa AG, Binningen
Contractor Huber & Straub AG, Basel

Extension to Göttibach School Building Thun (CH)

Project no. 1131
Realization 2000/01
Structural concept Stefan Bänziger
Project management Stefan Bänziger
Architecture Bauart Architekten und Planer AG, Berne
Client Building Department of the City of Thun
Contractor A. Soltermann AG Bauunternehmung, Unterkulm

Extension to Dorfmatt Secondary School Oberdorf (CH)

Project no. 1134 (competition)
Realization 2007–2009
Engineering consortium WGG Schnetzer Puskas Ingenieure AG, Basel / Sutter Ingenieur- und Planungsbüro AG, Arboldswil
Structural concept Heinrich Schnetzer
Project management Patrick Grieder
Architecture Werk 1 Architekten und Planer AG, Olten
Client Municipality of Oberdorf
Contractors Implenia AG / Thommen + Pergher AG

Emme Bridge Derendingen-Inkwil (CH)

Project no. 1138
Realization 2001/02
Engineering consortium WGG Schnetzer Puskas Ingenieure AG / BSB + Partner
Structural concept Heinrich Schnetzer
Project management Heinrich Schnetzer, Stefan von Ah
Client SBB AG, Berne
General contractor consortium Züblin Schlittler Spaltenstein Bau AG

Glass Pavilion, Sommerhaus de Vigier Solothurn (CH)

Project no. 1140
Realization 2000/01
Structural concept Tivadar Puskas
Project management Tivadar Puskas
Architecture Ern + Heinzl Gesellschaft von Architekten mbH, Solothurn
Client Bill de Vigier Foundation, Solothurn
Contractor Kurth Glas + Spiegel AG, Zuchwil

Forum 2004 Building and Plaza Barcelona (ES)

Project no. 1145 (competition)
Realization 2001–2004
Structural concept Heinrich Schnetzer
Project management Heinrich Schnetzer, Tivadar Puskas, Patrick Grieder, Mark Eitel
Local engineers Robert Brufau, Barcelona (ES)
Architecture Herzog & de Meuron Architekten AG, Basel
Client Ayuntamiento de Barcelona (ES)
Contractor Dragados S.A., Barcelona (ES) + Ferrovial S.A., Madrid (ES)

1108: Ruedi Walti, Basel; 1122: Maria Gambino, Basel; 1130: Ameropa, Binningen; 1131: Bauart Arch., Berne; 1134: René Rötheli, Baden; 1138: Foto Grossen, Biberist; 1140: unknown; 1145: José Miguel Hernandez, Spain

Dreispitz S-Bahn Station Basel

Project no. 1147a
Realization 2005/06
Structural concept Tivadar Puskas, Stefan Bänziger
Project management Stefan Bänziger, Christoph Raz (construction management)
Architecture Steinmann & Schmid Architekten AG, Basel
Client SBB AG, Olten; Canton of Basel-Land; Canton of Basel-Stadt
Contractor Batigroup AG Bau, Basel

Weihermatten Apartment Complex Binningen (CH)

Project no. 1153
Realization 2006/07
Structural concept Tivadar Puskas
Project management Ralf Zappe, Mark Eitel
Architecture Diener & Diener Architekten AG, Basel
Client/General contractor Glanzmann Service AG, Binningen

Extension to Gotthelf School Building Thun (CH)

Project no. 1171
Realization 2002
Structural concept Stefan Bänziger
Project management Stefan Bänziger
Architecture Bauart Architekten und Planer AG, Berne
Client Einwohnergemeinde Thun
Timber construction Wenger Holzbau AG, Steffisburg

Semi-detached House on Bischoffhöhe Riehen (CH)

Project no. 1179
Realization 2003/04
Structural concept Tivadar Puskas
Project management Mark Eitel
Architecture Morger & Degelo Architekten AG, Basel
Client private
Contractor Wenk AG, Basel

Semi-detached House on Bischoffweg Riehen (CH)

Project no. 1180
Realization 2003/04
Structural concept Tivadar Puskas
Project management Tivadar Puskas
Architecture Morger & Degelo Architekten AG, Basel
Client private
Contractor Glanzmann AG, Basel

CaixaForum Madrid Madrid (ES)

Project no. 1181
Realization 2003–2008
Structural concept Heinrich Schnetzer
Project management Heinrich Schnetzer, Manuel Alvarez
Engineering subcontractor: NB35 S. L. Ingenieros, Madrid (ES)
Architecture Herzog & de Meuron Architekten AG, Basel
Client Obra Social "laCaixa", Madrid (ES); Caixa d'Estalvis i Pensions de Barcelona, Barcelona (ES)
Contractor Ferrovial S.A., Madrid (ES)

Rheinpark Residential Development Birsfelden (CH)

Project no. 1185 (competition)
Realization 2002–2004
Structural concept Tivadar Puskas
Project management Patrick Grieder
Architecture Morger & Degelo Architekten AG, Basel
Client Helvetia Patria Versicherungen, Basel
Contractor Batigroup AG, Basel
General contractor Mobag AG, Basel

Redevelopment of Visp Railway Station Visp (CH)

Project no. 1187 (competition)
Realization 2004–2007
Engineering consortium Teysseire & Candolfi Ingenieure AG, Visp / WGG Schnetzer Puskas Ingenieure AG, Basel / BIAG Beratende Ingenieure AG, Visp
Structural concept Tivadar Puskas
Project management Nicola Benzoni, Judith Luible
Architecture Steinmann & Schmid Architekten AG, Basel
Client Gemeinde Visp SBB AG, Berne; Matterhorn Gotthard Bahn, Brig
Contractor Frutiger AG, GU, Thun; U. Imboden AG, Visp

1147a: Ruedi Walti, Basel; 1153: Ruedi Walti, Basel; 1171: Croci&du Fresne Photographie, Worblaufen; 1179: Ruedi Walti, Basel; 1180: Ruedi Walti, Basel; 1185: Morger&Degelo, Basel; 1187: Ruedi Walti, Basel

Renovation of SBB Bridge and Construction of New Waldenburgerbahn Bridge Liestal (CH)

Project no. 1195
Realization 2005/06
Engineering consortium WGG Schnetzer Puskas Ingenieure AG, Basel / Pfirter, Nyfeler + Partner AG, Basel
Structural concept Heinrich Schnetzer, Rudolf Gisi
Project management Rudolf Gisi
Client SBB AG, Olten
Contractor Meier + Jäggi AG, Zofingen

Renovation and New Build of Office / Apartment Building at Bäumleingasse 14 Basel

Project no. 1206
Realization 2005
Structural concept Tivadar Puskas
Project management Rolf Nachbur
Architecture Diener & Diener Architekten AG, Basel
Client Edwin Faeh, Basel
Contractor Knecht Bauunternehmung AG, Münchenstein

Extension and Renovation of Blumenrain Residential / Care Home for the Elderly Therwil (CH)

Project no. 1212 (competition)
Realization 2004–2007
Structural concept Tivadar Puskas
Project management Nicola Benzoni, Rolf Nachbur
Architecture Flubacher, Nyfeler + Partner Architekten AG, Basel
Client Stiftung Alterssiedlung Blumenrain, Therwil
General contractor Gribi Theurillat AG, Basel

Refurbishment of Rhine Promenade, Mittlere Brücke Basel

Project no. 1214 (competition)
Realization 2003/04
Structural concept Heinrich Schnetzer
Project management Patrick Grieder
Architecture Nussbaumer Trüssel Architekten und Gestalter, Basel
Client Building and Planning Department, Basel-Stadt
Contractor Engeli + Berger AG Bauunternehmung, Riehen

Renovation and New Build of Office Building at Spitalstrasse 8/12 Basel

Project no. 1217
Realization 2003/04
Structural concept Tivadar Puskas
Project management Tivadar Puskas, Stefan Bänziger
Architecture Morger & Degelo Architekten AG, Basel
Client Senn/Co. AG Bandfabrikation, Basel
General contractor Gribi Theurillat AG, Basel

Auenpark Residential Development Rohr (CH)

Project no. 1218
Realization 2002–2004
Structural concept Tivadar Puskas
Project management Patrick Grieder
Architecture Frei Architekten AG, Olten
Client Credit Suisse Asset Management, Zurich
General contractor Zschokke Generalunternehmung AG, Aarau

"Guido A. Zäch Institute," Paraplegic Center Nottwil (CH)

Project no. 1219
Realization 2003–2005
Structural concept Heinrich Schnetzer
Project management Stefan von Ah
Architecture Wilfrid & Katharina Steib, dipl. Architekten ETH BSA/SIA, Basel
Client Swiss Paraplegic Foundation, Nottwil
Contractor Aregger AG, Buttisholz

Refurbishment of St. Jakobshalle Basel

Project no. 1222
Realization 2006/07
Structural concept René Guillod
Project management Rolf Nachbur
Architecture Berrel Architekten BSA SIA SWB, Basel
Client Building and Planning Department, Basel-Stadt
Contractor Huber Straub AG, Basel

1206: Christian Richters, Berlin (DE); 1212: Tom Bisig, Basel; 1214: Ruedi Walti, Basel; 1217: Ruedi Walti, Basel; 1219: Walter Eggenberger, Schweizer Paraplegiker-Stiftung, Nottwil; 1222: Atelier Fontana, Basel

Detached House, Sonnenweg Otelfingen (CH)

Project no. 1236
Realization 2003/04
Structural concept Heinrich Schnetzer
Project management Mark Eitel
Architecture Rüedi & Pfluger Architekten, Basel
Client private
Contractor K. Eicher Bauunternehmung AG, Regensdorf

Wiesenkreisel Art Object Basel

Project no. 1237 (competition)
Realization 2004/05
Structural concept Tivadar Puskas
Project management Mark Eitel, Christoph Raz (construction management)
Architecture Harald König dipl. Architekt ETH, Zurich; Lucia Schnüriger (artist), Zurich
Client Building Department, Basel-Stadt
Steel construction Jakem AG Stahl- und Fassadenbau, Münchwilen

Lido Ice Skating Arena Rapperswil-Jona (CH)

Project no. 1239
Realization 2005/06
Engineering consortium WGG Schnetzer Puskas Ingenieure AG, Basel / Walter Böhler AG, Rapperswil-Jona
Structural concept Heinrich Schnetzer
Project management Mark Eitel, Stefan Bänziger
Architecture Zwimpfer Partner Architekten SIA, Basel
Client Municipality of Rapperswil; Municipality of Jona
General contractor Unirenova AG, Zurich

Elbphilharmonie Hamburg Hamburg (DE)

Project no. 1244
Realization 2007–2017
Structural concept Heinrich Schnetzer
Project management Heinrich Schnetzer, Mark Eitel, Kevin M. Rahner
Engineering subcontractor Rohwer Ingenieure VBI GmbH, Handewitt (Jarplund) (DE)
Architecture Herzog & de Meuron Architekten AG, Basel
Client ReGe Hamburg Projekt-Realizationsgesellschaft mbH, Hamburg (DE)
General contractor HOCHTIEF Construction AG, Hamburg (DE)

Expo 2005 – Swiss Pavilion at AICHI Aichi (JP)

Project no. 1248
Realization 2004/05
Structural concept Tivadar Puskas
Project management Manuel Alvarez
Architecture sab Architekten, Basel
Design Panorama 2000 group, Basel
Client Swiss Presence, Federal Department of Foreign Affairs (EDA)
General contractor Nüssli (Schweiz) AG, Hüttwilen

St. Jakob Tower Basel

Project no. 1251
Realization 2006–2008
Structural concept Tivadar Puskas
Project management Salomé Hug-Meier, Jan Stebler, Ralf Zappe
Architecture Herzog & de Meuron Architekten AG, Basel
Client UBS Fund Management, Basel; Kestenholz Basel AG, Basel; Musfeld AG, Basel
General contractor Implenia Generalunternehmung AG, Basel

Detached House Neuwiller (CH)

Project no. 1257
Realization 2004/05
Structural concept Tivadar Puskas
Project management Nicola Benzoni
Architecture sab Architekten AG, Basel
Client private
Contractor Bleyer Constructions, Gildwiller (FR)

Martinpark Kindergarten Thun (CH)

Project no. 1266
Realization 2003/04
Structural concept Tivadar Puskas
Project management Tivadar Puskas
Architecture Bauart Architekten AG, Berne
Client Municipal Properties Department of the City of Thun
Timber construction Wenger Holzbau + Schreinerei AG, Steffisburg

1237: Sarah Kreuter, Zurich; 1239: Johannes Marburg, Geneva; 1244: Foto-Werk, Michael Fritschi, Basel; 1248: Tobias Madörin, Basel; 1251: Ruedi Walti, Basel; 1257: Atelier Fontana, Basel

Holbeinpark Development Basel

Project no. 1269
Realization 2003–2006
Structural concept Tivadar Puskas
Project management Rolf Nachbur
Architecture Fischer Art AG Architekten, Basel
Client Schweizerische National Leben AG, Bottmingen
General contractor Batigroup AG Generalunternehmung, Basel

Perron 1 Office / Retail / Apartment Building Solothurn (CH)

Project no. 1278 (competition)
Realization 2008–2010
Structural concept Tivadar Puskas
Project management Jan Stebler
Architecture Morger + Dettli Architekten AG, Basel
Client Credit Suisse Asset Management, Zurich
General contractor Frutiger AG, Thun

New Monte Rosa Hut Zermatt (CH)

Project no. 1284
Realization 2008/09
Structural concept Tivadar Puskas
Project management Jan Stebler
Architecture ETH Studio Monte Rosa, Zurich; Bearth Deplazes Architekten AG, Chur
Client Swiss Alpine Club (SAC), Berne
Contractor Holzbau AG, Mörel
Timber construction engineers Holzbaubüro Reusser GmbH, Winterthur

Freienhof – Change of Use Basel

Project no. 1295
Realization 2004/05
Structural concept Tivadar Puskas
Project management Patrick Grieder
Architecture Zwimpfer Partner Architekten AG, Basel / Jeker Blanckarts Architekten SIA, Basel
Client Basler Lebensversicherungs-Gesellschaft, Basel
Contractor BBG Basler Baugesellschaft AG, Basel

Spirgarten Senior Citizen's Residential Home Zurich

Project no. 1304 (competition)
Realization 2005/06
Structural concept Tivadar Puskas
Project management Stefan Bänziger
Architecture Miller & Maranta dipl. Architekten ETH BSA SIA, Basel
Client Stiftung Spirgarten Zürich, Zurich
General contractor Mobag AG, Zurich

OASIS, Boldern Conference Center Männedorf (CH)

Project no. 1307 (competition)
Realization 2005/06
Structural concept Tivadar Puskas
Project management Stefan Bänziger
Architecture Vehovar + Jauslin Architektur AG, Zurich
Client Evangelisches Tagungs- und Studienzentrum Boldern, Männedorf
Contractor Männag AG Hoch- und Tiefbau, Männedorf

Renovation of Hotel Les Trois Rois Basel

Project no. 1310
Realization 2004–2006
Structural concept Tivadar Puskas
Project management Patrick Grieder, Leo Lanz (construction management)
Architecture Villa Nova Architekten AG, Basel
Client Hotel Les Trois Rois, Basel
Contractors Wenk AG / Glanzmann AG

Claramatte Park Facelift Basel

Project no. 1311 (competition)
Realization 2005/06
Structural concept Tivadar Puskas
Project management Thomas Gutzwiller (construction management)
Architecture Raderschall Partner AG Landschaftsarchitektur BSLA SIA, Meilen
Client Christoph-Merian-Stiftung CMS, Basel; Building Department of the Canton of Basel-Stadt, Municipal Parks Department, Department of Works, Building and Planning Department
Contractor Glanzmann AG, Basel

1269: unknown; 1278: Ruedi Walti, Basel; 1284: Tonatiuh Ambrosetti, Lausanne; 1295: Lilli Kehl, Basel; 1304: Ruedi Walti, Basel; 1307: Niklaus Spoerri Fotografie, Zurich; 1310: Grand Hotel Les Trois Rois, Basel; 1311: Serge Hasenböhler, Basel

Refurbishment of Trimbach Bridge Olten (CH)

Project no. 1312 (competition)
Realization 2004
Project management Heinrich Schnetzer, Mark Eitel
Client Canton of Solothurn, Department of Transport and Public Works, Solothurn
General contractor consortium Vogt Strassenbau AG (Astrada AG), Olten / Züblin Schlittler Spaltenstein Bau AG / Olten / WGG Schnetzer Puskas Ingenieure AG, Basel

Altmarkt Special Needs School Liestal (CH)

Project no. 1313 (competition)
Realization 2006/07
Structural concept Tivadar Puskas
Project management Ralf Zappe
Architecture sab Architekten AG, Basel
Client Insieme Baselland, Liestal
Contractor Glanzmann AG, Birsfelden

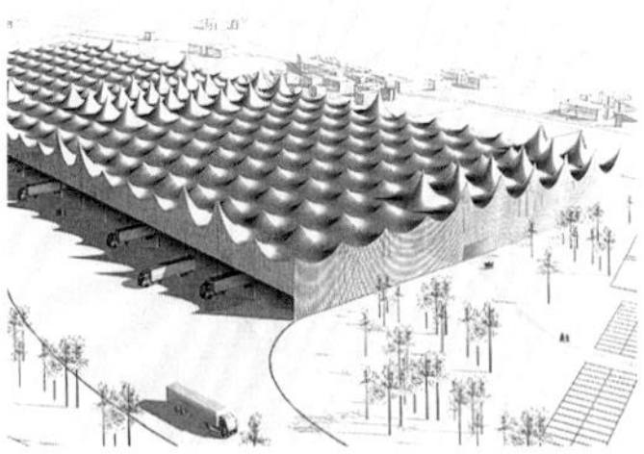

Prada Levanella, Warehouse and Distribution Center Montevarchi, Arezzo (IT)

Project no. 1319 (not built)
Conception 2004
Structural concept Heinrich Schnetzer
Project management Heinrich Schnetzer
Architecture Herzog & de Meuron Architekten, Basel
Client Prada Industrial SpA, San Zeno, Arezzo (IT)

Detached House Maisprach (CH)

Project no. 1325
Realization 2004/05
Structural concept Tivadar Puskas
Project management Salomé Hug-Meier
Architecture Weber Buess Architekten HTL, Basel
Client private
Contractor Hans Graf AG Bauunternehmung, Maisprach

Uno Office Building Liestal (CH)

Project no. 1328
Realization 2009–2011
Structural concept Tivadar Puskas
Project management Daniel Küpfer, Hansueli Küng
Architecture Christ & Gantenbein Architekten ETH SIA BSA, Basel
Client Basellandschaftliche Kantonalbank, Liestal
General contractor S + B Baumanagement AG, Olten

University Children's Hospital of Basel City and Basel Land (UKBB) Basel

Project no. 1335 (competition)
Realization 2007–2010
Structural concept Tivadar Puskas
Project management Patrick Grieder, Christoph Raz (construction management)
Architecture Stump & Schibli Architekten BSA AG, Basel
Client Cantons of Basel-Stadt and Basel-Landschaft
Contractor Erne AG Bauunternehmung, Laufenburg

Jinhua Architectural Park: Structure 1, Cube Jinhua (CN)

Project no. 1337
Realization 2004–2006
Structural concept Tivadar Puskas
Project management Tivadar Puskas
Architecture Herzog & de Meuron Architekten AG, Basel
Client Jindong New District Construction Headquarters of Jinhua City (CN)

Airport Hotel Basel

Project no. 1339 (competition)
Realization 2005–2007
Structural concept Tivadar Puskas
Project management Daniel Küpfer
Architecture Fankhauser Architektur AG, Reinach
Client Euroairport Hotel Basel AG, Basel
Contractors Marti AG / Glanzmann AG

1313: Georg Aerni, Zurich; 1319: Herzog & de Meuron, Basel; 1325: Weber Buess Arch., Basel; 1328: Roman Keller, Zurich; 1335: Juri Weiss, Basel; 1337: Iwan Baan, Amsterdam (NL); 1339: Fankhauser Architektur, Reinach

Im Viadukt Market Hall, Change of Use of Viaduct Arches Zurich

Project no. 1346 (competition)
Realization 2010
Structural concept Stefan Bänziger
Project management Stefan Bänziger, Florian Riebel
Architecture EM2N Architekten AG, Zurich
Client PWG foundation, Zurich
Construction management b+p Bau-Realization AG, Zurich
Contractor Piatti Bürgin Bau AG, Dietlikon

Laboratory Building WSJ 352, Novartis Campus Basel

Project no. 1356
Realization 2006–2010
Structural concept Tivadar Puskas
Project management Salomé Hug-Meier
Architecture Tadao Ando Architects & Associates Ltd., Osaka (JP)
Client Novartis Pharma AG, Basel
General planning Burckhardt & Partner AG, Basel
Contractor Implenia Bau AG, Basel

Extension to Mülimatt Retirement and Care Home Sissach (CH)

Project no. 1357 (competition)
Realization 2006–2008
Structural concept Tivadar Puskas
Project management Manuel Alvarez, Patrick Grieder
Architecture Ackermann Architekt BSA SIA AG, Basel
Client Stiftung Regionales Alters- und Pflegeheim Mülimatt, Sissach
General contractor Implenia Generalunternehmung AG, Basel

"Tree Pot" Leutschenpark Zurich

Project no. 1358 (competition)
Realization 2007/08
Structural concept Stefan Bänziger
Project management Stefan Bänziger
Architecture Dipol Landschaftsarchitekten GmbH, Basel; Christopher T. Hunziker GmbH, Zurich
Client City of Zurich, represented by Green City Zürich
Contractor Brunner Erben AG, Zurich

Modernization and Extension of Detached House near Basel

Project no. 1359
Realization 2005/06
Structural concept Heinrich Schnetzer
Project management Kevin M. Rahner
Architecture Herzog & de Meuron Architekten AG, Basel
Client private
Contractor Huber & Straub AG, Basel; Kaufmann Holz und Bau AG, Wallbach

Syngenta Works, Reception Building, Chemistry Building and Seed Center Stein (CH)

Project no. 1361 (competition)
Realization 2005–2007
Structural concept Heinrich Schnetzer
Project management Patrick Grieder
Architecture Burckhardt & Partner AG, Basel
Client Syngenta Crop Protection AG, Stein
Contractor ERNE AG Bauunternehmung, Laufenburg

Donau City Tower 2 Vienna (AT)

Project no. 1367 (not built)
Conception 2004/05
Structural concept Tivadar Puskas
Architecture Herzog & de Meuron Architekten AG, Basel
Client WED Wiener Entwicklungsgesellschaft für den Donauraum AG, Vienna (AT)

Detached House Inning am Ammersee (DE)

Project no. 1390
Realization 2006/07
Structural concept Tivadar Puskas
Project management Jan Stebler
Engineering subcontractor Zilch + Müller Ingenieure GmbH, Munich (DE)
Architecture Herzog & de Meuron Architekten AG, Basel / Muck Petzet Architekten, Munich (DE)
Client private
Contractor Bauunternehmen Assner, Waal (DE)
Timber construction Holzbau Schmid KG, Trostberg (DE)

1346: Roger Frei, Zurich; 1356: Johannes Marburg, Geneva; 1357: Roger Frei, Zurich; 1358: Christopher T. Hunziker, Zurich; 1361: Ruedi Walti, Basel; 1367: Herzog & de Meuron, Basel

International School Basel Aesch (CH)

Project no. 1395
Realization 2005–2007
Structural concept Tivadar Puskas
Project management Daniel Küpfer
Architecture sab Architekten AG, Basel
Client International School Basel, Reinach
General contractor HRS Hauser Rutishauser Suter AG, Basel
Contractor Rofra Bau AG, Basel

Actelion Business Center Allschwil (CH)

Project no. 1397
Realization 2007–2010
Structural concept Heinrich Schnetzer
Project management Hansueli Küng
Architecture GP Actelion: Herzog & de Meuron AG, Basel / Proplaning AG, Basel
Client Actelion Pharmaceuticals Ltd., Allschwil
Steel construction Winterhalter GmbH, Freiburg (DE)
Contractor Anliker AG Bauunternehmung, Emmenbrücke

Thurgau State Archive Frauenfeld (CH)

Project no. 1405 (competition)
Realization 2008–2010
Engineering consortium WGG Schnetzer Puskas Ingenieure AG / Rolf Soller AG, Kreuzlingen
Structural concept Tivadar Puskas
Project management Stefan Bänziger
Architecture Jessen Vollenweider Architektur GmbH, Basel
Client Thurgau Cantonal Building Department
Contractor Lerch AG Bauunternehmung, Winterthur

Refurbishment / Extension of Eichholz Camping Site Berne

Project no. 1426 (competition)
Realization 2007/08
Structural concept Tivadar Puskas
Project management Stefan Bänziger
Architecture Bauart Architekten und Planer AG, Berne
Client Stadtbauten Bern StaBe
Contractor Büchi Bauunternehmung AG, Berne

Dynamo Youth Culture House Zurich

Project no. 1438
Realization 2008/09
Structural concept Stefan Bänziger
Project management Stefan Bänziger
Architecture Phalt Architekten GmbH, Zurich
Client City of Zurich, represented by the Municipal Building Department
Steel construction Kaufmann Spenglerei & Sanitär AG, Zurich

Hinterwenkenweg Apartment House Riehen (CH)

Project no. 1446 (competition)
Realization 2009/10
Structural concept Heinrich Schnetzer
Project management Patrick Grieder
Architecture Pedrocchi Meier Architekten, Basel
Client private
General contractor Sulzer+Buzzi Baumanagement AG, Olten

Conversion of Abadia de Retuerta Monastery into a Hotel Valladolid (ES)

Project no. 1449
Realization 2008–2011
Structural concept Tivadar Puskas
Project management Tivadar Puskas
Architecture Burckhardt & Partner AG, Basel; Marco Serra Architekt, Basel
Client Novartis Pharma AG, Basel
Contractor VOLCONSA Construcción y Desarrollo de Servicios S.A., Valladolid (ES)

Im Kaspar, Row of Semi-detached Homes Münchenstein (CH)

Project no. 1453 (competition)
Realization 2007/08
Structural concept Tivadar Puskas
Project management Salomé Hug-Meier
Architecture Schmid Kuepfer Architekten AG, Basel
Client Christoph-Merian-Stiftung CMS, Basel
General contractor Schmid Kuepfer Architekten AG, Basel
Contractor Glanzmann AG, Birsfelden

1395: Foto-Werk, Michael Fritschi, Basel; 1397: Actelion Pharmaceuticals Ltd., Allschwil; 1405: Maximilian Meisse, Berlin (DE); 1426: unknown; 1438: Dominique Marc Wehrli, Zurich; 1446: Ruedi Walti, Basel; 1449: Burckhardt&Partner, Basel; 1453: Schmid Kuepfer Arch., Basel

ETH Zurich, Hönggerberg Campus, HPL Life Science Platform Zurich

Project no. 1454 (competition)
Realization 2009–2013
Structural concept Heinrich Schnetzer, Stefan Bänziger
Project management Stefan Bänziger
Architecture Burckhardt & Partner AG, Zurich
Client ETH Zürich Immobilien, Zurich
General planning Burckhardt & Partner AG, Zurich
Contractor Erne AG Bauunternehmung, Laufenburg

Markthalle Tower Basel

Project no. 1464 (competition)
Realization 2009–2012
Structural concept Tivadar Puskas
Project management Daniel Küpfer
Architecture Diener & Diener Architekten AG, Basel
Client Allreal Markthalle AG, Zurich
General contractor Allreal Generalunternehmung AG, Zurich
Contractor Marti AG, Basel

Technology Training Center Frauenfeld (CH)

Project no. 1471 (competition)
Realization 2009–2011
Structural concept Stefan Bänziger
Project management Stefan Bänziger, Judith Luible
Architecture Jessen Vollenweider Architektur GmbH, Basel
Client Thurgau Cantonal Building Department, Frauenfeld
Contractor Ernst Herzog AG Bauunternehmung, Frauenfeld

Expansion of the Swiss National Museum Zurich

Project no. 1472
Realization 2013–2016
Structural concept Heinrich Schnetzer
Project management Salomé Hug-Meier, Kevin M. Rahner
Architecture Christ & Gantenbein Architekten ETH SIA BSA, Basel
Client Federal Office for Buildings and Logistics (BBL)
Contractor Huber & Straub AG, Basel

Detached House Binningen (CH)

Project no. 1473
Realization 2007/08
Structural concept Heinrich Schnetzer
Project management Kevin M. Rahner
Architecture Buchner Bründler Architekten, Basel
Client private
Contractor Obrist Bauunternehmung AG, Wallbach

Air Service Basel

Project no. 1478 (competition)
Realization 2008–2010
Structural concept Heinrich Schnetzer
Project management Kevin M. Rahner
Engineering subcontractor CTE Mulhouse, Riedisheim (FR)
Architecture Zwimpfer Partner Architekten SIA, Basel
Client Air Service Basel GmbH, Basel
Steel construction Baumert Constructions Métalliques, Erstein (FR)

Zur Weid Sheltered Housing with Occupational Facilities Mettmenstetten (CH)

Project no. 1483 (competition)
Realization 2008/09
Structural concept Stefan Bänziger
Project management Stefan Bänziger
Architecture Bhend Klammer Architekten / Ramser Schmid Architekten
Client City of Zurich
General contractor Werubau AG Generalunternehmung, Meilen

Detached House Schaffhausen (CH)

Project no. 1488
Realization 2009–2011
Structural concept Tivadar Puskas
Project management Rolf Nachbur
Architecture btob architects, Basel; Henning König, Alexander Thomas, Basel
Client private
Contractor Gasser AG Bauunternehmung, Feuerthalen

1454: René C. Dürr, Zurich; 1464: Juri Weiss, Basel; 1471: Maximilian Meisse, Berlin (DE); 1472: Christ&Gantenbein Arch., Basel; 1473: Dominique Marc Wehrli, Zurich; 1478: Johannes Marburg, Geneva; 1483: Lucas Peters, Zurich; 1488: Roger Frei, Zurich

Gärtnerei Residential Development, im Forster Complex Zurich

Project no. 1490
Realization 2009–2011
Structural concept Stefan Bänziger
Project management Jochem Hilberink
Architecture EM2N Architekten AG, Zurich
Client Bös Fulen AG, represented by Immopoly GmbH, Zurich
Contractor Brunner Erben AG, Zurich

Firehouse 10, AZA Training Center, Andelfingen (CH)

Project no. 1492
Realization 2008–2013
Structural concept Stefan Bänziger
Project management Stefan Bänziger, Johannes Dudli
Architecture Atelier M Architekten GmbH, Zurich
Client Building Department of the Canton of Zurich
Contractor Landolt + Co. AG Bauunternehmung, Kleinandelfingen

City Sky Art Project Basel

Project no. 1506
Realization 2008
Structural concept Tivadar Puskas
Project management Tivadar Puskas
Design Littmann Kulturprojekte, Basel
Client Littmann Kulturprojekte, Basel
Contractor Glanzmann AG, Basel

Expansion Town Hall Regensdorf (CH)

Project no. 1514
Realization 2009/10
Structural concept Stefan Bänziger
Project management Stefan Bänziger, Florian Riebel
Architecture Phalt Architekten AG, Zurich
Client Municipality of Regensdorf
Contractor K. Eicher Bauunternehmung AG, Regensdorf

Densa Park Residential Development Basel

Project no. 1516
Realization 2009–2011
Structural concept Tivadar Puskas
Project management Patrick Grieder
Architecture Luca Selva AG Architekt ETH BSA SIA, Basel
Client Densa Immobilien AG, Basel
General contractor Priora AG, Basel
Contractor Marti AG Bauunternehmung, Basel

Renovation and Extension of Cinémathèque Suisse Penthaz (CH)

Project no. 1519 (competition)
Realization 2010–2013
Structural concept Stefan Bänziger
Project management Florian Riebel
Architecture EM2N Architekten AG, Zurich
Client Federal Office for Buildings and Logistics (BBL)
Contractor M. Rusconi & Cie S.A., Lausanne

ABR 5 Administrative Building, Roche Diagnostics Rotkreuz (CH)

Project no. 1521 (competition)
Realization 2009–2011
Structural concept Tivadar Puskas
Project management Daniel Küpfer
Architecture Burckhardt & Partner AG, Basel
Client Roche Diagnostics AG, Rotkreuz
Contractor Implenia Bau AG, Basel

Conversion of TP2 Freilager into School of Art and Design (HGK) FHNW Münchenstein (CH)

Project no. 1525
Realization 2010–2013
Structural concept Tivadar Puskas
Project management Rolf Nachbur
Architecture Müller Sigrist Architekten AG, Basel
Client Immobilien Basel-Stadt, represented by the Canton of Basel-Stadt – Building and Transport Department
Contractor Marti AG, Basel

1490: Roger Frei, Zurich; 1492: AMA Ackerknecht Minder Arch., Zurich; 1506: Klaus Littmann, Basel; 1514: Dominique Marc Wehrli, Zurich; 1516: Ruedi Walti, Basel; 1519: Roger Frei, Zurich; 1521: Thomas Jantscher, Colombier; 1525: Foto-Werk, Michael Fritschi, Basel

Avantage – Living in Wettstein Park Basel

Project no. 1530
Realization 2010–2012
Structural concept Tivadar Puskas
Project management Eva Feuling
Architecture Jessen & Vollenweider Architektur ETH SIA BSA, Basel
Client Proplaning AG, Basel
Contractor Implenia AG, Basel

Extension for Gerber Vogt AG Allschwil (CH)

Project no. 1532
Realization 2009
Structural concept Tivadar Puskas
Project management Hansueli Küng
Architecture FS Architekten, Magden
Client Gerber Vogt AG, Allschwil
Contractor Glanzmann AG Bauunternehmung, Birsfelden

Panoramica Apartment House Bottmingen (CH)

Project no. 1535
Realization 2010/11
Structural concept Tivadar Puskas
Project management Judith Luible, Rémy Jabas
Architecture Architect André Kaufmann SIA, Basel
Client Baugesellschaft Stallenstrasse, Bottmingen
Contractor Implenia Bau AG, Basel

Roche Opal Grenzach (DE)

Project no. 1546
Realization 2010/11
Structural concept Tivadar Puskas
Project management Wolfgang Herrmann, Salomé Hug-Meier
Architecture Christ & Gantenbein Architekten ETH SIA BSA, Basel
Client F. Hoffmann – La Roche AG, Basel
Contractor Glanzmann AG Bauunternehmung, Birsfelden

Change of Use into Jazz Campus, Utengasse 15/17 Basel

Project no. 1570
Realization 2010–2013
Structural concept Tivadar Puskas
Project management Patrick Grieder
Architecture Buol & Zünd Architekten BSA, Basel
Client Habitat Foundation, Basel
Contractor Knecht Bauunternehmung AG, Münchenstein

Refurbishment of Jakobsberg Substation, IWB Basel

Project no. 1575
Realization 2010–2014
Structural concept Rolf Nachbur
Project management Rolf Nachbur, Thomas Gutzwiller (construction management)
Architecture muellermueller Architekten BSA SIA, Basel
Client Industrielle Werke Basel (IWB)
Contractor Stamm Bau AG, Binningen

Apartment/Office Building, Birsigstrasse Basel

Project no. 1577
Realization 2010/11
Structural concept Tivadar Puskas
Project management Burkhard Schnabel
Architecture Kägi Schnabel Architekten ETH BSA SIA, Basel
Client Baugemeinschaft Klein Rickhoff, Basel
Contractor Obrist Bauunternehmung AG, Wallbach

Change of Use – Andlauerhof Basel

Project no. 1580
Realization 2011–2013
Structural concept Tivadar Puskas
Project management Patrick Grieder
Architecture Villa Nova Architekten AG, Basel
Client Andlauerhof AG, Feldbrunnen
Contractor Erne AG Bauunternehmung, Laufenburg

1530: Maximilian Meisse, Berlin (DE); 1532: Gerber Vogt, Allschwil; 1535: Architekt André Kaufmann, Basel; 1546: Walter Mair, Basel; 1570: Foto-Werk, Michael Fritschi, Basel; 1577: Kägi Schnabel Arch., Basel; 1580: Villa Nova Arch., Basel

Courtyard Development, Hegenheimerstrasse 137 Basel

Project no. 1582
Realization 2010/11
Structural concept Tivadar Puskas
Project management Judith Luible
Architecture Luca Selva AG Architekt ETH BSA SIA, Basel
Client Wohngenossenschaft Hegenheimerstrasse, Basel
Contractor BBL Baugesellschaft BL AG, Reinach

Seidentor Residential Park Arlesheim (CH)

Project no. 3003
Realization 2009–2011
Structural concept Tivadar Puskas
Project management Jan Stebler, Stefan Birk
Architecture Fankhauser Architektur AG, Reinach
Client/general contractor HRS Real Estate AG, Basel
Contractor Wenk AG, Füllinsdorf

Aquila Commercial and Residential Building Pratteln (CH)

Project no. 3006
Planning 2009–2012
Structural concept Heinrich Schnetzer
Project management Daniel Küpfer, Rémy Jabas
Architecture Christ & Gantenbein Architekten ETH SIA BSA, Basel
Client Immobilienfonds UBS Sima, Basel, represented by Stokar + Partner AG, Basel
Contractor Implenia Schweiz AG, Basel

Detached House, Maison Mosch Arlesheim (CH)

Project no. 3011
Realization 2010/11
Structural concept Tivadar Puskas
Project management Gabriel Ackermann
Architecture Dorenbach AG Architekten ETH SIA, Basel
Client private
Contractor Implenia Bau AG, Basel

Residential Development on Zellweger Development Site Uster (CH)

Project no. 3013
Realization 2011/12
Structural concept Stefan Bänziger
Project management Johannes Dudli
Architecture Annette Gigon / Mike Guyer Architekten, Zurich
Client Immopoly GmbH, Zurich
Contractor Marti AG, Zurich

School Building Zinzikon (CH)

Project no. 3017
Planning 2010–2012
Structural concept Stefan Bänziger
Project management Florian Riebel, Andreas Hirschi
Architecture Adrian Streich Architekten AG, Zurich
Construction management GMS Partner AG, Zürich-Flughafen
Client City of Winterthur, Building Department
Contractor Baltensperger AG, Seuzach

Renovation of ZKB Headquarters Zurich

Project no. 3018 (competition)
Realization 2012–2014
Structural concept Stefan Bänziger
Project management Florian Riebel
Architecture Jessen & Vollenweider Architektur ETH SIA BSA, Basel
Client Zürcher Kantonalbank, Zürich
Contractor Barizzi AG Bauunternehmung, Bertschikon

Doldertal Home for the Elderly – Renovation and Addition of Floor Zurich

Project no. 3021
Realization 2010
Structural concept Stefan Bänziger
Project management Stefan Bänziger, Jochem Hilberink
Architecture Ramser Schmid Architekten, Zurich
Client Property Administration of City of Zurich, represented by City of Zurich, Municipal Building Department
Contractor Abeco Bau AG, Zurich

1582: Ruedi Walti, Basel; 3003: Fankhauser Architektur, Reinach; 3006: Christ & Gantenbein Arch., Basel; 3011: Tom Bisig, Basel; 3013: Annette Gigon / Mike Guyer Arch., Zurich; 3017: Adrian Streich Arch., Zurich; 3021: Roger Frei, Zurich

Hermes Stand, BaselWorld 2013 Basel

Project no. 3024Hermes
Realization 2013
Structural concept SAPS Co., Ltd., Tokio (JP)
Executive engineer Tivadar Puskas
Project management Ayumi Isozaki
Architecture Toyo Ito & Associates, Architects, Tokyo (JP)
Construction management Christen et Cie., Vandœuvre
Client La Montre Hermès SA, Brügg b. Biel
Steel construction Jakem AG, Münchwilen
Timber construction Daniel Fournier Interior Concepts S.A.S, Paris (FR)

Conversion of Former ABB Site to Stamm Bau AG Competence Center Arlesheim (CH)

Project no. 3028a
Realization 2012/13
Structural concept Tivadar Puskas (competition)
Project management Hansueli Küng, Olivier Lichtenthaler
Architecture Fankhauser Architektur AG, Reinach
Client Stamm Bau AG, Binningen
Contractor Stamm Bau AG, Binningen

Residential Development, Via Suot Chesas Champfèr (CH)

Project no. 3041
Realization 2011/12
Structural concept Tivadar Puskas
Project management Salomé Hug-Meier
Architecture Diener & Diener Architekten, Basel
Client private
Contractor Walter Dietsche Baumanagement, Chur
Timber construction Bachofner GmbH, Frümsen

Kestenholz Truck Center Pratteln (CH)

Project no. 3042
Realization 2012/13
Structural concept Tivadar Puskas
Project management Hansueli Küng
Architecture Steinmann & Schmid Architekten AG, Basel
Client Zurimmo "B" Immobilien AG, Zurich, represented by UBS Fund Management, Basel
Contractor Implenia Generalunternehmung AG, Basel

Villa Zum Bannhübel Arlesheim (CH)

Project no. 3049
Realization 2010/11
Structural concept Tivadar Puskas
Project management Patrick Grieder
Architecture Dorenbach AG Architekten ETH SIA, Basel
Client private
Contractor Huber Straub AG Bauunternehmung, Basel

Detached House, Kahlstrasse Basel

Project no. 3050
Realization 2011/12
Structural concept Tivadar Puskas
Project management Giotto Messi
Architecture Buchner Bründler AG Architekten BSA, Basel
Client private
Contractor Morath & Crottaz AG, Basel

RIVA Residential Development on Schaffhauserrheinweg Basel

Project no. 3055 (competition)
RIVA engineering consortium: Schnetzer Puskas Ingenieure AG, Basel / Rothpletz, Lienhard + Cie AG, Berne
Realization 2012–2014
Structural concept Tivadar Puskas
Project management Patrick Grieder
Architecture Jessen & Vollenweider Architektur ETH SIA BSA, Basel
Client Sarasin Anlagestiftung Nachhaltigkeit Schweiz, Berne, represented by Dr. Meyer Asset Management AG, Berne
Contractor Marti AG, Basel

Detached House St. Gallen (CH)

Project no. 3056
Realization 2011/12
Structural concept Tivadar Puskas
Project management Giotto Messi
Local engineers Borgogno Eggenberger + Partner AG, St. Gallen
Architecture Marques AG, Lucerne
Client private
Contractor Stutz AG Hatswil, Immobilien & Baudienste

3024: Toyo Ito & Associates, Tokyo (JP); 3028a: Stamm Bau, Binningen; 3042: Steinmann & Schmid Arch., Basel; 3049: Tom Bisig, Basel; 3055: Jessen & Vollenweider Architektur, Basel

Basel Zoo, Outdoor Primates Facility
Basel

Project no. 3059
Realization 2011/12
Structural concept Conzett Bronzini Gartmann AG, Chur
Project management Rolf Nachbur (Schnetzer Puskas Ingenieure AG as engineering consultants)
Architecture Peter Stiner Architekt ETH BSA, Basel
Client Basel Zoo
Contractor Huber Straub AG Bauunternehmung, Basel
Mesh design Conzett Bronzini Gartmann AG, Chur; Pfeifer Ingenieure GmbH, Konstanz (DE)

Detached House
Winterthur (CH)

Project no. 3063
Realization 2010/11
Structural concept Stefan Bänziger
Project management Johannes Dudli
Architecture Maria del Carmen Martin Pastor & Partner dipl. Architektin ETH, Zurich
Client private
Contractor A. Blatter AG, Seuzach

Dreispitz Transitlager
Basel

Project no. 3070
Planning since 2014
Structural concept Tivadar Puskas
Project management Eva Feuling
Architecture/General planning services consortium BIG (Bjarke Ingels Group), Copenhagen (DK) / Proplaning AG, Basel
Client UBS Fund Management (Schweiz) AG, Basel

Letzibach Development Site, Zurich, Sub-project C, High-rise Development and Outdoor Spaces, Construction Plots C and E
Zurich

Project no. 3078 (competition)
Realization 2013–2015
Structural concept Stefan Bänziger
Project management Stefan Bänziger, Timothy Hafen
Architecture Adrian Streich AG / Loeliger Strub Architekten GmbH
Client Swiss Federal Railways SBB Immobilien Development, Zurich
General contractor Implenia Generalunternehmung AG, Basel

Residential Development, Faidostrasse
Basel

Project no. 3083
Realization 2012/13
Structural concept Mark Eitel, Heinrich Schnetzer
Project management Mark Eitel
Architecture Haberstroh Schneider Architekten / Proplaning AG
Client Varem AG, Basel
Contractor Marti AG, Basel

Laboratory Building WSJ 182, Novartis Campus
Basel

Project no. 3090
Realization 2012/13
Structural concept Tivadar Puskas
Project management Salomé Hug-Meier, Dominik Mölder
Architecture RMA Architects LLC, Brookline (USA)
Client Novartis Pharma AG, Basel
General planning Rapp Arcoplan AG, Basel
Contractors Morath & Crottaz AG, Basel / Huber + Straub AG, Basel

High-rise Blocks, Baden Nord, Brown Boveri Parking
Baden (CH)

Project no. 3092
Planning since 2010
Structural concept Stefan Bänziger
Project management Stefan Bänziger
Architecture Pool Architekten / Michael Meier & Marius Hug Architekten
Client ABB Immobilien AG, Baden

Northern Extension and New Ward Wing, Cantonal Hospital Frauenfeld
Frauenfeld (CH)

Project no. 3093 (competition)
Planning since 2010
Structural concept Tivadar Puskas
Project management Eva Feuling
Engineering subcontractor BHA Team Ingenieure AG, Frauenfeld
Architecture Schneider & Schneider Architekten ETH BSA SIA AG, Aarau
Client Cantonal Building Department, Frauenfeld

3059: Zoo Basel, Basel; 3063: Radek Brunecky, Zurich; 3070: BIG Architects, Copenhagen (DK) 3078: Adrian Streich Arch., Zurich; 3083: Foto-Werk, Michael Fritschi, Basel; 3092: Pool Arch., Zurich; 3093: Schneider & Schneider Arch., Aarau

Renovation of Kügeliloo Kindergarten Zurich

Project no. 3096
Realization 2012/13
Structural concept Stefan Bänziger
Project management Florian Riebel
Architecture Frei & Saarinen Architekten GmbH, Zurich
Client City of Zurich, Municipal Building Department
Contractor BWT Bau AG, Zurich

Detached House Kilchberg (CH)

Project no. 3101
Realization 2011/12
Structural concept Stefan Bänziger
Project management Florian Riebel
Architecture Frei & Saarinen Architekten GmbH, Zurich
Client private
Contractor Brönnimann Hoch- und Tiefbau AG, Rüschlikon

Residential Development, Scheffelstrasse Zurich

Project no. 3103
Realization 2012/13
Structural concept Stefan Bänziger
Project management Stefan Bänziger, Jochem Hilberink
Architecture Jakob Steib Architekten AG, Zurich
Client BSZ Immobilien AG, Zurich
Contractor Gautschi Bau AG, Affoltern am Albis

Renovation of Two Villas Zurich

Project no. 3105
Realization 2011/12
Structural concept Mark Eitel
Project management Mark Eitel
Architecture Miller & Maranta dipl. Architekten ETH BSA SIA, Basel
Client Jacobs Foundation, Zurich
Contractor Rieke AG Bauunternehmung, Zurich

Bächtelen Residential Development, Construction Plot A Köniz (CH)

Project no. 3107
Planning since 2010
Structural concept Heinrich Schnetzer
Project management Heinrich Schnetzer, Rémy Jabas
Architecture Buchner Bründler Architekten, Basel
Client Widmer Hans Management AG, Oberwil-Lieli, represented by Priora Generalunternehmung AG, Basel
General contractor Priora Generalunternehmung AG, Berne

Claraturm Basel

Project no. 3111
Planning since 2011
Structural concept Tivadar Puskas
Project management Giotto Messi
Architecture Morger + Dettli Architekten AG BSA/SIA, Basel
Client Balintra AG c/o UBS Fund Management AG, Basel

Hertenstein House Weggis (CH)

Project no. 3118
Realization 2011/12
Structural concept Tivadar Puskas
Project management Salomé Hug-Meier
Architecture Buchner Bründler Architekten, Basel
Client private
Contractor Catenazzi Bau AG, Weggis

Laufental Senior Citizens' Center Laufen (CH)

Project no. 3122 (competition)
Realization 2013–2015
Structural concept Tivadar Puskas
Project management Rolf Nachbur
Architecture/general planning Ackermann Architekt BSA SIA AG, Basel
Client Stiftung Seniorenzentrum Rosengarten, Laufental
Contractor Rofra Bau AG, Aesch

3096: Frei&Saarinen Arch., Zurich; 3101: Stefan Wülser, Zurich; 3103: Jakob Steib Arch., Zurich; 3105: Miller&Maranta dipl. Arch., Basel; 3107: Buchner Bründler Arch., Basel; 3111: Meyer Dudesek Arch., Zurich; 3118: Ruedi Walti, Basel; 3122: Ackermann Architekt, Basel

Detached House Kreuzlingen-Alp (CH)

Project no. 3128
Planning since 2011
Structural concept Tivadar Puskas
Project management Patrick Grieder
Architecture Buchner Bründler Architekten, Basel
Client private
Contractor Landolt + Co. AG Bauunternehmung, Kleinandelfingen

Apartment House, Neufrankengasse Zurich

Project no. 3139 (competition)
Realization 2011–2013
Structural concept Stefan Bänziger
Project management Jochem Hilberink
Architecture EM2N Architekten AG, Zurich
Client SBB Immobilien AG, Zurich
General contractor Halter AG Generalunternehmung, Zurich
Contractor Brunner Erben AG, Zurich

Detached House Galgenen (CH)

Project no. 3141
Realization 2012/13
Structural concept Stefan Bänziger
Project management Johannes Dudli, Reinhard Oppikofer
Architecture Kit Architects ETH SIA GmbH, Zurich
Client private
Contractor Implenia Bau AG, Siebnen

Mattenhof Residential Development Zurich

Project no. 3147
Planning since 2012
Structural concept Stefan Bänziger
Project management Johannes Dudli
Architecture Bachelard Wagner Architekten, Basel
Client Siedlungsgenossenschaft Sunnige Hof, Zurich

University of Applied Sciences and Arts Northwestern Switzerland Muttenz (CH)

Project no. 3149 (competition)
Planning since 2011
Structural concept Stefan Bänziger
Project management Stefan Bänziger
Architecture/general planning: Pool Architekten, Zurich
Client Building Department of the Canton of Basel-Landschaft; Building and Environmental Directorate, Liestal
Construction management Perolini Baumanagement AG, Zurich

Residential Development, Grimselstrasse Zurich

Project no. 3151 (competition)
Planning since 2011
Structural concept Tivadar Puskas
Project management Stefan Bänziger
Architecture Jessen & Vollenweider Architektur / Proplaning AG
Client Immobiliengesellschaft Turintra AG, Zurich, represented by UBS Fund Management, Zurich

Humanitas Residential and Care Home Riehen (CH)

Project no. 3157 (competition)
Planning since 2012
Structural concept Tivadar Puskas
Project management Patrick Grieder
Architecture Bachelard Wagner Architekten, Basel
Client Immobilien Basel-Stadt; Genossenschaft Humanitas, Riehen

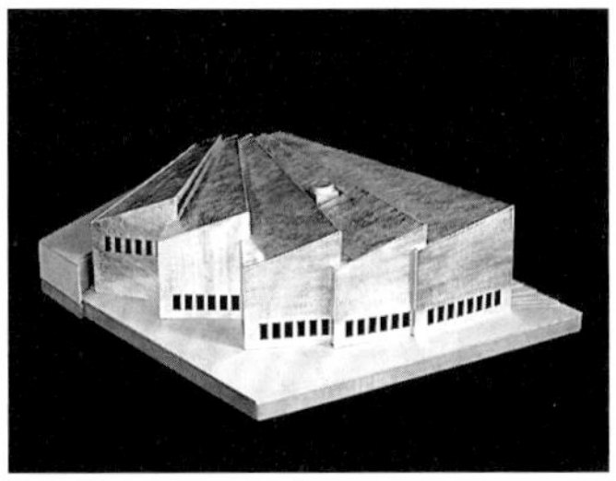

List Customer Center Arisdorf (CH)

Project no. 3158 (competition)
Planning 2011–2013
Structural concept Heinrich Schnetzer
Project management Rémy Jabas
Architecture Christ & Gantenbein Architekten / Aebli Zimmermann AG
Client List AG, Arisdorf
Contractor Huber Straub AG, Basel

3128: Buchner Bründler Arch., Basel; 3139: Simon Menges, Berlin (DE); 3141: Kit Architects, Zurich; 3147: Bachelard Wagner Arch., Basel; 3149: Pool Arch., Zurich; 3151: Jessen & Vollenweider Architektur, Basel; 3157: Bachelard Wagner Arch., Basel; 3158: Christian Kahl, Berlin (DE)

Animal Shelter Basel

Project no. 3159
Planning since 2011
Structural concept Heinrich Schnetzer
Project management Christoph Sterr
Architecture Schmid Kuepfer Architekten AG, Basel
Client Tierschutz beider Basel, Basel

Funicular Railway Le Locle (CH)

Project no. 3172
Planning since 2012
Structural concept Tivadar Puskas
Project management Giotto Messi
Local engineers AJS Ingénieurs civils SA, Neuchâtel
Architecture Diener & Diener Architekten, Basel
Client Ville du Locle
Contractor F. Bernasconi & Cie SA, Les Geneveys-sur-Coffrane

Ricola Herb Center Laufen (CH)

Project no. 3182z02
Realization 2013/14
Structural concept Heinrich Schnetzer
Project management Elisabeth Wuttke
Architecture Herzog & de Meuron Basel Ltd., Basel
Client Ricola AG, Laufen
General contractor Priora Generalunternehmung AG, Basel

Headquarters and Production Building for Swatch und Omega AG Biel (CH)

Project no. 3183 (competition)
Planning since 2011
Structural concept Tivadar Puskas, Jan Stebler
Project management Jan Stebler
Design architect Shigeru Ban Architects, Japan
General planning Itten + Brechbühl AG, Berne
Client The Swatch Group AG, Biel

WSL Plant Protection Laboratory Birmensdorf (CH)

Project no. 3186 (competition)
Realization 2013/14
Structural concept Tivadar Puskas
Project management Hansueli Küng, Dominik Häring
Architecture/general planning Burckhardt & Partner AG, Basel
Client Eidgenössische Forschungsanstalt WSL, Birmensdorf
Contractor Anliker AG Bauunternehmung, Thalwil

Balloon for the St. Prex Classics St. Prex (CH)

Project no. 3193
Realization 2012
Structural concept Heinrich Schnetzer
Project management Christoph Sterr
Architecture EPFL / ENAC / IA / ALICE (Atelier de la conception de l'espace), Lausanne
Client Association St. Prex Festival, St. Prex
Project management Anlag AG, Olsberg
Aluminum construction FHS Frech-Hoch AG, Sissach

GastroSocial High-rise / Office Building Aarau (CH)

Project no. 3205 (competition)
Realization 2013/14
Structural concept Tivadar Puskas (competition)
Project management Giotto Messi
Architecture Schneider & Schneider Architekten ETH BSA SIA AG, Aarau
Client GastroSocial, Aarau

Four Apartment Buildings, Jona-Garten Housing Development Wald (CH)

Project no. 3215
Planning since 2012
Structural concept Stefan Bänziger
Project management Stefan Bänziger
Architecture huggenbergerfries Architekten AG ETH SIA BSA, Zurich
Client Otto & Joh. Honegger AG, Wald, represented by Odinga und Hagen AG, Zurich

3159: Schmid Kuepfer Arch., Basel; 3172: Diener&Diener Arch., Basel; 3182z02: Herzog&de Meuron, Basel; 3183: Blaser Arch., Basel; 3186: Burckhardt&Partner, Basel; 3193: Joel Tettamanti, Zurich; 3205: Schneider&Schneider Arch., Aarau; 3215: huggenbergerfries Arch., Zurich

Apartment House, Rütiring Riehen (CH)

Project no. 3216
Realization 2013/14
Structural concept Tivadar Puskas
Project management Patrick Grieder
Architecture Staehelin Meyer Architekten ETH SIA, Basel
Client private
Contractor Huber Straub AG, Basel

Housing for the Elderly Im Büel Cham (CH)

Project no. 3219
Realization 2013–2015
Structural concept Tivadar Puskas
Project management Stefan Birk
Architecture Schneider & Schneider Architekten ETH BSA SIA AG, Aarau; Oeschger Architekten BSA SIA AG, Hausen/Brugg
Client Stiftung Wohnen im Alter, Cham
Contractor Saredi AG, Küssnacht am Rigi

Im Stückler Housing Development Zurich

Project no. 3220
Planning since 2012
Structural concept Stefan Bänziger
Project management Johannes Dudli
Architecture/general planning Adrian Streich Architekten AG / GMS Partner AG
Construction management GMS Partner AG, Zürich-Flughafen
Client Baugenossenschaft Halde, Zurich

Maison Davidoff Basel

Project no. 3222
Planning since 2012
Structural concept Tivadar Puskas
Project management Salomé Hug-Meier, Caterina Chilovi
Architecture Diener & Diener Architekten, Basel
Client Oettinger Davidoff Group, Basel

New-build Development, Sempacherstrasse 51 and 53 Basel

Project no. 3237 (competition)
Planning since 2012
Structural concept Heinrich Schnetzer
Project management Elisabeth Wuttke
Architecture/general planning: Miller & Maranta dipl. Architekten ETH BSA SIA, Basel
Client Stiftung SKB 1809, Basel

Heuried Sports Center Zurich

Project no. 3238 (competition)
Planning since 2012
Structural concept Stefan Bänziger
Project management Johannes Dudli
Architecture EM2N Architekten AG, Zurich
Client City of Zurich, Municipal Building Department
Timber construction Pirmin Jung Ingenieure für Holzbau AG, Rain

Residential Development, Rosentalstrasse 9–13 Basel

Project no. 3242
Planning since 2012
Structural concept Tivadar Puskas
Project management Tivadar Puskas, Caterina Chilovi
Architecture Morger + Dettli Architekten AG BSA/SIA, Basel
Client Zurich Imre AG, Zurich

Gartenhof School Buildings Allschwil (CH)

Project no. 3246
Planning since 2012
Structural concept Stefan Bänziger
Project management Stefan Bänziger, Salomé Hug-Meier
Architecture Birchmeier Uhlmann Architekten GmbH, Zurich / Carlos Rabinovich, Zurich
Client Einwohnergemeinde Allschwil, represented by Dietziker Partner Baumanagement AG, Basel

3216: Staehelin Meyer Arch., Basel; 3219: Schneider & Schneider Arch., Aarau; 3220: Adrian Streich Arch., Zurich; 3222: Diener & Diener Arch., Basel; 3237: Miller & Maranta Arch., Basel; 3238: EM2N Arch., Zurich; 3242: Meyer Dudesek Arch., Zurich; 3246: Birchmeier Uhlmann Arch., Zurich

Refurbishment of Lido Reinach (CH)

Project no. 3254
Planning since 2012
Structural concept Tivadar Puskas
Project management Patrick Grieder
Architecture Morger + Dettli Architekten AG BSA/SIA, Basel
Client Municipality of Reinach

Schachen Sports Hall Aarau (CH)

Project no. 3256 (competition)
Planning since 2012
Structural concept Tivadar Puskas, Giotto Messi
Project management Jan Stebler
Architecture Rolf Mühlethaler Architekt BSA SIA, Berne
Client Einwohnergemeinde Aarau, represented by Aarau Town Council and Municipal Building Department

Competition for Hardturm Stadium Zurich

Project no. WE407 (competition, 2nd place)
Planned 2012
Structural concept Tivadar Puskas, Giotto Messi
Architecture Graber Pulver Architekten AG
Client City of Zurich, Municipal Building Department

Competition Emmequerung Neumatt Ey (CH)

Project no. WE408 (competition, 2nd place)
Planned 2011
Structural concept Heinrich Schnetzer
Project management Heinrich Schnetzer
Architecture Marcel Baumgartner, Dipl. Architekt ETH SIA, Zurich
Client City of Burgdorf, Building Department
Timber construction Brawand Zimmerei AG, Grindelwald

Competition Linthsteg Weesen (CH)

Project no. WE445 (competition, not placed)
Planned 2011
Structural concept Heinrich Schnetzer
Project management Heinrich Schnetzer
Architecture Marcel Baumgartner, Dipl. Architekten ETH SIA, Zurich
Client Municipality of Weesen, Glarus

Refurbishment and Modernization of St. Jakobshalle Basel

Project no. WE492 (competition, first place)
Planning 2013–2015
Structural concept Tivadar Puskas
Project management Tivadar Puskas
Architecture Degelo / Berrel Berrel Kräutler, Basel
Client Building and Transport Department of the Canton of Basel-Stadt, Building Department

3254: bgrw, Basel (Nord GmbH, Basel); 3256: Rolf Mühlethaler Architekt, Berne; WE407: Graber Pulver Arch., Zurich; WE408: Marcel Baumgartner, Zurich; WE445: Marcel Baumgartner / nightnurse images, Zurich; WE492: Arch.gemeinschaft Degelo / Berrel Berrel Kräutler, Basel

Publications

2011

Tivadar Puskas / Daniel Küpfer (2011). "Raute, Ornament und Tragwerk" *Tec21* 137/36, pp. 12f.

Heinrich Schnetzer/ Aurelio Muttoni / Joseph Schwartz / Aita Flury (2011). "Starke Strukturen." In: Aita Flury (ed.). *Kooperation. Zur Zusammenarbeit von Ingenieur und Architekt.* Basel: pp. 193–206.

"Büroturm im Minergiestandard – innovative Doppelfassade mit dezentraler Fassadenlüftung," *Detail* 51/4, pp. 404–412.

2010

Evelyn C. Frisch (2010). "Wohlgeformte Raumerweiterung," *Steeldoc* 3/4, pp. 24–29.

Heinrich Schnetzer (2010). "Architekt und Ingenieur im Entwurfsprozess." In: Aita Flury (ed.). *Dialog der Konstrukteure.* Sulgen, pp. 108–112.

Heinrich Schnetzer (2010). "Handarbeit mit Beton," *Tec21* 136/38, pp. 20–25.

Heinrich Schnetzer (2010). "Wirkungsvoll kombiniert," *Tec21* 136/3–4, pp. 2–6.

Rahel Hartmann Schweizer (2010). "Transformation," *Tec21* 136/35, pp. 65–67.

2009

Heinrich Schnetzer (2009). "Der Lehrer." In: Thomas Vogel / Peter Marti (ed.). *Christian Menn. Brückenbauer.* Zurich, pp. 23–26.

2008

Herzog & de Meuron Architekten AG, Basel (2008). "St. Jakob-Turm, Basel," *Bauen Heute* 22/11–12, pp. 33–37.

Tivadar Puskas (2008). "Der Kräftefluss von der Spitze in den Fuss," *BAU Info* 14/11–12, pp. 61–68.

Heinrich Schnetzer (2008). "Dreibein, Korsett und Regenschirme," *Tec21* 134/36, pp. 24–28.

Heinrich Schnetzer / Mark Eitel / Kurt Andresen (2008). "Elbphilharmonie Hamburg – Grundsätzliche Überlegungen zur Zusammenarbeit Ingenieur – Architekt," *Bautechnik* 85/3, pp. 157–166.

2007

Catherine A. Cardno (2007). "Glass and Steel 'Cloud' redefines Hamburg Harbour," *Civil Engineering* 77/6, pp. 12f.

Maria Vittoria Corazza (2007). "The Fair looks at the sky / La Fiera guarda il cielo," *iic, l'Industria italiana del Cemento* 12, pp. 812–827.

Tivadar Puskas (2007). "Ingenieure im Planungswettbewerb," *Tec21* 133/12, pp. 16–18.

Tivadar Puskas / Stefan Bänziger (2007). "Grösste Sport- und Eventstätte der Schweiz," in: *Stahlbau* 76 (2007), H. 2, pp. 28–35.

2006

Tivadar Puskas / Stefan Bänziger (2006). "Stadionnutzung: Die grösste Sport- und Eventstätte," *Handelszeitung, Spezial Planungsunternehmer* 28. 3. 2006, p. 61.

2003

Tivadar Puskas / Stefan Bänziger (2003). "Schnell und hoch," *Tec21* 129/3–4, pp. 18–21.

2002

Tivadar Puskas (2002). "Der Messeturm in Basel," *Stahlbau* 71 (2002), H. 8, pp. 619–624.

Awards

2012

Courtyard development, Hegenheimerstrasse 137, Basel: 2012 Architecture Prize of the Swiss Heritage Society, Basel

2011

Actelion Business Center, Allschwil: Prix Acier Swiss Steel Construction Prize 2011

Mercedes Benz showroom, Kestenholz , Basel: Prix Acier Swiss Steel Construction Prize 2011 – Commendation

Im Viadukt Market Hall, change of use of viaduct arches: Brunel Award

2010

Im Viadukt Market Hall, change of use of viaduct arches: Public's choice – Good Buildings Award, Zurich

New Monte Rosa Hut, Zermatt: Swiss Solar Prize Category B: "New Buildings"

2009

New Monte Rosa Hut, Zermatt: "Pepper Symbol" award, Category: "Tourism", Milestone Special Prize for Sustainability

Metal Workshop, Dynamo Youth Culture House, Zurich: Prix Acier Swiss Steel Construction Prize 2009 – Commendation

2008

New Monte Rosa Hut, Zermatt: Holcim Award Europe Bronze

Dreispitz S-Bahn Station, Basel: Brunel Award 2008, Basel Heritage Prize

Redesign and Upgrade of Hardbrücke Railway Station, Zurich: Brunel Award – Commendation (Category: Architecture)

Claramatte Park Facelift, Basel: Good Buildings Award of the Cantons of Basel-Landschaft and Basel-Stadt

Altmarkt Special Needs School, Liestal: Good Buildings Award of the Cantons of Basel-Landschaft and Basel-Stadt

Messeturm Trade Fair Tower, Basel: Good Buildings Award of the Cantons of Basel-Landschaft and Basel-Stadt

St. Jakob Tower, Basel: Good Buildings Award of the Cantons of Basel-Landschaft and Basel-Stadt

Renovation and New Build of Office/Apartment Building, Bäumleingasse 14, Basel: Good Buildings Award of the Cantons of Basel-Landschaft and Basel-Stadt

2007

Redevelopment of Visp Railway Station: FLUX - Golden Transport Hub Prize

2005

Prada Aoyama, Tokyo (JP): Architectural Institute of Japan Prize –Architectural Design Division

2004

Detached House, Staufen: Häuser Award 2004, House of the Year

2002

School for Guide Dogs, Allschwil: Sarnafil Competition, 1st Prize

Apartment Building on Engelgasse, Basel: Good Buildings Award of the Cantons of Basel-Landschaft and Basel-Stadt

Rectory and Accommodation, Basel, Amerbachstrasse 9/11: Good Buildings Award of the Cantons of Basel-Landschaft and Basel-Stadt

IWB Warehouse Buildings, Basel, Neuhausstrasse 31: Good Buildings Award of the Cantons of Basel-Landschaft and Basel-Stadt

Schnetzer Puskas Engineers

Company portrait

The long history of Schnetzer Puskas Engineers began in 1953 with the founding of an engineering consultancy in Basel by Heinz Hossdorf. In 1972, Heinz Hossdorf and Kilian Weiss founded the general partnership Hossdorf und Weiss Ingenieure. The circle of partners was expanded in 1980: Kilian Weiss, René Guillod, and Rudolf Gisi took over the general partnership, operating under the new name Weiss Guillod Gisi Ingenieure. At this point Heinz Hossdorf retired as a partner.

In 1992, **Heinrich Schnetzer** (born 1958) became a partner and member of the management board. After training as a steel construction draughtsman, he initially studied structural engineering at the HTL in Muttenz. He completed his postgraduate studies at the ETH Zurich in 1988. He then worked as Research Assistant under Prof. Dr. Christian Menn and Research Associate under Prof. Dr. Peter Marti at the Institute of Structural Engineering at the ETH Zurich. The subject of his dissertation was "Stochastic Building Material Moduli for Concrete." In addition to his main occupation, Heinrich Schnetzer lectures regularly at various Swiss academic institutions and in 2012 served as a guest critic at the EPFL under Prof. Harry Gugger.

In 1996, the general partnership was transformed into the public limited company WGG Ingenieure AG. The same year, **Tivadar Puskas** (born 1960) became a partner and member of the management board. After training as a civil engineering draughtsman with Emch + Berger AG in Solothurn, Puskas studied structural engineering at the HTL in Burgdorf. Following a short period of practical experience at Balzari und Schudel Ingenieure in Berne he decided to further his studies at the ETH Zurich, from which he graduated in 1988. Since 2005, Tivadar Puskas has served as a guest critic at various academic institutions and has been involved in teaching activities such as lecturing on architecture subjects. In 2006, he became a partner and member of the management board.

In 2000, the company name was changed to WGG Schnetzer Puskas Ingenieure AG. The Zurich office was opened in 2004 with a staff of around ten under the direction of **Stefan Bänziger** (born 1972). Bänziger became a partner and a member of the management board in 2006. After leaving school he studied structural engineering at the ETH Zurich, specializing in the areas of structural design and materials. Immediately following graduation in 1997 he joined WGG Schnetzer Puskas Ingenieure. In 2011, he accepted a lecture in connection with the visiting professorship of Pool Architekten at the ETH Zurich.

Between 2006 and 2008, the three partners, Kilian Weiss, René Guillod and Rudolf Gisi, retired from the firm. In 2008, Mark Eitel, Patrick Grieder, Salomé Hug-Meier, Daniel Küpfer, Rolf Nachbur and Kevin M. Rahner were appointed Associates.

In 2009, Schnetzer Puskas Ingenieure AG was founded, with its head office in Basel and a branch office in Zurich. In 2012, Jan Stebler was appointed Associate and took over the running of the new branch office in Berne. In 2013, the management board was expanded to include Kevin M. Rahner and Jan Stebler. Johannes Dudli and Rémy Jabas were appointed Associates.

Team

This book relates to projects dating from the year 2000 onwards. Current and former team members since 2000 are mentioned below.

Members of the management board Stefan Bänziger, Tivadar Puskas, Dr. Heinrich Schnetzer

Associates Johannes Dudli, Mark Eitel, Patrick Grieder, Salomé Hug-Meier, Rémy Jabas, Rolf Nachbur, Kevin M. Rahner, Jan Stebler

Employees Gabriel Ackermann, Stefan Birk, Cyril Cerny, Caterina Chilovi, Matteo Della Giacoma, Christine Diebold, Eva Feuling, Menco Furter, Iván Garcia Ruíz, Markus Grieder, Thomas Gutzwiller, Timothy Hafen, Patrick Hänni, Dominik Häring, Christian Heinzel, Kathrin Hess, Jochem Hilberink, Andreas Hirschi, Ayumi Isozaki, Olivier Lichtenthaler, Stephan Luder, Enrica Mauro, Giotto Messi, Miran Misic, Dominik Mölder, Nicole Moritz, Dominik Musfeld, Reinhard Oppikofer, Nadine Pereira de Vasconcelos, Aleksandar Radovanovic, Dr. Florian Riebel, Roland Rolli, Elena Rosenke, Angelo Rosina, Michael Schiess, Sven Schläfli, Martina Schmid, Simon Spalt, Christoph Sterr, Christian Utzinger, Pedro Vitorino, Elisabeth Wuttke

Trainees Timon Braun, Mirko Henry, Ivona Samardzija, Davide Tagliente, Samuel Zekar

Former employees since 2000

Members of the management board Rudolf Gisi, René Guillod, Kilian Weiss

Associates Dr. Manuel Alvarez, Daniel Küpfer, Stefan von Ah

Employees Andrea Anastasi, Sandra Baumgartner, Nicola Benzoni, Christine Bergman, Susanne Binder, Andrea Blaser, Maurizio Brunone, Remo Bürgin, Rolf Buser, Marco Caflisch, Alessandro Castelli, Dr. Annette Detzel, Jessica Do Norte, Laetitia Eggenspiller-Kohler, Marco Fähndrich, Aleksandar Faltak, Debora Frey, Milan Gajic, Simon Gloor, Milorad Golubovic, Wolfgang Herrmann, Sven Humbel, Claudia Kliemt, Roman Köppli, Hansueli Küng, Béla Lamoth, Leo Lanz, Javier Leon, Judith Luible-Massonne, Michael Luternauer, Yves Manser, Ivana Maric, Tanja Matter, Birgit Merz-Schumacher, Simone Meyer, Heinrich Münch, Marcello Pizzichetta, Rozina Popovic, Christoph Raz, Petra Schafroth, Stefan Schär, Gabriele Schlittler, Burkhard Schnabel, Kurt Schönholzer, Miriam Schütt, Harry Skinner, Serafin Stauber, Henrik Stern, Michael Stoll, Tony Stucki, Jörg Sütterlin, Zoran Tanasic, Marc Tschopp, Cumhur Ürel, Elsbeth Vetter, Tina Visentin, Reto von Salis , Mirko Winkler, Ralf Zappe

Trainees Lina Dai, Christine Diebold, Stephanie Koeltzsch, Yves Manser, Ivana Maric, Tanja Matter, Harry Skinner, Serafin Stauber, Marc Tschopp, Pedro Vitorino, Mirko Winkler, Kilian Zeugin

Interns Kim Fröhlich, Andreas Hirschi, Christoph Jost, Jana Kaufmann, Jochua Reinle, Henrik Stern, Ken Zumstein

Authors

Marcel Baumgartner

Marcel Baumgartner studied architecture at the ETH Zurich and at the Graduate School of Design at Harvard University in Cambridge. He spent internships working in architectural practices in Berlin and New York. After completing his studies, he worked from 2001 to 2003 at Bearth & Deplazes Architekten in Chur. From 2003 to 2007 he was Research Assistant and from 2007 to 2011 Research Associate at the Chair of Architecture and Construction at the ETH Zurich under Professor Andrea Deplazes. Among other things, he headed the ETH Studio Monte Rosa from 2003 to 2009 and was responsible, as project-managing architect, for the construction of the SAC's new Monte Rosa hut. Since 2010, he has had his own architectural practice in Zurich, has participated in architectural competitions and realized first projects for public and private clients.

Jürg Conzett

Jürg Conzett is a partner in the engineering firm Conzett Bronzini Gartmann AG in Chur. The firm has twenty employees and specializes in bridge building and structural engineering projects. An important aspect of this involves work on existing structures. In 2010, Jürg Conzett curated the Swiss contribution "Landscape and Engineering Structures" for the Venice Biennale.

Aita Flury

Aita Flury is curator of the exhibition "Constructors' Dialogue. About the collaboration of civil engineer and architect in Switzerland." From 2006 to 2010 she was Associate Professor in Design at the HTW in Chur, since 2008 she has had her own architectural practice in Zurich. The interplay between practical and research activities as well as the relationship between space and structure play an important role in her work. Her most recent publication is *Cooperation. The Engineer and the Architect* (Basel: Birkhäuser Verlag 2012).

Harry Gugger

In 2010, Harry Gugger founded the Harry Gugger Studio following a twenty-year partnership with Herzog & de Meuron. His last projects with Herzog & de Meuron include the CaixaForum Madrid (2001–2008) and the extension to the Tate Modern, London (since 2004). With his own studio, Harry Gugger is involved with projects on various scales, with the focus on urban planning studies. In this work he benefits from the results of the research conducted under his professorship at the EPFL. In 2005, Harry Gugger was appointed full professor at the EPF Lausanne, where he founded the Laboratoire pour la Production d'Architecture (lapa). Since autumn 2011, he has in this capacity headed a laboratory for urban planning in Basel (Laboratoire Bâle, laba) and, in collaboration with ETH Zurich's Studio Basel, founded the "Cerberus" competence center for architecture and urban planning. Since 2010, Harry Gugger has also been a member of the board of the Holcim Foundation for Sustainable Construction.

Quintus Miller

Quintus Miller, born in 1961 in Aarau, studied architecture at the ETH Zurich. Between 1990 and 1994 he was design assistant at the EPF Lausanne and the ETH Zurich. Since 1994, he has been working in partnership with Paola Maranta in Basel. From 2000/01 he was Professeur invité at the EPF Lausanne, from 2007/08 visiting professor at the Accademia di Architettura at the Università della Svizzera Italiana, and from 2008 to 2010 visiting lecturer at the ETH Zurich. Since 2009 he has been a full professor of architecture at the Accademia di Architettura at the Università della Svizzera italiana. From 2004 to 2008, he was a member of the Urban Planning Committee of the city of Lucerne, and from 2005 to 2010 a member of the Fine Arts Committee of the municipality of Riehen. Since 2005 he has been a member of the Committee for the Preservation of Historical Monuments of the city of Zurich, and since 2011 a member of the Monuments Board of the Canton of Basel-Stadt. Since 2010 he has been chairman of the foundation board of the Archivio del Moderno at the Accademia di Architettura in Mendrisio.

Christian Menn

Christian Menn completed his structural engineering studies at the ETH Zurich in 1950 and, after working for three years as an engineer at Elektrowatt and Losinger & Cie AG, in 1956 he received a doctorate, also from the ETH Zurich, under Prof. Pierre Lardy. Before founding his own engineering practice in 1957 he worked as an engineer in Paris and Berne. Between 1957 and 1971, over eighty bridges throughout Switzerland were built under his direction, including the Rhine bridge in Tamins and the Nanin and Cascella bridges at the San Bernardino pass near Mesocco. From 1971 to 1992, he was professor at the ETH Zurich. Highlights of his career include the spectacular Ganter bridge on the Simplonstrasse, the Biaschina autobahn bridge in Giornico (concept for the competition project), the Charles River Bridge in Boston, and the Sunniberg bridge in Klosters. In 1996, he received an honorary doctorate from the University of Stuttgart, and in 2008 he received an honorary doctorate from the EPF Lausanne. In 2007, he was awarded the Bridge Design Award. To this day, Christian Menn participates as a consulting engineer in international projects.

Daniel Meyer

While studying structural engineering at the ETH Zurich, Daniel Meyer worked in architectural and engineering practices, on construction sites, and at the Institute for Hydrology + Hydraulics at the ETH Zurich. From 1989 to 1993, he worked as a project engineer at Wenaweser + Wolfensberger AG in Zurich. He was then project manager at Dr. Bollinger + Grohman in Frankfurt am Main until 1994. In 1995, he founded the firm Dr. Lüchinger + Meyer Bauingenieure AG together with Dr. Paul Lüchinger. His field of activity includes structural engineering and consulting. Since 2009, Daniel Meyer has been a member of the management board of the SIA, since 2010 as chairman of the SIA 144 regulation committee.

Image Credits

Actelion Pharmaceuticals Ltd., Allschwil: p. 82/3
Tadao Ando, Osaka (JP): p. 38 (3)
Iwan Baan, Amsterdam (NL): p. 257
Christ & Gantenbein Architekten, Basel: p. 55 (t.), 213
EM2N Architekten AG, Zurich: p. 143, 144 (b.)
Aita Flury, Zurich: p. 275
Foto-Werk GmbH, Michael Fritschi, Basel: p. 245–247, 251, 296/7, 353
Roger Frei, Zurich: p. 147
Graber Pulver Architekten AG, Zurich: p. 149, 151 (b.)
José Miguel Hernandez, Spain: p. 91
Herzog & de Meuron, Basel: p. 78, 86 (t.), 87, 140 (b.), 171, 172 (t.), 174 (t.), 238, 253, 256, 317
Thomas Jantscher, Colombier: p. 186 (t.), 187
Josef Meyer Stahl und Metall AG, Emmen: p. 208, 209 (t. / l.)
Prof. Vittorio Magnago Lampugnani, Zurich: p. 38 (2)
Johannes Marburg, Geneva: p. 43, 46/7, 77, 81, 263
Marcel Baumgartner dipl. Architekten ETH SIA, Zurich: p. 283
Marcel Baumgartner dipl. Architekten ETH SIA, Zurich / nightnurse images GmbH, Zurich: p. 287
Marcel Baumgartner dipl. Architekten ETH SIA, Zurich / Schnetzer Puskas Engineers, Basel: p. 284/5, 288/9
Miller & Maranta Architekten, Basel: p. 34 (t.), 35
Morger & Dettli Architekten AG, Basel: p. 219
Nacása + Partners, Tokyo (JP): p. 174 (b.), 175
NB 35 S. L. Ingenieros, Madrid (ES) / Schnetzer Puskas Engineers, Basel: p. 88/9
Hans Pfaffen, Chur: p. 299, 301 (b.), 302/3
Paolo Poloni, Zurich: film stills within the conversations
Pool Architekten, Zurich: p. 115–117, 119
ARGE Rheinbrücke, Frutiger AG Engineering, Thun: p. 266/7
ARGE Schneider & Schneider Architekten AG, Aarau / Oeschger Architekten AG, Hausen: p. 121
Schnetzer Puskas Engineers, Basel: p. 22/3, 39, 41/2, 44/5, 49/50, 51 (t.), 53/4, 55 (b.), 58/9, 72–75, 79/80, 85, 90, 93/4, 95 (t.), 98/9, 106, 109/10, 112/3, 118, 122/3, 126/7, 134, 137–139, 140 (t.), 141 (t.), 144 (t.), 145/6, 150, 151 (t.), 154/5, 166, 169/70, 172 (b.), 173, 177–181, 184/5, 201, 203, 205–207, 209 (r.), 214–217, 220–223, 226/7, 241, 242 (t.), 248–250, 254/5, 259–261, 262 (t.), 274, 280, 291–295, 300, 306/7, 314, 315 (b.), 318/9, 321–323, 326/7
Christopher Scholz, Frankfurt (DE): p. 325
Daniel Spehr, Basel: p. 183, 186 (b.)
Toyo Ito & Associates, Architects, Tokyo (JP): p. 315 (t.)
Wacker Ingenieure, Birkenfeld (DE): p. 190/1
Ruedi Walti, Basel: p. 34 (b.), 51 (b.), 95 (b.), 111, 210/1
Juri Weiss, Basel: p. 141 (b.)
Timothee Zurbuchen, Lausanne: p. 328/9
Zwimpfer Partner Architekten, Basel: p. 262 (b.)

Acknowledgements

We wish to express our heartfelt thanks to the Institute for the History and Theory of Architecture (gta) at the ETH Zurich for the unique opportunity to publish a book on the subject of structural design and to curate the associated exhibition.

We wish to thank all of the authors who contributed to the book, as well as the people, institutions, companies, and sponsors who made it possible for the work to be published in the first place.

We would like to express our particular thanks to Philippe Carrard, whose patient enquiries and insistence encouraged us to write a publication about our work.

A big thank-you is due to Aita Flury, the publisher of our book. During the course of conversations with us, the engineers, she was able, time and again, to bring our thoughts and formulations into focus by discussing the development of engineering structures and through critical questioning.

We also wish to express our heartfelt thanks to all outside authors for their inspired texts which have made it possible to include important external perspectives on our work.

Thanks go to Clementine van Rooden for her assistance with our own text contributions. We would like to thank Tiziana De Filippo of the Institute for the History and Theory of Architecture for the precise and tenacious copy-editing of the book. Heartfelt thanks go to Dorothee Wettstein of design studio 1kilo who displayed considerable verve in giving the book a fitting visual format. We would like to thank Nicole Moritz for the commitment with which she attended to internal coordination within the firm. We also thank Ayumi Isozaki for the careful revision of all drawings and the coordination within the firm.

We would like to express our heartfelt thanks to all those persons, institutions, and companies involved with the book who, as sponsors, made the publication of the book possible. The realization of the structures presented in this book was only made possible by our clients – architects, developers, departments and authorities, institutional awarding authorities, general contractors, and contractors – who have placed their trust in us over the years.

Lastly but most importantly, we would like to thank our team. Without our colleagues there would be no collective body of work. Their untiring interest in the planning and realization of engineering structures is the foundation for our achievements and the impetus for the further development of our firm.

Heinrich Schnetzer, Tivadar Puskas, Stefan Bänziger

Imprint

Project direction Philippe Carrard
Concept and editorial direction Aita Flury, Heinrich Schnetzer, Tivadar Puskas, Stefan Bänziger
Editorial collaboration Clementine van Rooden
Picture editing Nicole Moritz-Wittwer
Productions of plans Ayumi Isozaki and team Schnetzer Puskas Engineers
Editorial assistance Tiziana De Filippo, collaboration: Sandra Gomez
Translation David Koralek, ArchiTrans | architectural translations, Berlin; Translation Agency Antoinette Aichele-Platen, Munich (Prologue and Annex)
English editing Marion Elmer
Graphic concept and design 1kilo
Lithography, printing and binding freiburger graphische betriebe

ISBN 978-3-85676-332-9

The Deutsche Nationalbibliothek lists this publication in the Deutsche Nationalbibliografie; detailed bibliographic data are available in the Internet at http://dnb.d-nb.de.

This book appears on the occasion of the exhibition *Schnetzer Puskas Ingenieure. Entwurf Struktur Erfahrungen* from 31 october to 11 december at the ETH Zurich.

This book is available in a German edition: ISBN 978-3-85676-321-3

Support
Ackermann Architekt BSA SIA AG, Basel
Adrian Streich Architekten AG, Zurich
Amt für Hochbauten der Stadt Zürich, Zurich
b+p baurealisation ag, Zurich
dany waldner ag, Zurich
Egli Rohr Partner AG, Baden
Gerber-Vogt AG, Fenster- und Fassadenbau, Allschwil
Graber Pulver Architekten AG, Zurich/Berne
Herzog & de Meuron Basel Ltd., Basel
Implenia Schweiz AG, Buildings Nordwest, Generalunternehmung, Basel
ITECO Ingenieurunternehmung AG, Affoltern am Albis
Kellerhals Anwälte, Basel
KNECHT Bauunternehmung AG, Münchenstein
Marti AG Basel, Basel
MCH Group AG, Basel
:mlzd, Biel
Montana Bausysteme AG, Villmergen
Morger + Dettli Architekten AG, Basel
MTR Tief- und Rückbau AG, Pratteln
Perolini Baumanagement AG, Zurich
phalt Architekten AG, Zurich
pool Architekten, Zurich
Priora Generalunternehmung AG, Basel
SBB Immobilien, Zurich
Schweizerischer Ingenieur- und Architektenverein SIA, Zurich
Spannverbund Bausysteme GmbH, Bülach
Stalder & Buol Architektur GmbH, Zurich
Stokar+Partner AG, Basel
Swiss Prime Site AG, Olten
Theo Hotz Partner AG, Zurich
Thomas Glanzmann GmbH, Basel
UBS Fund Management (Switzerland) AG, Basel
usic – Schweizerische Vereinigung Beratender Ingenieurunternehmungen, Berne
Walt+Galmarini AG, Zurich
Zwimpfer Partner Architekten, Basel/Zurich

burckhardtpartner
Burckhardt+Partner AG, Basel

GHELMA SPEZIALTIEFBAU
Ghelma AG Spezialtiefbau, Meiringen

HS HUBER STRAUB Bauunternehmung
HUBER STRAUB AG, Basel

ETH
Eidgenössische Technische Hochschule Zürich
Swiss Federal Institute of Technology Zurich

DARCH gta
Institute for the History and Theory of Architecture
Faculty of Architecture
Swiss Federal Institute of Technology Zurich

In partnership with

sia
schweizerischer ingenieur- und architektenverein
swiss society of engineers and architects

SCHWEIZER BAUMUSTER-CENTRALE ZÜRICH
Schweizer Baumuster-Centrale, Zurich

In cooperation with

BAUKOMA
Baukoma AG, Kerzers

elco heating solutions
Elcotherm AG, Vilters

Eternit®
Eternit (Schweiz) AG, Niederurnen

Holcim
Holcim (Schweiz) AG, Zurich

PLOT FACTORY
PlotFactory AG, Weisslingen

Schweizer
Ernst Schweizer AG, Hedingen

VELUX®
VELUX Schweiz AG, Trimbach

zzwancor
ZZ Wancor AG, Regensdorf